GEOINFORMATION

Photogrammetry, Remote Sensing, GIS and GPS

The Author

Dr. N.V. Prasad, M.Tech., M.Sc., M.Ed., LLB, Ph.D., is an Associate Professor of GIS and Remote Sensing and working/worked at several Universities in India and abroad and has been teaching Geoinformatics-related courses since 2000. He was the recipient of *several* distinguished achievement awards of academic and research from various renowned organizations and Institutions. He received E.P. Odum Gold Medal, 2009 (The International Society for Ecological Communication); Leading Scientists of the World (Silver Medal) 2011 and 2013 (IBC, Cambridge, England); Life time achievement award, 2011 (WCASC, Great Britain); and Rashtriya Gaurav Award 2011 (IIFS, New Delhi) for his commendable academic and research activities in the field of science. He is the author of three books, *i.e.*, Remote Sensing of Mangrove Wetlands (ISBN: 978-3-639-35843-8), GIS and Remote Sensing for Sustainable Forest Management (ISBN: 978-3-639-36769-0), and Ecology of Estuarine Mangrove Zooplankton (ISBN: 978-3-639-35303-7). He published about 50 research papers in National and International Journals and presented papers of 12 national and 5 international conferences held at India and abroad. He is the Assistant Editor of Indian Journal of Environment and Ecoplanning and previously worked as Editorial member of National Journal of Life Science. He is also freelance Journalist/writer and published more than 200 article in various daily newspapers and magazines in the field of Science, Environment, History and Current issues.

GEOINFORMATION

Photogrammetry, Remote Sensing, GIS and GPS

Volume 1

N.V. Prasad

2023

Daya Publishing House®

A Division of

Astral International Pvt. Ltd.

New Delhi – 110 002

First Published, 2015
Reprinted, 2023

Cataloging in Publication Data--DK
Courtesy: D.K. Agencies (P) Ltd. <docinfo@dkagencies.com>

Prasad, N. V. *(Associate professor of geoinformatics),* **author.**
Geoinformation : photogrammetry, remote sensing, GIS and GPS / N.V. Prasad.
2 volumes ; cm
Includes bibliographical references (pages) and index.

ISBN 9789351246213 (Vol.1)
ISBN 9789351246183 (Set)

1. Photogrammetry. 2. Remote sensing. 3. Geographic information systems. 4. Global Positioning System. I. Title.

DDC 526.982 23

Published by : **Daya Publishing House®**
A Division of
Astral International Pvt. Ltd.
– ISO 9001:2015 Certified Company –
4736/23, Ansari Road, Darya Ganj
New Delhi-110 002
Ph. 011-43549197, 23278134
E-mail: info@astralint.com
Website: www.astralint.com

Printed at : **Replika Press Pvt. Ltd.**

Dedicated to All time legendary person...

Padma Shri Dr. *Nandamuri Taraka Rama* Rao

ANDHRA PRADESH STATE COUNCIL OF HIGHER EDUCATION
(A Statutory Body of the Government of A.P.)
Opp. Mahavir Hospital, Mahavir Marg, Masab Tank, Hyderabad - 500 028.
✆: (O) 040-23311879 (R) 23387773 Fax : 040-23311470
E-mail:l.venugopalreddy@gmail.com; chairman.apsche@gmail.com URL:www.apsche.org

Prof. L. Venugopal Reddy
CHAIRMAN

Foreword

There is no doubt that today, perhaps more than ever before, humanity faces a myriad of complex and demanding challenges. This has been propelled by the ever increasing global population and intense pressure being exerted on the Earth's resources. The resulting consequences are severe changes in land cover, diminishing biodiversity and natural habitats, dwindling fresh water supplies, and the degradation in the quality of the little that is available, and changing weather and climatic patterns, especially global warming with its associated predicted catastrophes such as rising sea level and increased numbers of extreme weather events.

These human-induced and natural impacts on the globe need to be well understood in order to develop informed policies, decisions, and remedial measures to mitigate current and future negative impacts. This can be achieved through continuous monitoring of the earth surface to acquire data that can be soundly and rigorously analyzed to provide information about the current state of the earth and its changing patterns, and to enable predictions of possible future impacts. Environmental and global monitoring techniques that may provide such information are under scrutiny from an increasingly environmentally conscious society that demands the efficient delivery of such information at a minimal cost.

In addition, it is the nature of changes that they vary both spatially and temporally, thereby putting pressure on traditional methods of data acquisition, some of which are very labour intensive. With these challenges, conventional monitoring techniques, particularly those that record spatial changes, call for more sophisticated approaches that deliver the necessary information at an affordable cost. Developing pragmatic and sustainable solutions to address these and many other similar challenges requires the use of geodata and the application of geoinformatics. Geoinformatics, defined by Ehlers (2003) as "the art, science or technology dealing with the acquisition, storage, processing, production, presentation and dissemination of geoinformation", is a multidisciplinary field. It has at its core different technologies that support the acquisition, analysis, and visualization of geodata. The geodata is usually acquired from Earth observation sensors as remotely sensed images, analyzed by geographic information systems (GIS), and visualized on paper or on computer screens. Furthermore, it combines geospatial analysis and modelling, development of geospatial databases, information systems design, human-computer interaction, and both wired and wireless networking technologies. Geoinformatics uses geocomputation and geovisualization for analyzing information. Typical branches of geoinformatics include: photogrammetry, remote sensing, GIS, global positioning system (GPS), cartography, geodesy, and web mapping.

This book presents the concepts, methods and applications of geoinformation. The book also features the state-of-the-art discussion of the advanced topics in GIS and remote sensing. The concepts and applications of geoinformation presented in this book will be of benefit to university students, researchers, and decision-makers across a wide range of fields.

I appreciate and congratulate Dr. Prasad for bringing out such a informative and qualitative book and I wish him all the success in all his feature endeavors.

Prof. L. Venu Gopal Reddy, Ph.D

Chairman, AP State Council of Higher Education
& Former Vice-Chancellor,
Andhra University, Visakhapatnam and
Acharya Nagarjuna University, Guntur

Preface

The need to understand our dynamic and changing planet perhaps has never been more important. However, in order to gather, analyze, and gain knowledge from the complex systems and forces which affect the earth, large amounts of data are required. Even the simple task of answering technologically, the fundamental question, "where am I?" requires the ability to generate, explore, describe, and use vast datasets. The study of Geoinformatics promises to serve the need to explore, to research and enhance humanity's grasp of the world.

Spatial data handling involves many disciplines. We can distinguish disciplines that develop spatial concepts, provide means for capturing and processing of spatial data, provide a formal and theoretical foundation, are application-oriented, and support spatial data handling in legal and management aspects. The term geoinformation/geoinformatics is used globally with varying connotations, though it has its origin and popular use in Europe. Geoinformation/geoinformatics is consider as a new discipline integrating elements of various disciplines that deal with geographic or geospatial data, *i.e.*, data linked to some locality on the surface of the Earth. These disciplines include, but are not limited to, surveying, remote sensing, geographic information systems (GIS), cartography, geosciences, informatics, computer science, statistics, and management. Surveying and mapping have recently undergone a transition from discipline-oriented technologies, such as geodesy, surveying, photogrammetry and cartography, into the methodology-oriented integrated discipline of geoinformation. This is based on GPS positioning, remote sensing, digital photography for data acquisition and GIS for data manipulation and data output.

Interest in Geoinformation has tremendously grown in recent years in many different ways. Researchers from a variety of academic disciplines are using spatial thinking and Geoinformation tools to develop spatially-explicit models. Broad public interest in this subject is being fuelled by popular applications like Google maps, personal navigation devices, MapQuest, etc. Web-based software developers are increasingly exploring "mash-ups" integrating different information sources to web-based maps, such as Google Earth and MS Virtual Earth. Therefore, there is a need to bring key GIS concepts and results to a diverse audience as current Geoinformation literature largely caters to either Geoinformation specialists or end-users of popular GIS software.

For remote sensing, the basic fundamentals are the properties of electromagnetic radiation and their interaction with matter. This radiation is received by sensors and platforms in analogue or digital form, and is subject to image processing. In photogrammetry, the stereo-concept is used for the location of information in 3D. With the advent of high-resolution satellite systems in stereo the theory of analytical photogrammetry restituting 2D image information into 3D is of increasing importance, merging the remote sensing approach with that of photogrammetry. The result of the restitution is a direct input into geographic information systems in vector or in raster form.

Fundamentally, GIS is an applications-led technology, yet successful applications need appropriate scientific foundations. GIS is a science, a technology, a discipline, and an applied problem solving methodology. GIS is of enduring importance because of its central coordinating *principles*, the specialist *techniques* that have been developed to handle spatial data, the special *analysis* methods that are key to spatial data, and because of the particular *management* issues presented by geographic information (GI) handling. Each section of this book investigates the unique, complex, and difficult problems that are posed by geographic information, and together they build into a holistic understanding of all that is important about GIS.

Effective use of GIS is impossible if they are simply seen as black boxes producing magic. GIS is applied rarely in controlled, laboratory-like conditions. Our messy, inconvenient, and apparently haphazard real world is the laboratory for GIS, and the science of real-world application is the difficult kind-it can rarely control for, or assume away, things that we would prefer were not there and that get in the way of almost any given application. Scientific understanding of the inherent uncertainties and imperfections in representing the world makes us able to judge whether the conclusions of our analysis are sustainable, and is essential for everything except the most trivial use of GIS. GIS is also founded on a search for understanding and predictive power in a world where human factors interact with those relating to the physical environment. Good science is also ethical and clearly communicated science, and thus the ways in which we analyze and depict geography also play an important role.

This book attempts to present the required basic background for Remote Sensing, Photogrammetry, GIS and GPS in the new Geoinformation/ Geoinformatics concept in which the different methodologies must be combined. The fundamentals of these are described in detail, with an emphasis on global, regional and local applications. This book also provides a comprehensive look at the challenges and discoveries made in this emerging field. It is a collection of informed and pertinent articles on all vital aspects of this field is divided in to 5 chapter as follows: (1) Introduction, (2) Photogrammetry, (3) Remote Sensing, (4) GIS, and (5) Global Positioning System (GPS). These chapters outline the possibilities and challenges faced in the field and represents areas in which it can expand and become further defined. In the context of data integration, a short introduction to the GPS satellite positioning system is provided. This book will appeal to a wide range of readers from advanced undergraduates to all professionals in the growing field of Geoinformation. The concepts and applications of Geoinformation presented in this book will be of benefit to decision makers across a wide range of fields including those working in environmental management, urban planning, tourism industry, in the emergency services, public health and epidemiology, crime mapping, market analysis and e-commerce, or mineral exploration, among many others. This book is the companion for everyone who desires a rich understanding of how Geoinformation is used in the real world.

Unlike many books on Geoinformation, this one isn't meant to keep you spellbound for days or weeks. Instead, you can use this book when you need to answer basic questions or figure out what questions to ask your Geoinformation–specialist friends. Think of this book as a reference you can use to find what you need when you need it.

Dr. N.V. Prasad

Contents

Volume 3

– *Chapter 1* –
INTRODUCTION

Geoinformation: An Overview

We are presently positioned at the beginning of twenty first century with the fast growing trends in computer technology information systems and virtual world to obtain data about the physical and cultural worlds, and to use these data to do research or to solve practical problems. The current digital and analog electronic devices facilitates the inventory of resources and the rapid execution of arithmetic or logical operations. These Information Systems are undergoing much improvement and they are able to create, manipulate, store and use spatial data much faster and at rapid rate as compared to conventional methods.

An Information System, a collection of data and tools for working with those data, contains data in analog form or digital form about the phenomena in the real world. Our Perception of the world through selection, generalization and synthesis give us information and the representation of this information that is, the data constitute a model of those phenomena. So the collection of data, the database is a physical repository of varied views of the real world representing our knowledge at one point in time. Information is derived from the individual data elements in a database, the information directly apparent, *i.e.*, information is produced from data by our thought processes, institution or whatever based on our knowledge. Therefore in a database context the terms data, information and knowledge are differentiated. It can be summarized that the data is very important and added value as we progress from data to information, to knowledge. The data, which has many origins and forms, may be any of the following:

- ☆ Real, for example the terrain conditions etc.
- ☆ Captured, *i.e.*, recorded digital data from remote sensing satellites or Arial photographs of any area.
- ☆ Interpreted, *i.e.*, landuse from remote sensing data.
- ☆ Encoded *i.e.*, recordings of rain-gauge data, depth of well data etc.
- ☆ Structured or organized such as tables about conditions of particular watershed.

1. Concepts of Space and time in Spatial Information Systems

Spatial information is always related to geographic space, *i.e.*, large-scale space. This is the space beyond the human body, space that represents the surrounding geographic world. Within such space, we constantly move around, we navigate in it, and we conceptualise it in different ways. Geographic space is the space of topographic, landuse/landcover, climatic, cadastral, and other features of the geographic world. Geographic information system technology is used to manipulate objects in geographic space, and to acquire knowledge from spatial facts.

Geographic space is distinct from small-scale space, or tabletop space. In other words, objects that are smaller than us, objects that can be moved around on a tabletop, belong to small-scale space and are not subject of our interest.

The human understanding of space, influenced by language and cultural background, plays an important role in how we design and use tools for the processing of spatial data. In the same way as spatial information is always related to geographic space, it relates to geographic time, the time whose effects we observe in the changing geographic world around us. We are less interested in pure philosophical or physical considerations about time or space-time, but more in the observable spatio-temporal effects that can be described, measured and stored in information systems.

1.1 Spatial Information Systems

The handling of spatial data usually involves processes of data acquisition, storage and maintenance, analysis and output. For many years, this has been done using analogue data sources, manual processing and the production of paper maps. The introduction of modern technologies has led to an increased use of computers and information technology in all aspects of spatial data handling. The software technology used in this domain is geographic information systems (GIS).

A general motivation for the use of GIS can be illustrated with the following example. For a planning task usually different maps and other data sources are needed. Assuming a conventional analogue procedure we would have to collect all the maps and documents needed before we can start the analysis. The first problem we encounter is that the maps and data have to be collected from different sources at different locations (*e.g.*, mapping agency, geological survey, soil survey, forest survey, census bureau, etc.), and that they are in different scales and projections. In order to combine data from maps they have to be converted into working documents of the same scale and projection. This has to be done manually, and it requires much time and money.

With the help of a GIS, the maps can be stored in digital form in a database in world co-ordinates (meters or feet). This makes scale transfor-mations unnecessary, and the conversion between map projections can be done easily with the software. The spatial analysis functions of the GIS are then applied to perform the planning tasks. This can speed up the process and allows for easy modifications to the analysis approach.

2. GIS, Spatial Information Theory, and the Geoinformatics Context

Spatial data handling involves many disciplines. We can distinguish disciplines that develop spatial concepts, provide means for capturing and processing of spatial data, provide a formal and theoretical foundation, are application-oriented, and support spatial data handling in legal and management aspects. The following Table 1.1 shows a classification of some of these disciplines. They are grouped according to how they deal with spatial information. The list is not meant to be exhaustive. The discipline that deals with all aspects of spatial data handling is called geoinformatics.

Table 1.1: Classification of Disciplines

Characteristics of Disciplines	*Sample Disciplines*
Development of spatial concepts	Geography
	Cognitive Science
	Linguistics
	Psychology
Means for capturing and processing spatial data	Remote Sensing
	Surveying Engineering
	Cartography
	Photogrammetry
Formal and theoretical foundation	Computer Science
	Expert Systems
	Mathematics
	Statistics
Applications	Archaeology
	Architecture
	Forestry
	Geo-Sciences
	Regional and Urban Planning
	Surveying
Support	Legal Sciences
	Economy

3. Meaning and Scope

Geoinformatics (or Geoinformation) has been described as "the science and technology dealing with the structure and character of spatial information, its capture, its classification and qualification, its storage, processing, portrayal and dissemination, including the infrastructure necessary to secure optimal use of this information" (Groot, 1989) or "the art, science or technology dealing with the acquisition, storage, processing production, presentation and dissemination of geoinformation" (Ehlers and Amer, 1991).

The term geoinformatics is used globally with varying connotations, though it has its origin and popular use in Europe. According to the International Institute of Geo-Information Science and Earth Observations (ITC, the Netherlands), technologies supporting the processes of acquiring, analyzing, and visualizing spatial data form the core of geoinformatics (ITC, 2001). Other researchers consider geoinformatics as a new discipline integrating elements of various disciplines that deal with geographic or geospatial data, *i.e.*, data linked to some locality on the surface of the Earth. These disciplines include, but are not limited to, surveying, remote sensing, geographic information systems (GIS), cartography, geosciences, informatics, computer science, statistics, and management (Virrantaus and Haggrén, 2000).

Other terms, Geo-information Science, and Spatial Information Science have much the same meaning. A related term that is sometimes used synonymously with geoinformatics is geomatics. Geomatics is encompasses geoinformatics, but geomatics also focuses on surveying. It was originally introduced in Canada, and became very popular in French speaking countries. Laurini and Thompson (1992) describe it as "the fusion of ideas from geo-sciences and informatics." The term geomatics, however, was never fully accepted in the United States where the term geographical information science is preferred. Goodchild (1992) defines GIS research as "research on the generic issues that surround the use of GIS technology, impede its successful implementation, or emerge from an understanding of its potential capabilities."

Geoinformatics has at its core the technologies supporting the processes of acquiring, analyzing and visualizing spatial data. Both geomatics and geoinformatics include and rely heavily upon the theory and practical implications of geodesy. Geoinformatics combines geospatial analysis and modelling, development of geospatial databases, information systems design, human-computer interaction and both wired and wireless networking technologies. Geoinformatics uses geocomputation and geovisualization for analyzing geoinformation.

Surveying and mapping has recently undergone a transition from discipline oriented technologies, such as geodesy, surveying, photogrammetry and cartography into a methodology oriented integrated discipline of geoinformation based on GPS positioning, remote sensing, digital photography for data acquisition and GIS for data manipulation and data output. For remote sensing the basic fundamentals are the properties of electromagnetic radiation and their interaction with matter. This radiation is received by sensors on platforms in analogue or digital form to result in images, which are subject to image processing. In photogrammetry the stereo-concept is used for the location of the information in three dimensions. With the advent of high resolution satellite systems in stereo, the theory of analytical photogrammetry, restituting two-dimensional image information into three dimensions, is of increasing importance merging the remote sensing approach with that of photogrammetry. The result of the restitution is a direct input into geographic information systems in vector or in raster form. The application of these is possible at the global, regional and local levels. Data integration is made possible by geocoding, in which the GPS satellite positioning system plays an increasing role.

Branches of geoinformatics include Photogrammetry, Cartography, Geodesy, Remote Sensing, Geographical Information Systems (GIS), Global Positioning

System (GPS) and Web mapping. The very nature of Earth system science and geoinformatics ties them together in a synergistic bond. Geoinformatics provides an environment for the integration of geospatial data, models and knowledge and provides extremely useful decision-making tools for the planning, monitoring, management and appropriate utilization of resources at the country, regional and global levels.

4. Historical Development

4.1 Geodesy

Geodesy is the science that determines the size and shape of the earth, the precise positions and elevations of points, and lengths and directions of lines on the earth's surface, and the variations of terrestrial gravity. In ancient Greece (Homer, 800 B.C.) the earth's surface was believed to be a disk surrounded by the oceans. But, not long after, Pythagoras (550 B.C.) and Aristotle (350 B.C.) postulated that the earth was a sphere. Eratosthenes, an Alexandrine scientist, (276-195 B.C.) is given credit for first measuring earth's size. His results were about 16 per cent too big. At Syene (today's Assuan) located at the Tropic of Cancer at a latitude of 23.5° the sun reflected from a deep well on June 21, while it would not do so in Alexandria at a latitude of 31.1°. Erathosthenes measured the distance between the two cities along the meridian by cart-wheel computing the earth's spherical radius as 5909 km. Poseidonius (135-50 B.C.) also determined earth's size using arc measurements between Rhodes and Alexandria (11 per cent too large). Arabian efforts about 827 A.D. near Baghdad by the caliph Abdullah al Mamun gave an answer only about 3.6 per cent too large.

Little is recorded in western literature, but see Smith (1986), about geodesy until after the Middle Ages. However with invention of the telescope in the early 1600's, publication of 14 place logarithms and use of triangulation for arc measurement, many advancements were made. Later developments also included the theory of gravity, differential and integral calculus, standardization of length and introduction of least squares adjustment.

In 1615 a Dutchman, Willebrord Snellius, measured an arc more than 80 miles long with a series of 33 triangles. His computation of the earth's circumference was too small by about 3.4 percent. In 1669-70 a Frenchman, Jean Picard, measured an arc on the meridian through Paris and obtained results too large by only 0.7 percent. An interesting note here is that Newton needed and used Picard's measurements in development of his theory of universal gravitation which he published in 1687. Newton also concluded the earth is flattened at the poles due to earth's rotation.

Picard's work was extended north to Dunkirk and south to Collioure by the Cassini brothers about 1700. The total arc was about 8° 20' but the arc was computed in two segments-the parts north and south of Paris. The radius of the earth determined from the northern portion was shorter than that computed from the southern portion. Hence the Cassini's concluded, in fact insisted, the earth is elongated at the poles. In 1735 and 1736 two geodetic surveying expeditions were sent by the French Academy of Science to Peru (Bouguer) and Lapland (Maupertuis) to make precise measurements to settle the dispute. Results showed quite conclusively the earth is flattened at the poles. Distances on the ellipsoid could consequently be determined by the astronomical observations of latitude and longitude at the respective points on the earth's surface.

Laplace (1802), C.F. Gauss (1828) and F.W. Bessel (1837), however, recognized that astronomic observations were influenced by the local gravity field due to mass irregularities of the earth's crust. This was confirmed among others by G. Everest observing huge deflections of the vertical in the Himalayas. This led to the establishment of local best-fitting reference ellipsoids for positional surveys of individual countries.

In the simplest case latitude and longitude was astronomically observed at a fundamental point, and an astronomical azimuth was measured to a second point in the triangulation network spanning a country. Within the triangulation network, at least one side was measured by distance measuring devices on the ground. For the determination of a best-fitting ellipsoid, several astronomic observation stations and several base lines were used. The coordinates of all triangulated and monumental points were calculated and least squares adjusted on the reference ellipsoid with chosen dimensions. On the chosen reference ellipsoid, the ellipsoidal latitudes and longitudes were obtained for all points of the first order triangulation network. This network was subsequently densified to second, third and fourth order by lower order triangulation. The survey accuracy of these triangulation networks of first to fourth order was relatively high, depending on the observational practices, but discrepancies between best fitting ellipsoids of neighbouring countries were in the order of tens of metres. For the purpose of mapping, the ellipsoidal coordinates were projected into projection coordinates.

Since the earth's gravity field influences the flow of water on the earth's surface, ellipsoidal coordinates without appropriate reductions cannot be used for practical height determinations. Height reference systems are therefore separate from position reference systems based on reference ellipsoids. An ideal reference surface would be the equipotential surface of the resting oceans, called 'the geoid'. Due to earth tides influenced by the moon and planets, due to ocean currents and winds influenced by climate and meteorology, this surface is never resting. For this reason the various countries engaged in mapping systems have created their own vertical reference systems by observing mean sea level tides at tidal benchmarks. Spirit levelling extended the elevations in level loops of first order over the mapping area of a country to monumented benchmarks. These level loop observations, corrected by at least normal gravity, could be densified by lower order levelling to second, third and fourth order. As is the case for positions, differences of several metres in height values may be the result of the different height reference systems of different countries.

The different reference systems for position and height still used for mapping in the countries of the world are in transition, changing into a new reference frame of three- or four-dimensional geodesy. This has become possible through the introduction of the US Navy Navstar Global Positioning Systems (GPS) in the 1980s.

It now consists of twenty-four orbiting satellites at an altitude of 20200 km. These orbit at an inclination of 55° for 12 hours, allowing a view, in a direct line of sight, of at least four of these satellites from any observation point on the earth's surface for 24 hours of the day. Each of the satellites transmits timed signals on two carrier waves with 19.05 cm and 24.45 cm wavelengths. The carrier waves are modulated with codes containing the particular satellite's ephemeris data with its corrections. When three satellites with known orbital positions transmit signals to a ground receiver, the observed distances permit an intersection of 3D coordinates on the earth's surface. Now GPS has become a new tool for detailed local surveys and its updating.

4.2 Surveying

Surveying, according to the American Congress on Surveying and Mapping (ACSM), is the science and art of making all essential measurements to determine the relative position of points and/or physical and cultural details above, on, or beneath the surface of the Earth, and to depict them in a usable form, or to establish the position of points and/or details. While geodesy's main goal was to determine the size and shape of the earth, and to provide control for the orientation of subsequent surveys on the earth's surface, it was the aim of survey technology to provide tools for the positioning of detailed objects.

The history of surveying traces back to the separation of farm tracts and the flooding of the Nile River in Egypt (3000 BC). As the river would flood every year, there grew a need for markers to delineate the individual parcels. Stones were placed by a rope stretcher, or surveyor, and geometric schemes devised so that if the river washed the markers away, they could be easily replaced. The nearly perfect squareness and north-south orientation of the Great Pyramid, built c. 2700 BC, affirm the Egyptians' command of surveying. A recent reassessment of Stonehenge (2500 BC) suggests that the monument was set out by prehistoric surveyors using peg and rope geometry. The Groma Surveying instrument was originated in Mesopotamia during early 1st millennium BC.

Under the Romans, land surveyors were established as a profession, and they established the basic measurements under which the Roman Empire was divided, such as a tax register of conquered lands (300 AD). The rise of the Caliphate led to extensive surveying throughout the Arab Empire. Arabic surveyors invented a variety of specialized instruments for surveying, including: Instruments for accurate levelling, a rotating alhidade, and a surveying astrolabe. In England, The Domesday Book was published by William the Conqueror (1086). This covered all England, contained names of the land owners, area, land quality, and specific information of the area's content and inhabitants.

In the 18th century in Europe triangulation was used to build a hierarchy of networks to allow point positioning within a country. Highest in the hierarchy were triangulation networks. These were densified into networks of traverses (polygons), into which local mapping surveying measurements, usually with measuring tape, corner prism and the familiar red and white poles, are tied. In the late 1780s, a team from the Ordnance Survey of Great Britain, originally under

General William Roy began the Principal Triangulation of Britain using the specially built Ramsden theodolite. The continental Europe's Cadastre was created in 1808. This was founded by Napoleon I (Bonaparte) and contained numbers of the parcels of land (or just land), land usage, names etc., and value of the land.

The first direct distance measurements have been in place since the Babylonians and the Romans. Direct distance observations by tapes and chains, as they have been used in former centuries, have made way to the preference of angular measurement by theodolites, which permitted the use of trigonometry to calculate distances in over determined angular triangulation networks. Even for detailed terrestrial topographic surveys, instruments were developed in the early 1900s, the so-called tacheometers, which were able to measure distances rapidly in polar mode by optical means, combining these with directional measurements.

In the 1950s, the direct measurement of distances became possible by the invention of electronic distance-measuring devices. The tellurometer, developed in South Africa, utilized microwaves as a carrier onto which measurement phases of different wavelengths were modulated. The Swedish geodimeter used light as a carrier and later most manufacturers used

Infrared carrier waves. These distance-measuring capabilities were soon combined with directional measurement devices, known from theodolites in the form of electronic tacheometers. The automatic coding of directional and distance measurements was perfected in the so-called 'total stations'.

Nowadays it is possible to integrate GPS receivers and PC graphic capabilities in a single instrument, the so-called 'Electronic field book', which can be used effectively for updating vector graphics or for orienting oneself in the field with the help of digitized and displayed images. For the survey of heights, simple route survey devices, such as the barometer, are considered to be too inaccurate according to today's standards. Spirit level devices have gradually been replaced by instruments with automatic compensators assuring a horizontal line of sight by the force of gravity. The optically observed level rods read by the operator have likewise been automated by digitally coded reading devices to permit more rapid levelling operations.

4.3 Photogrammetry

Photogrammetry is the technology to derive geometric information of objects with the help of image measurements. The development of photogrammetry clearly depends on the general development of science and technology. It is interesting to note that the four major phases of photogrammetry are directly related to the technological inventions of photography, airplanes, computers and electronics.

Figure 1.1 depicts the four generations of photogrammetry. Photogrammetry had its beginning with the invention of photography by Daguerre and Niepce in 1839. The first generation, from the middle to the end of last century, was very much a pioneering and experimental phase with remarkable achievements in terrestrial and balloon, The second generation, usually referred to as *analog photogrammetry*, is characterized by the invention of stereophotogrammetry by

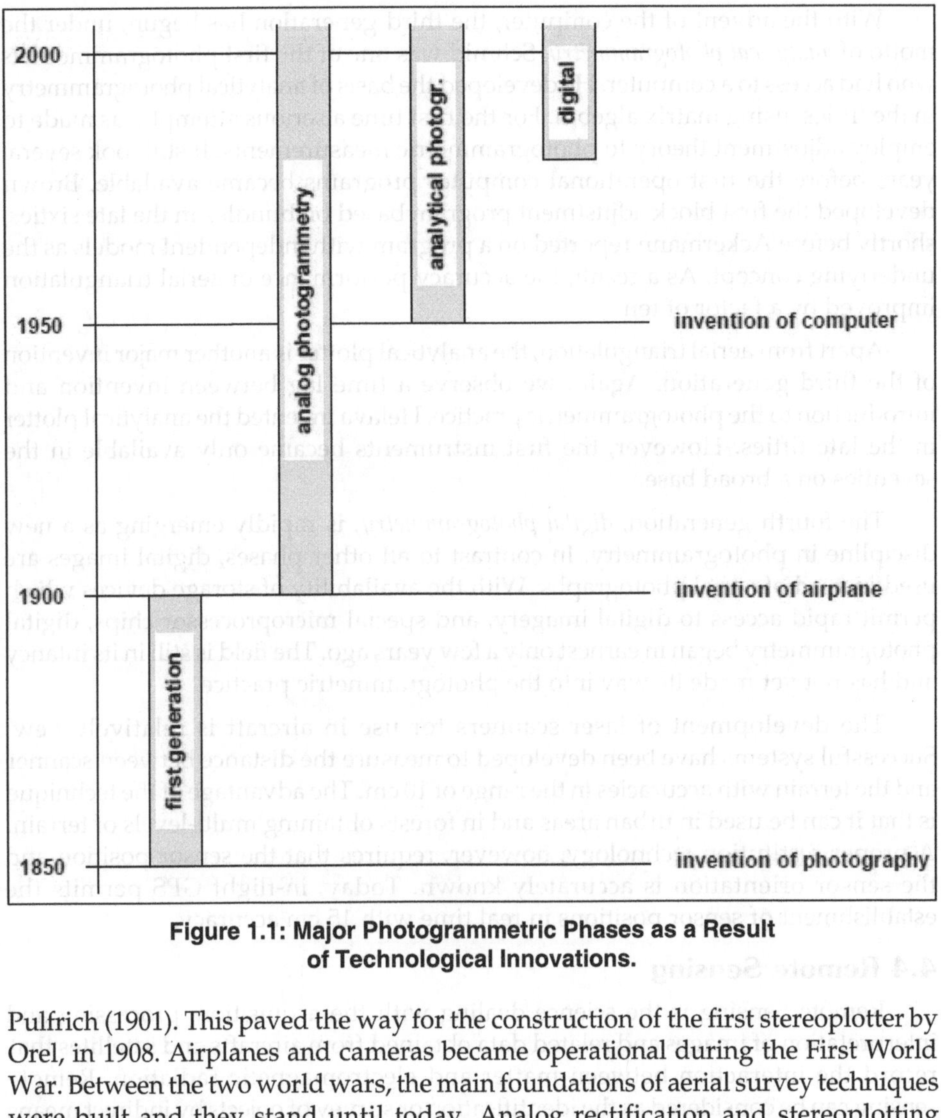

Figure 1.1: Major Photogrammetric Phases as a Result of Technological Innovations.

Pulfrich (1901). This paved the way for the construction of the first stereoplotter by Orel, in 1908. Airplanes and cameras became operational during the First World War. Between the two world wars, the main foundations of aerial survey techniques were built and they stand until today. Analog rectification and stereoplotting instruments, based on mechanical and optical technology, became widely available. Photogrammetry established itself as an efficient surveying and mapping method. The basic mathematical theory was known, but the amount of computation was prohibitive for numerical solutions and consequently all the efforts were aimed toward analog methods. Von Gruber is said to have called photogrammetry the art of avoiding computations.

After World War II, modifications and simplifications of the optical and mechanical plotting instruments developed in Switzerland, in Britain, in France, in Italy, in the USA and the USSR played an important role in meeting basic mapping demands all over the globe.

With the advent of the computer, the third generation has begun, under the motto of *analytical photogrammetry*. Schmid was one of the first photogrammetrists who had access to a computer. He developed the basis of analytical photogrammetry in the fifties, using matrix algebra. For the first time a serious attempt was made to employ adjustment theory to photogrammetric measurements. It still took several years before the first operational computer programs became available. Brown developed the first block adjustment program based on bundles in the late sixties, shortly before Ackermann reported on a program with independent models as the underlying concept. As a result, the accuracy performance of aerial triangulation improved by a factor of ten.

Apart from aerial triangulation, the analytical plotter is another major invention of the third generation. Again, we observe a time lag between invention and introduction to the photogrammetric practice. Helava invented the analytical plotter in the late fifties. However, the first instruments became only available in the seventies on a broad base.

The fourth generation, *digital photogrammetry*, is rapidly emerging as a new discipline in photogrammetry. In contrast to all other phases, digital images are used instead of aerial photographs. With the availability of storage devices which permit rapid access to digital imagery, and special microprocessor chips, digital photogrammetry began in earnest only a few years ago. The field is still in its infancy and has not yet made its way into the photogrammetric practice.

The development of laser scanners for use in aircraft is relatively new. Successful systems have been developed to measure the distance between scanner and the terrain with accuracies in the range of 10 cm. The advantage of the technique is that it can be used in urban areas and in forests obtaining multi-levels of terrain. A proper restitution technology, however, requires that the sensor position and the sensor orientation is accurately known. Today, in-flight GPS permits the establishment of sensor positions in real time with 15 cm accuracy.

4.4 Remote Sensing

Remote Sensing is the science dealing with the acquisition, processing and interpretation of images and related data obtained from aircrafts and satellites that record the interaction between matter and electromagnetic radiation. Remote sensing can be considered as the identification or survey of objects by indirect means using naturally existing or artificially created force fields. The technology of remote sensing originated in the science and technology of aerial photointerpretation (Silva, 1978). Interpretation of aerial photographs had been operational for much of the twentieth century, at least since the First World War. Then new imagery and sensing techniques quickly emerged based on various computer technologies, engineering designs, launch and deployment opportunities, and developments in sensor science (Figure 1.2).

The use of the term remote sensing began as a way of describing some of the new imagery and analysis techniques acquired alongside aerial photography during the height of the Cold War. The methods of analysis soon diverged-although, in

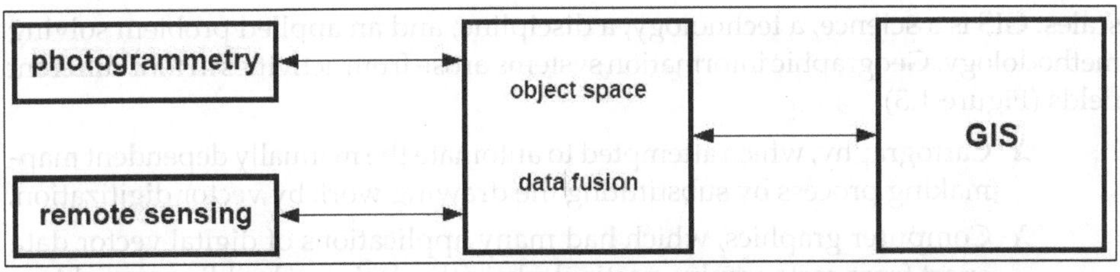

Figure 1.2: Relationship of Photogrammetry, Remote Sensing and GIS.

the view of many, including Buiten (1993) and Beaubien (1994), visual and digital analysis methods have recently reconverged, to the benefit of the overall remote sensing approach. Remote sensing has since developed into a scientific field that can stand on its own. In 1985, Curran suggested that rapid technological advancements (*e.g.*, computers) and improved sensor systems had propelled remote sensing into a stage of exponential growth.

Empirical evidence for the broader interpretation of the growth of remote sensing is found by considering the occurrence of milestones in the development of sensing and space technology, such as the development of radar in the 1940s, thermal imagery in the 1950s, the routine availability of meteorological satellite data in the 1960s, the launch of the ERTS and Landsat satellites in the 1970s, the continuation of medium- and coarse-scale monitoring of the globe in the 1980s, the advent of satellite high spatial detail imagery in the late 1990s, the rapid refinement and anticipated widespread application of lidar in the 2000s.

Higher spatial resolution was achieved by the French Spot satellites launched since 1986 with panchromatic pixel sizes of 10 m and multispectral resolution at 20 m pixels. The Indian satellites IRS 1C and 1D reached panchromatic pixel sizes of 6m in 1996. Since 1999, the US commercial satellite IKONOS 2 is in orbit, which produces digital panchromatic images with 1m pixels on the ground. This was surpassed by the US commercial satellite Quickbird, with 0.6 m pixels on the ground. In 1991, the European Space Agency (ESA) commenced a radar satellite programme ERS 1 and 2, which was followed by the Japanese radar sensor on JERS 1 in 1994, the Canadian Radarsat in 1995 and the Russian Almaz in 1995. The Space Shuttle Radar Topography mission SRTM in the year 2000 carried an American C band radar sensor and a German X band radar sensor by which large portions of the land mass of the earth were imaged at pixel sizes of 25 m to 30 m.

A new development is the construction of hyperspectral sensing devices, which, at lower resolutions, permit the scanning of the earth in more than 1000 narrow spectral bands to identify objects by their spectral signatures. In recent years, the focus has been increasingly on the development and methodology of digital remote sensing applications in a wide variety of settings, and at virtually every temporal and spatial scale imaginable.

4.5 Geographic Information Systems (GIS)

The field of geographic information systems (GIS) is concerned with the description, explanation, and prediction of patterns and processes at geographic scales. GIS is a science, a technology, a discipline, and an applied problem solving methodology. Geographic information systems arose from activities in four different fields (Figure 1.3).

- Cartography, which attempted to automate the manually dependent map-making process by substituting the drawing work by vector digitization.
- Computer graphics, which had many applications of digital vector data apart from cartography, particularly in the design of buildings, machines and facilities.
- Databases, which created a general mathematical structure according to which the problems of computer graphics and computer cartography could be handled.
- Remote sensing, which created immense amounts of digital image data in need of geocoded rectification and analysis.

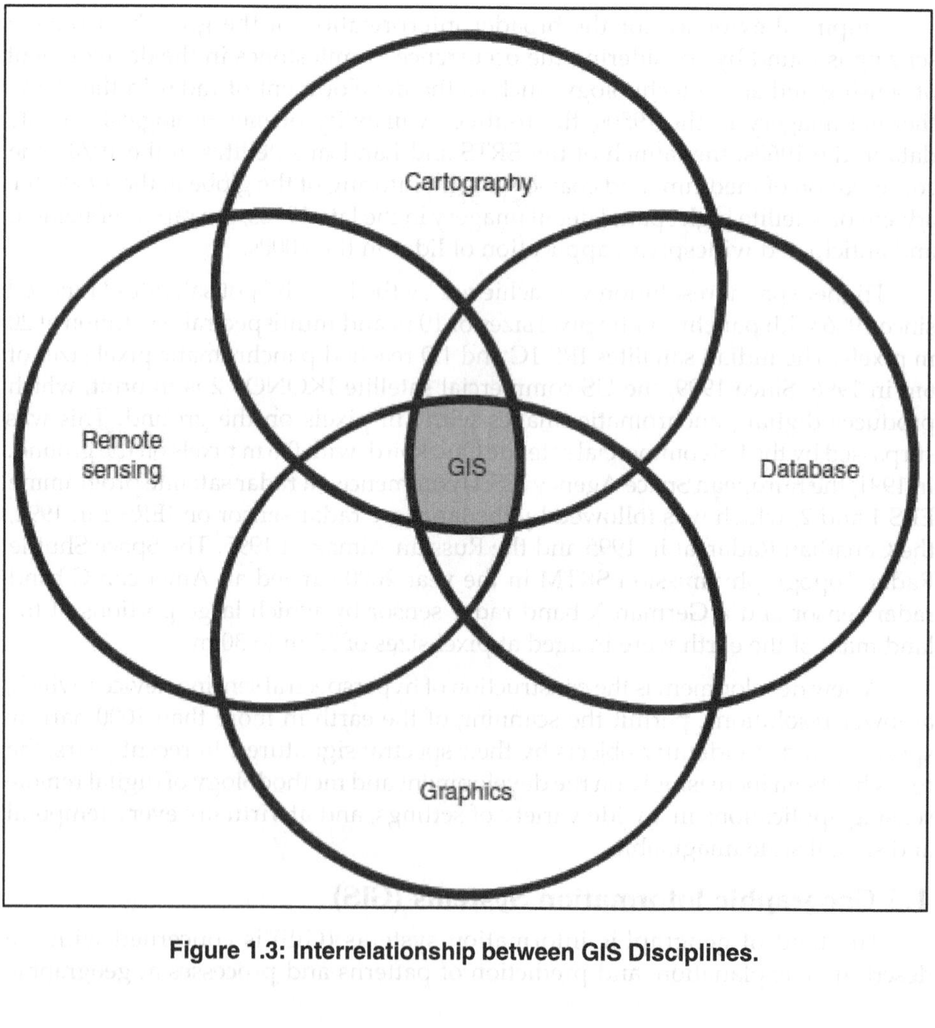

Figure 1.3: Interrelationship between GIS Disciplines.

GIS was developed parallel in North America, Europe, and Australia. The first theoretical and exploratory attempts for a GIS design started in the 1960s. The first GIS was the Canada Geographic Information System, designed in the mid-1960s as a computerized map measuring system. Later in late 1960s, Harvard University's Laboratory for Computer Graphics and Spatial Analysis developed a general-purpose GIS- a project that led eventually to the ODYSSEY GIS of the late 1970s. The UK Experimental Cartography Unit (ECU) pioneered high quality computer mapping in 1968. The first automated cartography developments occurred in the 1960s, and by the late 1970s most major cartographic agencies were already computerized to some degree. But the magnitude of the task ensured that it was not until 1995 that the first country (Great Britain) achieved complete digital map coverage in a database.

Remote sensing also played a part in the development of GIS, as a source of technology as well as a source of data. Many technical developments in GIS originated in the Cold War. The first military satellites of the 1950s were developed and deployed in great secrecy to gather intelligence, but the declassification of much of this material in recent years has provided interesting insights into the role played by the military and intelligence communities in the development of GIS.

As the developments in computing speed and computer storage grew rapidly in the 1980s and 1990s, it became obvious that an integration of industrial efforts for user requirements was mandatory. This led to the creation of the Open GIS Consortium (OGIS) which, along with the standards organizations ISO and CEN, led to the transferability of data created on different systems, to the integration of vector and raster data, to the establishment of meta databases and to the transmission of data in the Internet or Intranet. It was soon realized that data created and analysed for project purposes only was a duplication of effort, and that the tendency went into the creation of base data, which had to be continuously updated not only to provide an inventory of geospatial information, but to permit its crossreferenced analysis and to integrate the effort into a management system for sustainable development.

4.6 Global Positioning System (GPS)

The Global Positioning System (GPS) was developed by the U.S. Department of Defense (DOD), Ivan Getting, and the Massachusetts Institute of Technology (MIT). Originally consisting of 11 orbiting satellites, the GPS–official name NAVSTAR (Navigation System with Timing And Ranging)–was launched by DOD in 1978 strictly for military use. The idea for the system began much, much earlier. In fact the basic idea of GPS navigation can be traced back centuries to the first explorer who asked the question, "Where am I?" The technology of GPS history, however, first emerged in the 1950s.

In 1957, Russia launched a satellite (Sputnik) into space. After the launch, MIT researchers noticed that as Sputnik orbited the planet, its radio signal varied in strength. As it approached their position, the signal strength increased. When the satellite departed the MIT researchers' position, the signal strength decreased. From this increase and decrease of the radio signal, the MIT researchers could determine

Sputnik's exact orbit. This recognition that radio signals from a satellite, or "artificial star" could determine distinct positions on the ground was truly the launching pad of the GPS system.

Even before that, however, the seeds for the GPS technology were being planted. In 1951, Dr. Ivan Getting, a graduate of MIT student and a Rhodes Scholar from Oxford, put his Ph.D. in astrophysics to use at Raytheon. The Air Force requested a guidance system for a proposed ICBM travelling via railroad. In response to this request, Dr. Getting developed the first three-dimensional, position-finding system based on time difference of arrival. This system became the basis for the future GPS.

The deployment portion of the GPS history began in 1973 with the decision to develop a satellite navigation system based on existing technology of the U. S. Air Force and the U. S. Navy. The system went through extensive testing during the next three years. In 1977 the first transmitters were installed on the surface of the Earth and tested–even though no satellites have been launched yet. The transmitters were dubbed Pseudolites (pseudo satellites).

Between 1978 and 1985, eleven satellites were launched into space and put into position. In 1979, the decision was made to increase the number of satellites to 18. In 1980, the first Block I satellite was launched. This "bird' had sensors specifically designed to detect atomic explosions, and was placed in orbit as a means of monitoring the Soviet Union's compliance with the 1963 agreement with the United States to refrain from nuclear testing.

Also in 1980, the onboard atomic clocks, the most accurate timepieces in the world, were activated. Developed by physicists, atomic clocks measure time by the change in energy levels of electrons. These clocks are stable, continuous, and accurate to a nanosecond, or one-billionth of a second.

In 1983, the GPS ceased being solely a military system and was made available for public use. President Ronald Reagan declassified the GPS system in response to a tragedy involving a civilian airplane. Korean Airline flight 007 got lost over Soviet territory and was shot down by Soviet fighters. In 1986, the Challenger space shuttle tragedy halted space shuttle launches, and thereby delayed the GPS system, because shuttles were supposed to transport the new Block II satellites. Eventually, Delta rockets, the original transport, were put back into use to launch the satellites, and in 1988, the decision was made to increase the number of satellites to 24, because functionality with 18 satellites was still limited.

Though the GPS system was available for public use, the first Gulf War in 1990 saw temporary deactivation of use by the public, because the military needed more receivers. Public use returned in 1993 along with the decision that the GPS system would be available free of charge to the entire world. Full Operational Capacity (FOC) was achieved in July of 1995 with the placement and activation of the last of the 24 satellites.

Since 1995, GPS history has seen considerable technological advancement. More satellites have been put into orbit, increasing both availability and accuracy.

Previously GPS could locate a subject within 100 meters. That accuracy improved to finding subjects within 10-15 meters, and at present can be measured within centimeters in some cases. Availability improved from utilization in military and the position-required industries of aviation, nautical navigation, and land surveying, to include many personal applications. With the use of a fairly inexpensive receiver, employers use GPS to track their fleet of vehicles, parents use the system to keep track of small children and of disobedient teenagers, and pet lovers use it to keep track of their dogs.

5. Integration of Geoinformation Technologies

Figure 1.4 outlines the differences between the classical spatial information system approach and the new geospatial information concept.

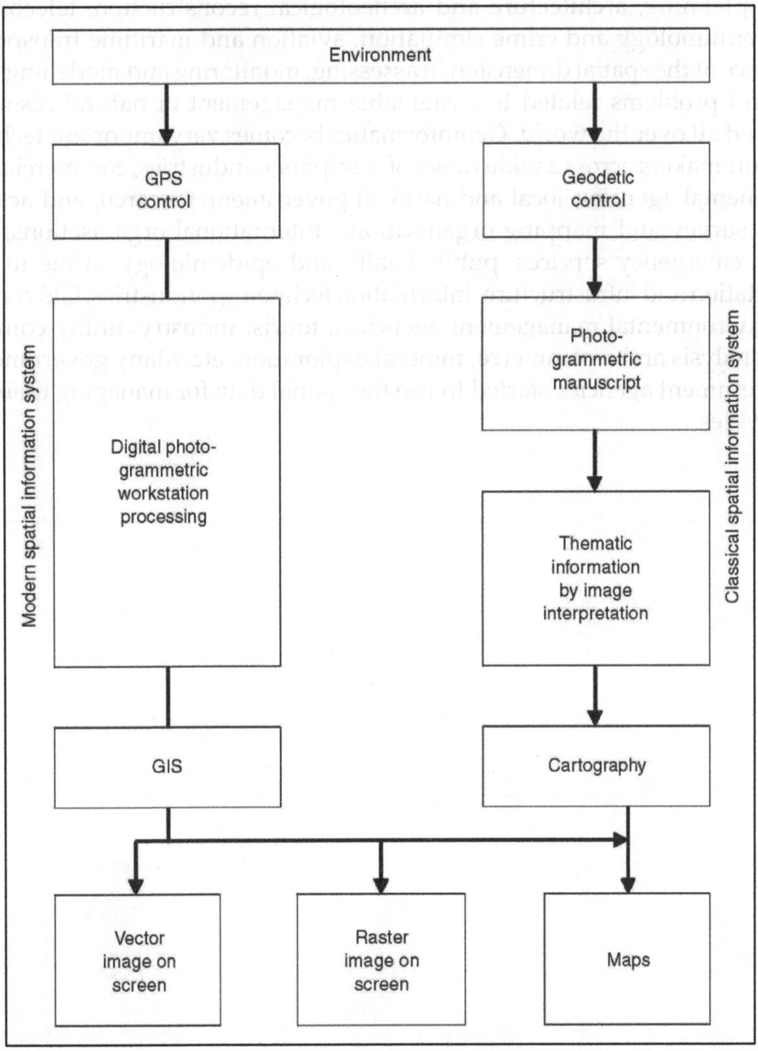

Figure 1.4: Classical and Modern Geospatial Information System.

In the past, the disciplines of geodesy, photogrammetry and cartography worked in an independent fashion to provide printable maps. In the new concept GPS, remote sensing, digital photogrammetry and GIS are able not only to produce printable maps, but to display raster and vector images on a computer screen and to analyse them in an interdisciplinary manner for the purposes of society.

6. Major Applications of Geoinformatics

Many fields benefit from geoinformatics, including urban planning and land use management, in-car navigation systems, virtual globes, public health, local and national gazetteer management, environmental modelling and analysis, military, transport network planning and management, agriculture, meteorology and climate change, oceanography and coupled ocean and atmosphere modelling, business location planning, architecture and archeological reconstruction, telecommunications, criminology and crime simulation, aviation and maritime transport. The importance of the spatial dimension in assessing, monitoring and modelling various issues and problems related to sustainable management of natural resources is recognized all over the world. Geoinformatics becomes very important technology to decision-makers across a wide range of disciplines, industries, commercial sector, environmental agencies, local and national government, research, and academia, national survey and mapping organisations, International organisations, United Nations, emergency services, public health and epidemiology, crime mapping, transportation and infrastructure, information technology industries, GIS consulting firms, environmental management agencies), tourist industry, utility companies, market analysis and e-commerce, mineral exploration, etc. Many government and non-government agencies started to use the spatial data for managing their day to day activities.

– *Chapter 2* –
PHOTOGRAMMETRY

Elements of Photogrammetry

Photogrammetry is the science of quantitative analysis of measurements from photographs. The word photogrammetry has been derived from three Greek words: ***Photos***-light; ***Gramma***-to drawn or written; ***Metron***-to measure. Photogrammetry is defined as the art, science and technology of obtaining reliable Information about physical objects and the environment through process of recording, measuring and interpreting photographic images and patterns of recorded radiant electromagnetic energy and other phenomena (American Society of Photogrammetry, 1998). As implied by its name, the science originally consisted of analyzing photographs. It is generally perceived as making measurements on photographs. It has been used for many years in mapping applications throughout the world. It provides an accurate and cost-effective alternative to conventional ground surveying techniques.

If one looks to a definition of surveying, it may state that surveying is the art and science of making measurements on, below, or above the surface of the earth. In this context, photogrammetry is a branch of surveying. Indeed, one can argue that photogrammetry, at one level, is a tool just like a Global Positioning System (GPS) receiver or a total stations.

Today, digital images are becoming the medium of choice for many photogrammetric applications. In fact, the distinction between remote sensing and photogrammetry has gradually become blurred as conventional photogrammetrist employ "non-conventional" imagery for mapping purposes. One of the main advantages of photogrammetry is that the photograph/image forms a permanent record of the situation at that instant in time.

Photogrammetry is a measurement technology in which the three-dimensional coordinates of points on an object are determined. This is achieved by stereoscopy, which is measurements made in two (or more) photographic images taken from different positions (different views). We say commonly photographs, but it may be also imagery stored electronically on tape or disk taken by video or CCD (Charge Coupled Device) cameras or radiation sensors such as scanners. The results can be:

- ☆ Coordinates of the required object-points
- ☆ Topographical and thematical maps
- ☆ Rectified photographs (orthophoto)

It's most important feature is the fact that the objects are measured without being touched. Therefore, the term "remote sensing" is often used instead of "photogrammetry". "Remote sensing" is a rather young term, which was originally confined to working with aerial photographs and satellite images. Today, it includes also photogrammetry, although it is still associated rather with "image interpretation". Principally, photogrammetry can be divided into range photogrammetry (mostly aerial photogrammetry) and close range photogrammetry (mostly terrestrial photogrammetry).

1. What is it Good for?

The applications of photogrammetry are widely spread. Principally, it is utilized for object interpretation (what is it? What's its type, quality, and quantity) and object measurement (where is it? What's its form and Size?). Aerial photogrammetry is mainly used to produce large scale topographical or thematical maps and digital terrain models (DTM), representing the terrain relief. It is also used to combine live action with computer generated imagery in movie post-production. Among the users of close-range photogrammetry are architects and civil engineers (supervising buildings, document their current state, deformations or damages), archaeologists, geophysicists, geodesists, surgeons or police departments (documentation of traffic accidents and crime scenes).

2. How Does it Work?

In principal, the 3D coordinates define the locations of object points in the 3D space. The image (photography) coordinates define the locations of the object points' images on the film or an electronic imaging device. The exterior orientation of a camera defines its location in space (translation) and its view direction (orientation). The inner orientation defines the geometric parameters of the imaging process. This is primarily the focal length of the lens, but can also include the description of lens distortions. Further additional observations play an important role in extracting these relations for 'solving' the images, such as: scale bars, basically a known distance of two points in space, and known fix points (anchor points), the connection to the basic measuring units is created.

Photography in its broadest sense is a process that converts the real 3-dimensional world into flat 2-dimensional images (Figure 2.1). The camera is the device that makes this transformation or mapping from 3-dimensions to 2-dimensions. Unfortunately, we cannot map the 3-dimensional world onto two dimensions completely so some information is lost (primarily the depth).

Whereas, photogrammetry in its broadest sense 'reverses' the photographic process described above. It converts, or maps, the flat 2-dieptmensional images back into the real 3-dimensional world. However, since information is lost in the photographic process, we cannot reconstruct the 3-dimensional world completely with just one photograph. As a minimum, we require two different photographs to reconstruct the 3-dimensional world. The more photographs we take and use the extra information in them we improve the process. The 3-dimensional coordinates we produce from the measurements of multiple photographs are the end result of photogrammetry.

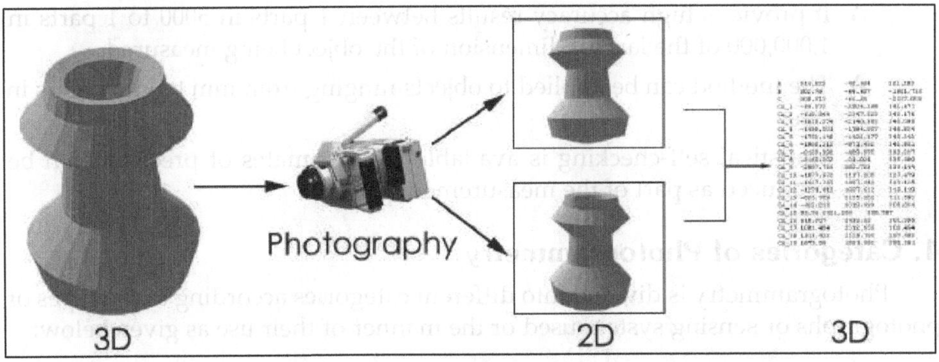

Figure 2.1: Concept of Photography and Photogrammetry.

Photogrammetry uses the basic principle of Triangulation, whereby intersecting lines in space are used to compute the location of a point in all three dimensions. However, in order to triangulate a set of points one must also know the camera position and aiming angles (together called the orientation) for all the pictures in the set. A process called Resection does this. Some camera calibration has to be done in order to define its errors and remove them from the process (some cameras are to self-calibrate). Usually these procedures are performed simultaneously in a process called the Bundle Adjustment.

The methodology is measuring common points that are identified on each image. A line of sight (or ray) can be 'constructed' from the spatial camera location (focal) to a point on the object. The intersection of these rays (aerial triangulation) that determines the three-dimensional location of each point measured. In principal, there are more sophisticated algorithms that can exploit other information about the scene that is known a priori (before the photographing procedure is done), for example symmetries, allowing the reconstruction of 3D coordinates from only one camera position (more rare). Mathematical algorithms for photogrammetry typically express the problem of minimizing the sum of the squares of a set of errors, mainly exist in the set of measurements (LSM-least squares method). Nowadays photogrammetric methods rely mainly on images produced from aerial or satellite imagery (or scans produced by digital cameras), and radar or laser produced data (*i.e.*, remote sensing).

3. What Does it do?

Besides conventional mapping, photogrammetry has been used for many different scientific tasks. It is an ideal technology when measuring features and phenomenon that are inaccessible for one reason or another. Additionally, the object to be measured may be too hot or cold, too soft or fragile, or too dangerous to measure my conventional means.

☆ Measures the 3-D location of points or features at one instant using between one and many cameras at the same time, or over a period of time using one camera.

- ☆ It provides high accuracy results between 1 parts in 5000 to 1 parts in 1,000,000 of the largest dimension of the object being measured.
- ☆ The method can be applied to objects ranging from mm to kilometers in size.
- ☆ Statistical self-checking is available and estimates of precision can be produced as part of the measurement process.

4. Categories of Photogrammetry

Photogrammetry is divided into different categories according to the types of photographs or sensing system used or the manner of their use as given below:

(i) On the Basis of Orientation of Camera Axis

- ☆ *Terrestrial or Ground photogrammetry:* When the photographs are obtained from the fixed and usually known position or near the ground with camera axis is horizontal or nearly horizontal. In terrestrial photogrammetry maps are prepared from terrestrial (or ground) photographs or terrestrial photogrammetry employees photograph taken form points on earth surface for measurement purposes. The terrestrial photographic surveying considered as the further development of plane table surveying.
- ☆ *Aerial photogrammetry:* Photographs of terrain in an area are taken by a precision photogrammetric camera mounted in an aircraft flying over an area. In **Aerial Photogrammetry** the camera is mounted in an aircraft and is usually pointed vertically towards the ground. Multiple overlapping photos of the ground are taken as the aircraft flies along a flight path. These photos are processed in a stereo-plotter (an instrument that lets an operator see two photos at once in a stereo view). These photos are also used in automated processing for Digital Elevation Model (DEM) creation.

(ii) On the Basis of Sensor System Used

Following names are popularly used to indicate type of sensor system used in recording imagery.

- ☆ Radargrammetry: Radar sensor
- ☆ X-ray photogrammetry: X-ray sensor
- ☆ Hologrammetry: Holographs
- ☆ Cine photogrammetry: motion pictures
- ☆ Infrared or colour photogrammetry: infrared or colour photographs

(iii) On the Basis of Principle of Recreating Geometry

When single photographs are used with the stereoscopic effect, if any, it is called monoscopic photogrammetry. If two overlapping photographs are used to generate three-dimensional view to create relief model, it is called stereophotogrammetry. It is the most popular and widely used form of photogrammetry.

(iv) On the Basis of Procedure Involved for Reducing the Data from Photographs

Three types of photogrammetry are possible under this classification:

- ☆ ***Instrumental or Analogue photogrammetry:*** In analog photogrammetry, optical or mechanical instruments were used to reconstruct three-dimensional geometry from two overlapping photographs (Figure 2.2). The main product during this phase was topographic maps.

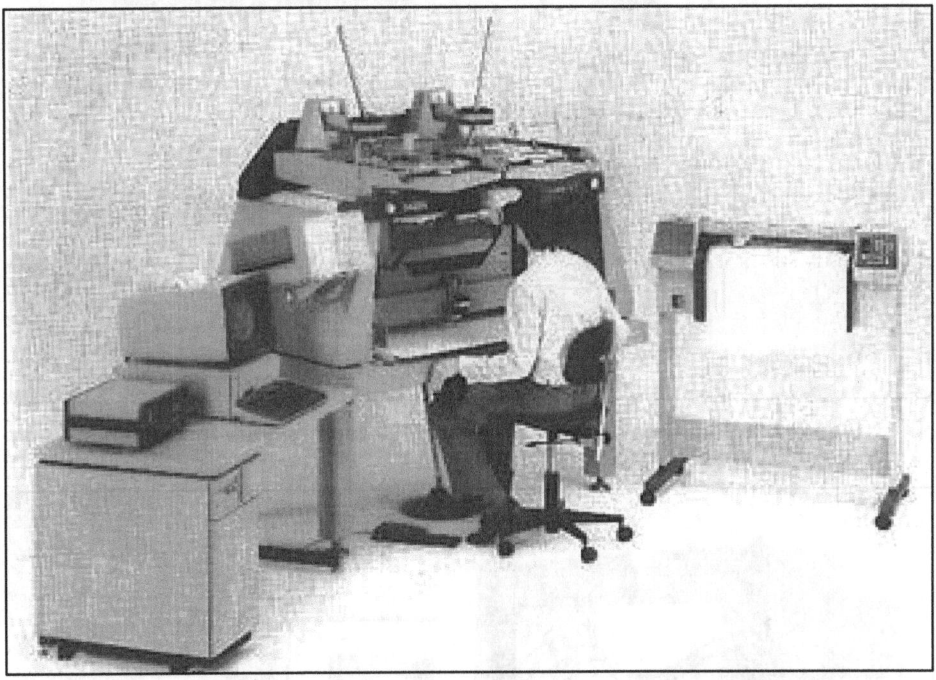

Figure 2.2: Analogue Photogrammetry Instrument.

- ☆ ***Semi-analytical or Analytical:*** Analytical photogrammetry solves problems by establishing mathematical relationship between coordinates on photographic image and real world objects (Figure 2.3). The computer replaces some expensive optical and mechanical components. The devices were analog/digital hybrids. Semi-analytical approach is hybrid approach using instrumental as well analytical principles. The 3D modelling is mathematical (not re-created) and measurements are made in the 2D images. Analytical aerotriangulation, analytical plotters, and orthophoto projectors were the main developments during this phase. The outputs of analytical photogrammetry can be topographic maps, but can also be digital products, such as digital maps and DEMs.
- ☆ ***Digital Photogrammetry or Softcopy photogrammetry:*** Digital photogrammetry is sometimes called softcopy photogrammetry. It uses digital image processing principle and analytical photogrammetry tools

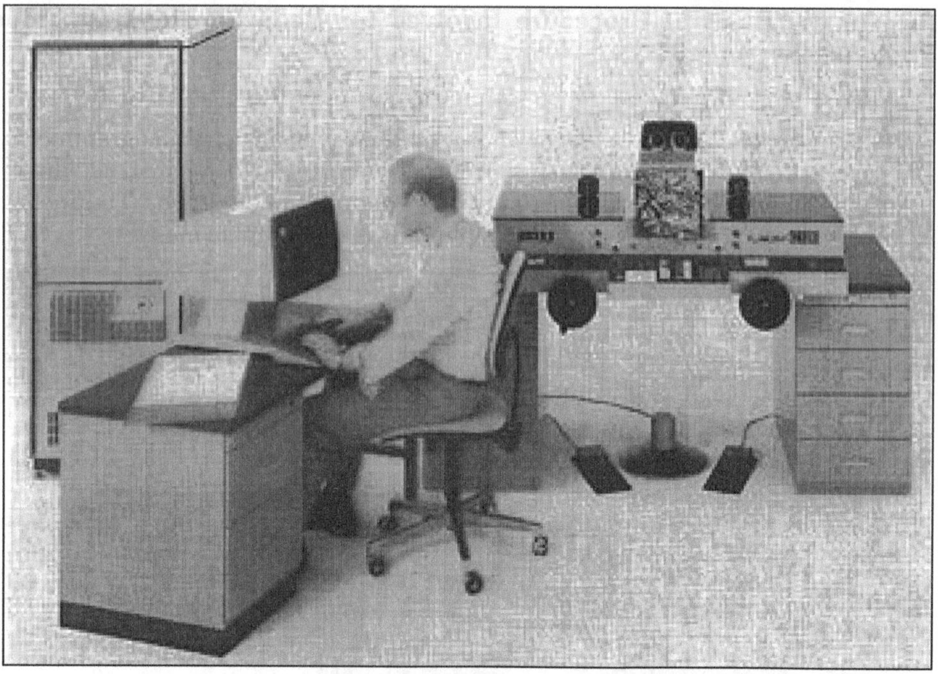

Figure 2.3: Semi-Analytical or Analytical Photogrammetry Instruments.

Figure 2.4. Digital photogrammetry.

to carry out photogrammetric operation on digital imagery. Digital photogrammetry is applied to digital images that are stored and processed on a computer (Figure 2.4). The output products are in digital form, such as digital maps, DEMs, and digital orthophotos saved on computer storage media.

(v) On the Basis of Platforms on which the Sensor is Mounted

If the sensing system is spaceborne, it is called space photogrammetry, satellite photogrammetry or extra-terrestrial photogrammetry. Satellite photogrammetry is a relatively new area of practice, especially among conventional photogrammetric companies. Historically, photogrammetry had been used to locate satellites in space in order to establish a more global network of control. With the advent of Global Positioning System (GPS) receivers, the use of photogrammetry with satellites has taken on a new role. Added to this is the emergence of very high-resolution imagery, on the order of one-meter or less. Thus, the future will see more utilization of this high-resolution 11 acre = 43,560 square feet imagery to both supplement and supplant conventional aerial imagery.

Out of various types of the photogrammetry, the most commonly used forms are stereophotogrammetry utilizing a pair of vertical aerial photographs (stereopair) or terrestrial photogrammetry using a terrestrial stereopair.

5. Photographing Devices

A photographic image is a "central perspective". This implies, that every light ray, which reached the film surface during exposure, passed through the camera lens (which is mathematically considered as a single point, the so called "perspective center"). In order to take measurements of objects from photographs, the ray bundle must be reconstructed. Therefore, the internal geometry of the used camera (which is defined by the focal length, the position of the principal point and the lens distortion) has to be precisely known. The focal length is called "principal distance", which is the distance of the projection center from the image plane's principal point. Depending on the availability of this knowledge, the photogrammetrist divides photographing devices into three main categories:

(i) Metric Cameras

These cameras have stable and precisely known internal geometries and very low lens distortions. Therefore, they are very expensive devices. The principal distance is constant, which means, that the lens cannot be sharpened when taking photographs. As a result, metric cameras are only usable within a limited range of distances towards the object. The image coordinate system is defined by (mostly) four fiducial marks, which are mounted on the frame of the camera. Terrestrial cameras can be combined with tripods and theodolites. Aerial metric cameras are built into aero planes mostly looking straight downwards. Today, all of them have an image format of 23 by 23 centimeters.

(ii) Stereometric Camera

If an object is photographed from two different positions, the line between the two projection centers is called "base". If both photographs have viewing directions, which are parallel to each other and in a right angle to the base (the so called "normal case"), then they have similar properties as the two images of our retinas. Therefore, the overlapping area of these two photographs (which are called a "stereo-pair") can be seen in 3D, simulating man's stereoscopic vision. In practice, a stereo-pair can be produced with a single camera from two positions or using a stereometric camera. A stereometric camera in principle consists of two metric cameras mounted at both ends of a bar, which has a precisely measured length (mostly 40 or 120 cm). This bar is functioning as the base, while both cameras have the same geometric properties. Since they are adjusted to the normal case, stereo-pairs are created easily.

(iii) 'Amateur' Cameras

The photogrammetrist speaks of an "amateur camera", when the internal geometry is not stable and unknown, as is the case with any commercial available camera. Photographing a test field with many control points and at a repeatable fixed distance setting (for example at infinity), a calibration of the camera can be calculated. In this case, the four corners of the camera frame function as fiducials. However, the precision will never reach that of metric cameras. Therefore, they can only be used for purposes where no high accuracy is demanded.

6. Photogrammetric Techniques

Depending on the available material (metric or non-metric camera, stereo-pairs, shape of recorded object, control information etc.) and the required results (2D or 3D, accuracy), different photogrammetric techniques can be applied. Depending on the number of photographs, three main categories can be distinguished:

(i) Mapping from a Single Photograph

This is only useful for plane (2D) objects. Obliquely photographed plane objects show perspective deformations which have to be rectified. The rectification can be neglected, only if the object is flat and the picture is made from a vertical position towards the object. In this case, the photograph will have a unique scale factor, which can be determined if the length of at least one distance at the object is known. There are some common techniques that can calculate this.

(ii) Stereo-photogrammetry

As the term already implies, stereo-pairs are the basic requirement here. These can be produced using stereometric cameras. If only a single camera is available, two photographs have to be made from different positions, trying to match the conditions of the "normal case", as in vertical aerial photographs. They are made using special metric cameras that are built into an aeroplane looking straight downwards. While taking the photographs, the aeroplane flies over a certain area in a way that the whole area is covered by overlapping photographs. The overlapping part of each stereo-pair can be viewed in 3D and consequently mapped in 3D using one of following techniques:

- ✰ ***Analogue***: the analogue method was mainly used until the 70ies. The method tries to 'convert' the recording procedure. Two projectors, which have the same geometric properties as the used camera (these can be set during the so called "inner orientation"), project the negatives of the stereo-pair. Their positions then have to be exactly rotated into the same relationship towards each other as at the moment of exposure ("relative orientation"). After this step, the projected bundle of light rays from both photographs intersect with each other forming a three dimensional optical model. At last, the scale of this model has to be related to its true dimensions and the rotations and shifts in relation to the mapping (world) coordinate system are to be determined. Therefore, at least three control points, which are not on one straight line, are required ("absolute orientation").

  The optical model is viewed by means of a stereoscope. The intersection of rays can then be measured point by point using a measuring mark. This consists of two marks, one on each photograph. When viewing the model, the two marks fuse into a 3D one, which can be moved and raised until the desired point of the 3D object is met. The movements of the mark are mechanically transmitted to a drawing device. In that way maps can be created (or actually – were created....).

- ✰ ***Analytical***: the first analytical plotters were introduced in 1957. From the 1970ies on they became commonly available on the market. The idea is still the same as with analogue instruments, but here, a computer manages the relationship between the image and the real-world coordinates. The restitution of the stereo-pair is done within three steps.

  After restoration of the "inner orientation", where the computer may now also correct for the distortion of the film, both pictures are relatively oriented. After this step, the pictures will be looked at in 3D.

  Then, the absolute orientation is performed, where the 3D model is transferred to the real-world coordinate system. Therefore, at least three control points are required. After completing these, any detail can be measured out of the stereo-model in 3D. Like in the analogue instrument, the model and a corresponding measuring mark are seen in 3D. The movements of the mark are under the control of the user. The main difference to the former analogue plotting process is that the plotter doesn't plot any more directly onto the map but onto the monitors screen or into the database of the computer. The analytical plotter uses the computer to calculate the real-world coordinates which can be stored as an ASCII file or transferred on-line into CAD-programs. In that way, 3D drawings are created, which can be stored digitally, combined with other data and plotted later at any scale.

- *Digital*: digital techniques have become widely available during the last decade. Here, the images are not on film but digitally stored on tape or disc. Each picture element (pixel) has its known position and measured intensity value (one for grayscale images, several for colour or multispectral images).

(iii) Mapping from Several Photographs

This kind of restitution, which can be done in 3D, has only become possible by analytical and digital photogrammetry. 3D objects are photographed from several positions, *i.e.*, more than two photographs are used. The photographs can be taken with different cameras and at different times (dependent that the objects photographed does not move). The technique is that first a bundle adjustment has to be calculated. Using control points and triangulation points the geometry of the whole block of photographs is reconstructed with high precision. Then the image coordinates of any desired object-point measured in at least two photographs can be intersected. The result is the coordinates of the required points. In that way, the whole 3D object can be digitally reconstructed.

7. Advantages of Photogrammetry

- Very precise
- Time effective
- Cost effective
- Based on well-established and tested algorithms.
- Less manual effort
- More geographic fidelity
- Corrects all sorts of distortions
- Provide a reasonable geometric modelling alternative when little is known about the geometric nature of the image data
- Provide an integrated solution for multiple images or photographs simultaneously
- Achieve a reasonable accuracy without a great number of GCPs
- Create a three-dimensional stereo model or to extract the elevation information

8. Applications

Photogrammetry is used in many different fields including topographic mapping, architecture, engineering, archeology, manufacturing, quality control, police traffic crash and crime scene investigations and geology, as well as meteorology. It is also used to combine live action with computer generated imagery in movie post-production. The new generation of 3D movies are good examples of the use of photogrammetry in film industry. The following description give an overview of various applications areas of photogrammetry (Rampal, 1982).

(i) **Geology:** Structural geology, investigation of water resources, analysis of thermal patterns on earth's surface, geomorphological studies including investigations of shore features.

- ☆ Engineering geology
- ☆ Stratigraphics studies
- ☆ General geologic applications
- ☆ Study of luminescence phenomenon
- ☆ Recording and analysis of catastrophic events
- ☆ Earthquakes, floods, and eruption.

(ii) **Forestry:** Timber inventories, cover maps, acreage studies.

(iii) **Agriculture:** Soil type, soil conservation, crop planting, crop disease, crop-acreage.

(iv) **Design and construction:** Data needed for site and route studies specifically for alternate schemes for photogrammetry. Used in design and construction of dams, bridges, transmission lines.

(v) **Planning of cities and highways:** New highway locations, detailed design of construction contracts, planning of civic improvements.

(vi) **Cadastre Cadastral problems** such as determination of land lines for assessment of taxes. Large scale cadastral maps are prepared for reapportionment of land.

(vii) **Environmental Studies** and Land-use studies.

(viii) **Exploration:** To identify and zero down to areas for various exploratory jobs such as oil or mineral exploration.

(ix) **Military intelligence:** Reconnaissance for deployment of forces, planning maneuvers, assessing effects of operation, initiating problems related to topography, terrain conditions or works.

(x) **Medicine and surgery:** Stereoscopic measurements on human body, X-ray photogrammetry in location of foreign material in body and location and examinations of fractures and grooves, biostereometrics.

(xi) **Miscellaneous:** Crime detection, traffic studies, oceanography, meteorological observation, Architectural and archaeological surveys, contouring beef cattle for animal husbandry etc.

9. Role of Photogrammetry in a GIS

One area of Geomatics that has had a major impact on the photogrammetry and remote sensing industry/profession has been the development of geographic information systems (GIS). A GIS is an information system designed to provide the decision-maker with the most current, accurate and complete data about phenomenon or events within an area. This is referred to as geoinformation when these events and phenomenon contain a spatial reference. There is growing importance in this technology as more and more decisions are made based upon this data. It is a part of the national, and in some cases international, infrastructure in an information society (Heipke, 2004).

The complexity of a GIS when compared to conventional information systems lies in the fact that we deal with geo-objects, which are objects that have a spatial and even temporal component. As Heipke (2004) points out, these objects are complex in that each has its own identity and can be described by different geometric, thematic, radiometric and temporal attributes. Furthermore, objects are controlled by both topological and geometrical relationships.

Heipke (2004) describes four major requirements for a modern geographic information system. Each of them draws heavily from photogrammetry and remote sensing in order to provide the system with the data needed to meet the user needs. These requirements are:

(a) The data should be topologically-based that uses an object model to define the features. It must also be able to accommodate changes in the data and to convey that change using multiple presentations. There are a couple of key issues in this requirement important from a photogrammetry and remote sensing point of view. First, topology pertains to relationships that exist between the data and many of these relationships are easily recognized and measured within imagery. The second issue is the method of describing the geospatial data. Two methods exist as described by Heipke [2004]: field-based and object based models. A field-based model consists of collecting continuous data over an area, such as contours. Normally, a digital terrain model (DTM) or digital elevation model (DEM) is obtained by observing the elevation, as an example, at discrete points over the site from which contours are derived using some type of interpolation process. The object based model will locate features and their attributes and then them into groups like roads, trees, buildings, etc. Imagery thus represents the major form of data acquisition within a GIS. In fact, Heipke (2004) has identified the role of imagery within a GIS as:

- ☆ Imagery is the prime means of geospatial data acquisition.
- ☆ Imagery is the ideal medium to use as a base or backdrop upon which map data can be overlaid to facilitate the user understands of the GIS data.
- ☆ As 3-D rendering increases, imagery is critical in providing a more realistic rendition of the conditions found at this site.

A third issue relates to the handling of change within an area. It is a well-documented fact that the geospatial data represents a major cost in GIS development and that this data needs to be maintained in order for the GIS to remain useful. Imagery is one of a number of strategies that can be used to monitor and assess change. A picture is s snapshot of what existed at an instant in time and thus images can be used as a base from which change can be monitored. Of course, there are also a number of other types of strategies that we utilize in the data maintenance area.

(b) Modern geographic information systems need to incorporate imagery within their software system and to provide the user with the tools to acquire data from the imagery and process that data as necessary. At one time, photogrammetry and remote sensing, and to some extent even GIS, were considered as separate disciplines. But today that paradigm is no longer valid and the boundaries used to define the extent of these fields are blurred. In fact, modern data acquisition involved the integration of a number of technologies including imagery (multi-temporal, multi-spectral, and multi-sensor), global positioning system positioning, inertial navigation, and collateral data contained in other geospatial databases.

(c) The workflow needs to be improved such that updating can be performed more efficiently through automation. Additionally, versioning needs to be incorporated in order to monitor change as it occurs over time. As we have seen, maintaining currency within the data is critical in the life cycle of a GIS. The goal is to be able to detect change, capture the data that describes this change and update the database in an automated fashion. While automation has begun to increase our productivity in this area, there is still a lot of work to be accomplished.

(d) There is a need for integration within our systems. Heipke (2004) talks about data integration or data fusion but integration also involves the means by which data are collected. Data fusion is a major problem because these data sets may employ different sensors, were produced for different purposes, and may contain data at different resolutions or accuracy levels. The goals, as Heipke (2004) points out, are:

- ☆ Being able to import data from one dataset to another from which useful information can be obtained,
- ☆ To be able to derive new thematic information by integrating two different datasets, and
- ☆ To automatically assess the quality of the data, make corrections to the data when necessary and to improve the overall quality of the database.

The other issue with integration is the bringing together of different data collection tools into a single system. There are many examples of this, one being laser scanning or lidar. Here, data are collected from GPS receivers, an inertial measurement unit and a laser scanner to obtain, as an example, terrain elevation data for a DTM. Our understanding of how each component works is essential in ascertaining the accuracy of the data. Equally important is to assess how the components work together as a system.

From the above, it is fairly evident that imagery collected for photogrammetry and remote sensing applications will continue to play an important and critical role in the development and enhancement of geographic information systems.

Evolution of Photogrammetry

Although photogrammetry employs photographs (or digital imagery today) for measurements, the concepts go back into history even earlier. In 1480, Leonardo da Vinci (Figure 2.5) wrote the following: "Perspective is nothing else than the seeing of an object behind a sheet of glass, smooth and quite transparent, on the surface of which all the things may be marked that are behind this glass. All things transmit their images to the eye by pyramidal lines, and these pyramids are cut by the said glass. The nearer to the eye these are intersected, the smaller the image of their cause will appear".

Figure 2.5: Leonardo da Vinci.

In 1492 he began working with perspective and central projections with his invention of the Magic Lantern (The Magic Lantern is the term coined to a device that acts very similar to the current day slide projector), although there is no evidence that he actually built a working model and some claim the device actually dates back to the early Greeks. The principles of perspective and projective geometry form the basis from which photogrammetric theory is developed. Many of da Vinci's artistic contemporaries contributed to the understanding of projective geometry from a graphical perspective.

Other scientists continued this work on projective geometry mathematically. For example, Albrecht Duerer, in 1525, using the laws of perspective, created an instrument that could be used to create a true perspective drawing. In 1759, Johan Heinrich Lambert, in a treatise "Perspectiva Liber", developed the mathematical principles of a perspective image using space resection to find a point in space from which a picture is made. The relationship between projective geometry and photogrammetry was first developed by R. Sturms and Guido Hauck in Germany in 1883.

The first photograph was obtained by Joseph Nicephone Niépce (1765-1833). The positive image Niépce required an eight-hour exposure. In 1837, Jacques Mandé Daguerre obtained the first "practical" photograph using a process called the Daguerreotype. Around 1840, the French geodesist Dominique François Jean Arago began to advocate the use of "photogrammetry", using the daguerreotype, in front of the French Arts and Science Academy.

1. Development Cycles

The developments in photogrammetry from around 1850, have followed four development cycles (Konency, 1985). Each of these periods extended about fifty years. These cycles include:

(a) Plane table photogrammetry (1850 to 1900),
(b) Analog photogrammetry (1900 to 1960),
(c) Analytical photogrammetry (1960 to present) and
(d) Digital photogrammetry, which is just beginning to be a presence in the photogrammetric Industry.

Konency's (1985) development cycles are based upon the economic theory of Kondratjew. Collier (2002) uses a different approach, dividing data collection developments in topographic mapping into 4 phases. The first phase was prior to World War I. This segment is best described by individualized experimentation into new techniques but with little change in the map making process. He also characterized this period as one of "considerable cross-national cooperation". The second phase comprises the period of World War I. Experimentation gave way to development of simple map making techniques geared towards mapping needs of the military. The 1920s comprise phase three. We see the introduction of the airplane as well as considerable technological advances in instrumentation. Mapping focus changes from military uses to civilian needs. As Collier points out, this period is also characterized as nationalistic with little international cooperation. Finally, phase four saw the beginnings of large-scale civil mapping. Photogrammetry matured from an experimental map making technology into a proven technique for accurate maps. While much has changed in the technology since the 1930s, Collier would probably argue that these are further maturation of what developed in the 1930s. He argues that the introduction of the slotted template and multiplex plotter changed the face of topographic mapping.

As Collier (2002) maintains, the status of topographic mapping was poor prior to 1900. The Ordnance Survey in Great Britain did have complete topographic map coverage of their country at a scale of 1:10,000. Belgium was similarly mapped that the same scale. Other countries had mixtures of scales and incomplete coverages. In the United States, only about 26 per cent of the country had topographic coverage. For towns and special projects the map scales were either 1:31,680 or 1:24,000 while settled rural areas had coverage at a scale of 1:62,500. The remaining portions of the U.S. with topographic map coverage had maps at a scale of 1:125,000 or 1:250,000. In Canada the situation was even worse.

Maps are critical elements for national development. When the East India Company obtained control of India, one of the first orders of business was to survey the country. Much of this need was driven by fiscal needs, although these topographic maps were used for other purposes. As the English Empire expanded into Africa, the necessity of maps became quickly apparent. The universal method of creating topographic maps in 1900 was by ground survey.

1.1 Plane Table Photogrammetry

In 1849, Aimé Laussedat (1819-1907) was the first person to use terrestrial photographs for topographic map compilation. He is referred to as the "Father of Photogrammetry". The process Laussedat used was called iconometry [*icon* (Greek) meaning image, *metry* (Greek) which is the art, process, or science of measuring]. In 1858, he experimented with aerial photography supported by a string of kites but abandoned it a couple of years later. In 1862, Laussedat's use of photography for mapping was officially accepted by the Science Academy in Madrid. He also tried balloon photography and is the first person to have captured an image from balloons, but deserted it because of the difficulty of obtaining a sufficient number of photographs to cover all of the area from one air station. At the Paris Exposition in 1867, Laussedat exhibited the first known phototheodolite and his plan of Paris derived from his photographic surveys. These maps were comparable to earlier maps compiled from conventional field surveys which showed that this new technology could be used for mapping.

Plane table photogrammetry is an extension of the conventional plane table surveying. Each exposure station was determined by resection and plotted on the plane table. The exposed photos were oriented on the plane table and the directions to the different objects were transferred onto the map sheets.

With the advent of photography and the ability to make exposures from the air, it was soon found that there were military applications to this technology. In 1855, Nadar (Gaspard Felix Tournachon) used a balloon at 80-meters to obtain the first aerial photograph. In 1859 the Emperor Napoleon ordered Nadir to obtain reconnaissance photography in preparation of the Battle of Solferino. In the Franco-Prussian War during the 1870s, the Prussian army installed a photo field detachment to obtain stereophotos, most notably of the fortification of Strasbourg.

Paulo Ignazio Pietro Porro (1801-1875) was an Italian geodesist and optical engineer. As a geodesist, he invented the first tacheometer (tachymeter) in 1839. In 1847 he was able to improve image quality of a lens system all the way to the edges by using three asymmetrical lens elements. He also developed an erecting lens imaging system in 1854. Porro developed a panoramic camera in 1858 that was equipped with a sighting telescope, compass, and level. The image was recorded on sensitized paper mounted on a cylinder. In 1865 he designed the photogoniometer. This development is significant in photogrammetry because of its application in removing lens distortion. His approach was to look at the image with a telescope through the camera lens. This concept was also independently considered by Carl Koppe (1884-1910). Therefore, this concept is called the Porro-Koppe Principle.

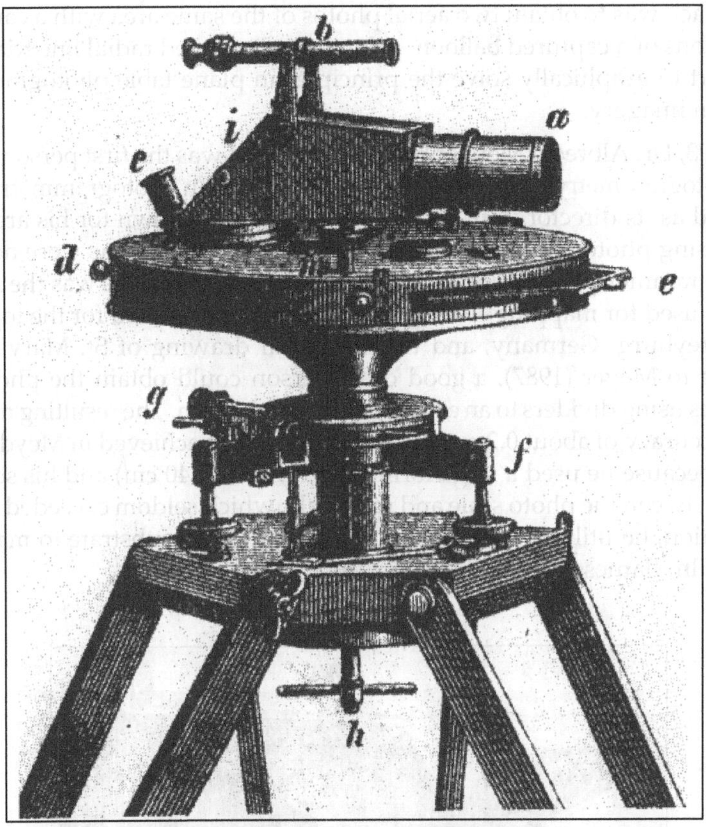

Figure 2.6: Chevallier's Photographic Plane Table.

In 1858, the Frenchman Chevallier developed a "photographic plane table" (Figure 2.6). This camera exposed the photo plate in a horizontal position. The light rays were deflected at right angles by using a prism attached to the sighting device that could be rotated.

Birdseye (1940) indicated that while pioneers were refining panoramic photography, there was no evidence that Porro, Chevallier, and others actually used these cameras for topographic surveys. Moessard, in 1884, not only developed a "cylindrographe" but also used it for mapping purposes. In 1885, George Eastman used nitrocellulose as a film base and later (1890) replaced the photographic dry plate for roll film. F. Stolze discovered the principle of the floating mark, which is used for stereoscopic measurements.

In the U.S., James Fairman was issued a patent for an "Apparatus for aerial photography" on August 2, 1887. This device used a camera supported by either a balloon or a kite with the shutter being operated with a clock mechanism. This was the first aerial photography captured from a balloon inthe U.S. On December 12, 1893, Cornele B. Adams was given a patent for his "Method of Photogrammetry".

His approach was to obtain two aerial photos of the same area with a camera from two positions of a captured balloon. Adams also invented radial line triangulation in an effort to graphically solve the principles in plane table photogrammetry to his balloon imagery.

In 1893, Dr. Albrecht Meydenbauer (1834-1921) was the first person to use the term "photogrammetry". He founded the Royal Prussian Photogrammetric Institute and served as its director until 1909. Meydenbauer is known for his architectural surveys using photogrammetry. Believing that current cameras were not suitable for photogrammetry, he designed his first camera in 1867. This was the first wide-angle lens used for mapping -105° Pantoshop lens. It was used for the topographic map of Freyburg, Germany, and the structural drawing of St. Mary's Church. According to Meyer (1987), a good draftsperson could obtain the photographic coordinates using dividers to an accuracy of about 0.1 mm. The resulting map would have an accuracy of about 0.2 mm. High accuracy was achieved in Meydenbauer's methods because he used a large format size (40 cm × 40 cm) and his selection of the ratio between the photo scale and mapscale, which seldom exceeded a factor of 2. In addition, he utilized mirror glass for the emulsion substrate to maintain the required film flatness during the exposure.

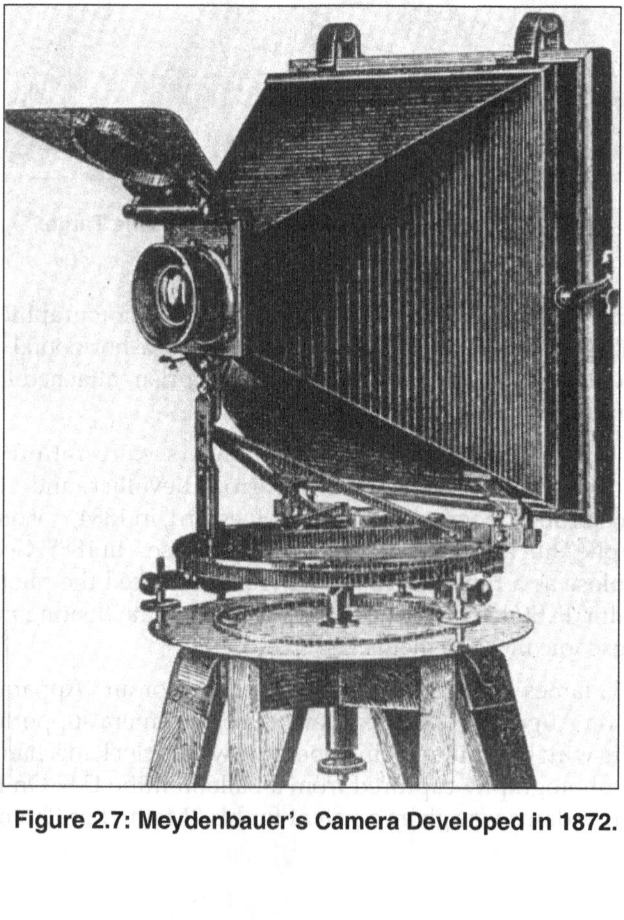

Figure 2.7: Meydenbauer's Camera Developed in 1872.

Meydenbaur continued to work on different camera designs. For example, in 1872 he developed two new cameras using a smaller 20 cm × 20 cm format (Figure 2.7). One had a focal length of 12 cm and the other 21 cm. The significance of this camera was that the plates were changed after each exposure. This was done using a light-tight container. To change the film, the camera was removed from the tripod and placed in the container.

1.2 Analog Photogrammetry

Two important developments were required to advance photogrammetry to its second phase - analog photogrammetry. First, stereoscopy was becoming widely used. Second, the development of the airplane by the Wright brothers in 1903. This provided a better camera platform than the terrestrial camera. Wilbur Wright has been identified as the first person to obtain aerial imagery from an airplane. He took movies of his flight in Centoci, Italy, on April 24, 1909. The first aerial photography captured from a plane for mapping purposes was done by Captain Cesare Tardivo for a 1:4,000 mosaic of Bengasi, Italy. He presented this survey in a paper delivered at the International Society of Photogrammetry meeting in Vienna on September 25, 1913.

In Canada, Edouard Deville (1849-1924), the Surveyor General of the Dominion, invented the first stereoscopic-plotting instrument called the Stereo-Planigraph in 1896 (Figure 2.8). While this was the first attempt to use stereo overlapping photos, the complexity of the instrument resulted in little use. Stereoscopy was achieved through a Wheatstone Stereoscope to create the stereomodel. Within the model, Deville used a tracer point to perform the mapping. He was very successful in the mapping of the Canadian Rocky Mountains and is referred to as "Canada's Father

Figure 2.8: Deville's Stereo Planigraph.

of Photogrammetry". Deville used a camera and theodolite mounted on the same tripod (non-stereoscopic photography). The theodolite was used to establish control while the camera was used for acquiring the photography for mapping. In order to transfer the imagery to the map, projective grids (called the Canadian Grid Method) were employed. In 1888, a single topographic party of four surveyors, under the chief J. J. McArthur, surveyed around 500 square miles of the Canadian Rockies.

The early pioneers like Meydenbaur advanced the ideas that were presented by Lausseat. Planimetric detail and contours were derived suing graphical methods directly from the photographs. The processes were difficult to perform and tedious. Thus, even though photogrammetry was being used for mapping purposes, they primarily found use in situations where conventional plane table mapping was either too difficult or impracticable to use. The U.S. Geological Survey began to use photogrammetry for topographic mapping in 1904. C.W. Wright and F.E. Wright took photos from the ground using a panoramic camera in their survey in Alaska. They used a commercial camera, modified by adding level bubbles and internal scales within the camera.

Theodor Scheimpflug (1865-1911), an Austrian, developed the theory of the double projector, which offered direct viewing of the projected images. He also presented a method of maintaining a sustained focus in rectification of photographs that is called the Scheimpflug Condition. In order to have sharp focus when the negative and easel (positive) planes are not parallel, the negative plane, easel plane and lens plane (plane perpendicular to the optical axis) must intersect along one line. This perspective transformation was used in the development of the "Perspektograph" in 1903. Scheimpflug introduced the concept of radial triangulation and is sometimes considered the initiator of aerial photogrammetry since he was the first to successfully use aerial photographs for practical mapping.

According to Birdseye (1940), Scheimpflug also worked with kites but moved to balloons and later to dirigible balloons. While Laussedat was unsuccessful with balloons, Scheimpflug developed a multi-lens camera in 1900 that consisted of a central vertical lens surrounded by seven oblique lenses. This gave him some very wide angle composite photography with which to work. One of the issues with balloons is the effect of wind on the flight line. The dirigible balloon gave the pilot better control over the flight direction. This led to the expansion of aerial photogrammetry in Europe.

Beginning in 1899, the German Sebastian Finsterwalder began to publish papers on analytical photogrammetry. When studying the resection of single and double points in space, he showed the existence of a critical surface for single-point resection. Using vector terminology, Finsterwalder showed the analytical conditions that must be met for rays to intersect. In 1899, Finsterwalder described the principles of modern double-image photogrammetry and the methodology of relative and absolute orientation. In addition, he introduced the necessity of redundant rays to recreate the proper geometry and used least squares theory to describe the relationship of the vectors between corresponding rays.

Figure 2.9: Dr. Carl Pulfrich.

In 1901, Dr. Carl Pulfrich (1858-1929) (Figure 2.9), a German physicist, designed the first stereocomparator employing x and y coordinate scales and presented the results at the 73rd Conference of Natural Science and Physicians in Hamburg. This was the first photogrammetric instrument manufactured by Zeiss. Pulfrich is sometimes referred to as the "Father of Stereophotogrammetry".

At about the same time, Dr. Henry George Fourcade (1865-1948), from South Africa, independently developed a similar stereocomparator. The main difference is that Fourcade utilized grid plates instead of x and y coordinates. He was also the first to discuss the need for reseau in a surveying camera for use with the measuring stereoscope. Because of the independent development, many refer to the stereocomparator as the Pulfrich-Fourcade stereocomparator.

In Germany, Eduard von Orel (1877-1941), in 1908, developed the first stereoautograph. This plotter was significant because its construction principles made terrestrial photogrammetry practical in mountainous areas. It allowed the operator to trace elevation contours directly. The first prototype, using the Pulfrich stereocomparator, was built by the Zeiss Works and employed the geometrical linkage referred to as the "Zeiss parallelogram". The "Orel-Zeiss Stereoautograph", developed by Pfeiffer and Bauersfeld provided continuous operation without the need for computations required of other plotters. W. Bauersfeld, the chief engineer at Zeiss Works, produced the first stereoplanigraph, which is a universal direct projection stereo instrument. Numerous other individuals made significant contributions to photogrammetric theory and instrumentation. These include Max Gasser who built a double projection plotter for vertical photography (this is the forerunner of the Multiplex plotter) and the Italian Umberto Nistri (1895-1962) who created a double projectionplotter using alternating image projection.

Frederick Vivian Thompson also developed his stereoplotter in 1908. Thompson's stereoplotter received weak acceptance in Great Britain, although it was employed by the Survey of India for the Karakorum Range mapping. Both plotters suffered from the fact that they could not be adapted for aerial photography since each instrument required coplanarphotography, a situation that rarely existed in aerial photography.

Professor Reinhard Hugershoff (1882-1941) was a surveyor and photogrammetrist who contributed much to the development of surveying and mapping instrumentation. He created the first analog plotter in 1921 called the Hugershoff Autocartograph. It was a very complex mechanical plotter that incorporated two photogoniometers. The instrument could be used to map both planimetric features and contours. It was universal in that it could be used for terrestrial, vertical aerial, oblique, and convergent photography. He also developed an aerial camera utilizing glass plates that was used to obtain oblique photography by hand or by attaching the camera on the side of the aircraft. When cameras began

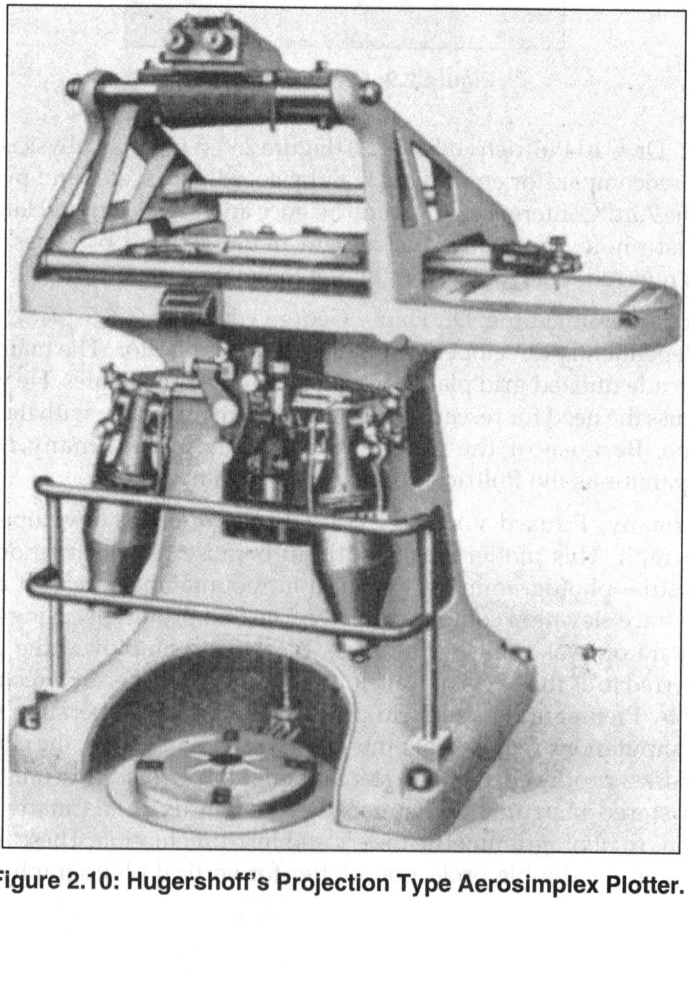

Figure 2.10: Hugershoff's Projection Type Aerosimplex Plotter.

to be employed through a hole in the fuselage, Hugershoff developed a ring mount that could be corrected for drift. Hugershoff also developed a projection plotter where the imagery was projected from below on a light table (Figure 2.10). Stereoscopy was achieved with a mirror stereoscope.

Many of Hugershoff"s instruments were used in major engineering projects. In an attempt to solve the problems of size in his autocartograph, Hugershoff developed the Aerocartograph in 1926. This instrument used space rods instead of the complex mechanical system used in his earlier instrument. It also had the capability of changing the optical path allowing the operator to view the left photograph with the right eye and right photography with the left eye. This gave the photogrammetrist the ability to perform analog aerotriangulation. Hugershoff. Later in life he worked with Zeiss Works in developing other photogrammetric equipment.

Otto von Gruber (1884-1942) derived, in 1924, the projective equations and their differentials, which are fundamental to analytical photogrammetry. At the time of this development, the process was very complex and cumbersome. His method of relative orientation of a stereoplotter makes the process of orientation easier and quicker. This procedure is still in use today and the six model points where parallax is cleared in the model are often referred to as the von Gruber points.

In a similar vein, Earl Church (1890-1956) also contributed to the theory of analytical photogrammetry. He developed the analytical solutions to space resection, orientation, intersection, rectification, and control extension using direction cosines. There are two main criticisms of Church's approach to analytical photogrammetry. His approach is based on unique solutions were there were no redundant control. Hence, one bad control point would invalidate all of the subsequent calculations. Additionally, he never performed any error analysis to his procedure. Because of this, problems occurred with real photography. Church, a professor at Syracuse University and one of the founding members of the American Society of Photogrammetry, is referred to as the "American Father of Photogrammetry".

Dr. Bertil Hallert from Sweden is best known for his investigation into errors, stereoplotter orientation procedures, and standards for plotter calibration. During the early part of the twentieth century, many of the figures in analog stereoplotter manufacturing began to develop their unique brand of instrument. Heinrich Wild, who had already made significant advances in surveying instrumentation, developed the "Autograph". This instrument, which was used for terrestrial mapping, was modified in 1926 to accommodate aerial photography.

In Italy, Professor Santoni, who was at the Officine Galileo, developed the Autoreductor in 1920. This was followed by the Stereocartograph in 1925. In the ensuring years, numerous stereoplotters were introduced into the photogrammetric industry. The production of the C4 by Zeiss in 1930 is important because it is the first Zeiss plotter that could be used with both terrestrial and aerial photography. Stereoplotters were also being developed by the French. G.J. Poivilliers designed the Stereotopograph in 1919. This began a series of stereoplotters that were used by the French Central Mapping Agency. In 1933, R. Feber developed a direct projection plotter that used alternating image projection.

In the early days of aerial mapping, the photographer would simply point the camera toward the area to be photograph and snapped the picture. Simple pointing mechanisms were attached to the camera for aiming purposes to ensure that the desired area was captured on film. It was quickly learned that vertical photography was better than oblique. Thus, mounts were attached to the sides of the aircraft to fix the orientation of the camera.

Early cameras used glass plates and dark slides to obtain an exposure. To make the process simpler, magazines were developed so that the photographer could take a series of pictures during the flight. Early American photogrammetric pioneers included the brothers Arthur Brock Jr. and Norman H. Brock who, along with Edward H. Cahill, developed aerial cameras and plotting instruments. They were the first to create an aerial camera that was mounted in the plane instead of holding the camera over the side. This camera was very unique and was geared to meet the military needs in war. The camera used the ferrotype in order to provide field commanders with imagery as soon as possible. An electric motor was employed to drive the camera and a gyroscope was incorporated to keep the lens axis in a vertical direction. The motor moved the ferrotype from the magazine to the focal plane. Once the exposure was made the ferrotype was placed in a magazine that contained developer. The idea behind this design was that upon landing the pilot would take the exposed ferrotypes and place them into fresh water to stop the developing. The camera utilized a 12" focal length with an f-stop of 4.5 with a 4" × 5" image format. The final product did not include the gyroscope since it did not accurately hold the camera in a vertical direction. Moreover, the ferrotypes were replaced with cut film.

In 1916 they developed another camera utilizing roll film. It could take 100 pictures per roll (4 ½" wide × about 560" long) (Figure 2.11). To hold the lens in a vertical direction, it was pendulously held in a gimbal and dampened to minimize oscillations. The camera was linked to the instrument panel for the pilot to operate. They could take anywhere between 3 and 10 photographs per minute. One of the significant events associated with this camera is that it was the first war pictures taken by the U.S. Army (war with Mexico).

One of the true giants in photogrammetry in the American private sector was Sherman Mills Fairchild (April 7, 1896–March 28, 1971). He was an entrepreneur with a wide range of interests who became a very successful businessman (Doyle, 1980). His inventiveness was demonstrated during World War I when he discovered the problems of image distortion on aerial cameras that were in use at the time. He developed a between-the-lens shutter using a rotary blade. This improved the quality of the image because of the sharper definition without the resulting distortions. This development so impressed Billy Mitchell that he used it to record his battleship bombing tests.

But, as the war ends, the military loses interest in the camera. This left a lasting impression on Fairchild. In the future he will create a business organization as he begins any new technological development.

Figure 2.11: USC and GS 9-lens Camera.

In 1920, Sherman Fairchild formed the Fairchild Aerial Camera Corporation where he developed the K-3 camera and its successors. To demonstrate the advantages of this camera, he produced an aerial map of Manhattan Island by mosaicking 100 aerial photos. The camera was mounted in the airplane at a time when aerial photos were being acquired at the side of the aircraft. Soon after this, he created Fairchild Aerial Surveys to provide photogrammetric services. Fairchild cameras represented an industry standard for a number of years. They were used in the Explorer II balloon flights and in the Apollo missions for mapping the moon's surface. Fairchild soon recognized that the biplanes being used in aerial mapping were lacking in their design for accurate photogrammetric applications. He then forms the Fairchild Airplane Manufacturing Corporation. The result is a monoplane that offers better visibility for mapping along with a heated cabin. The FC-1 had a high wing over the cabin and also employed folding wings for easy storage. The production model, called the FC-2, made Fairchild the leading monoplane manufacturer known for its incorporation of the newest technologies available.

Another pioneer aviator and photogrammetrist was Talbert "Ted" Abrams (1896-1990). As a young man, he joined the U.S. Marine Aviation Section. Upon discharge, he and his wife Leota formed the ABC Airborne Corporation, which was later renamed Abrams Aerial Survey Corporation. Over the years the company grew to be one of the leading photogrammetric firms in the country. Recognizing that airplanes were not designed with aerial mapping in mind, Ted Abrams developed the Explorer in 1937 (Figure 2.12). The plane consisted of a large "glass nose, twin tail booms, a pusher engine, and a tricycle landing gear". In 1937, the Abrams Instrument Corporation was created to manufacture camera parts and accessories. In addition, Ted Abrams served as an advisor in military and created the Abrams School of Aerial Surveying and Photo Interpretation.

In 1936, Robert Ferver, from France, was awarded a U.S. patent for the Gallus-Ferber Photorestituteur which was the first orthophoto production instrument although it was not used much because it was not economical. This plotter that was used to raise or lower one of the projectors. The movable projector did not have a filter attached to it.

In the U.S., Russel Kerr Bean made significant contributions in photogrammetric instrumentation. In 1956 he was awarded a patent for an "Ellipsoidal Reflector Projector for Stereo-Photogrammetric Map Plotting" known as the ER-55 and it was used by the US Geological Survey (USGS) during the 1950s and 1960s (Figure 2.13). This plotter accepted stereoscopic imagery not only from vertical photography but also convergent low oblique and transverse oblique photography. This plotter was later manufactured by Bausch and Lomb Optical

Figure 2.12: Ted Abrams Standing Next to the Explorer Airplane he Developed.

Figure 2.13: ER-55 Plotter Developed by Bean at the USGS.

Co. as the Balplex plotter. In 1959 he was awarded a patent for an orthoscope. This instrument produced photography at the same level of accuracy as a map at a fraction of the cost. Bean's orthophotoscope was different from Farber's instruments in that it employed a movable film surface. The surface had a slot on it, which allowed for movement–the X and Y directions. Bean also developed a camera calibration system consisting of a bank of collimators that also projected resolution targets onto the film.

Harry T. Kelsh (1889-1979) made an important contribution to photogrammetric instrumentation in the development of the Kelsh stereoplotter in 1945. This optical projection plotter offered private photogrammetrists the opportunity to perform accurate mapping without the expense that was required for the European stereoplotters.

Instrument makers have had significant influences on photogrammetric developments. In 1819, Kern of Aarau, Switzerland, was founded and began manufacturing precision surveying and mapping instruments. Kern introduced the highly popular PG2 analog stereoplotter. Over 700 of these instruments were sold worldwide. In 1980, Kern introduced the DSR1 analytical stereoplotter.

One of the early leading photogrammetry manufacturers was Zeiss. It's dominance in the photogrammetry industry in the early part of the 20th century is due to the fact that many of the early pioneers were employed by the company. These individuals include von Orel, Pulfrich, Walter Bauersfeld, Willi Sandor, and von Gruber. Some of the milestones at Zeiss in photogrammetry include:

- ☆ 1901 Zeiss' first photogrammetric instrument, the Stereocomparator, made using the design by Pulfrich.
- ☆ 1921 Stereoplanigraph C1 produced
- ☆ 1930 C4 went into production

Wild Heerbrugg was founded in 1921. They became a world leader in the manufacture of accurate surveying and mapping instruments. Their A8 and B8 Aviograph stereoplotters were very successful analog instruments with over 2,000 sold worldwide.

In 1988, Kern and Wild merged and eventually formed Leica in 1990. Using the expertise from both companies, the SD 2000 analytical plotter was launched in 1991. In 2001, Leica acquired Azimuth Corporation, ERDAS, and LH Systems giving Leica the capabilities of offering clients LIDAR scanning systems, remote sensing/ image processing software packages, and digital stereoplotter capabilities. Some of the major milestones for Wild in photogrammetry are given as follows:

- ☆ 1921–Wild is founded by three individuals: Jacob Schmidheiny, Robert Helbling and Heinrich Wild.
- ☆ 1922–First phototheodolite.
- ☆ 1925–First Universal Autograph is manufactured.
- ☆ 1926–Wild begins production of the A2 Autograph.
- ☆ 1927–First C1 aerial camera manufactured.
- ☆ 1928–Begin production of the C2 and C3 aerial cameras.
- ☆ 1932–Wild manufactures theE1 Rectifier, A3 Autograph and a mirror stereoscope.
- ☆ 1933–C12 Stereometric Camera and A4 Autograph being to be built.
- ☆ 1937–Wild introduces the A5 Autograph and RC Automatic Aerial Camera.
- ☆ 1938–Wild manufactures the A6 Second-Order Stereoplotter and S3 stereoscope. It also reigns to design the BC2 Ballistic Camera.
- ☆ 1942–The RC5 Automatic Aerial Camera and ST1 and ST2 mirror stereoscopes are produced.
- ☆ 1948–The Aviotar high-precision photogrammetric lens is produced.
- ☆ 1949–Wild produces the A7 Autograph.
- ☆ 1950–The A8 Autograph and RC7 aerial camera with Aviotar lens are manufactured.
- ☆ 1952–The BC4 Ballistic Camera and the Aviogon lens are produced.

- ☆ 1954–The RC8 Aerial Camera with the Aviogon lens is produced.
- ☆ 1955–Wild produces the STKL Precision Stereocomparator that is used in analytical photogrammetry.
- ☆ 1957–The A9 Autograph, RC9 Aerial Camera, U3 Diapositive Printer and VG1 Enlarger are produced. The RC9 is a super wide-angle camera and the A9 is the plotter designed to accommodate the super wide-angle photography.
- ☆ 1958–Wild begins manufacture of the B8 and B9 Aviograph stereoplotters and the E3 Rectifier.
- ☆ 1962–The BS Stereomat, C120 and C40 Stereometric camera, U9 Fixed-Ratio Printer, E4 Rectifier-Enlarger and the U4 Diapositive Printer enter the photogrammetric marketplace.
- ☆ 1963–Wild introduces the Universal-Aviogon lens which is corrected for visible and infrared light wavelengths.
- ☆ 1964–The A40 Autograph is rolled out for the ISP Congress.
- ☆ 1968–Wild introduces the A2000 Stereomat at the ISP Congress and begins production on the EK8 Coordinate Printer, RC 10 Universal Film Camera with Super-Aviogon 11 lens, A10 Autograph and ST10 Strip Stereoscope.
- ☆ 1971–The B8S Aviograph, P32 Terrestrial Camera and APK1 Panorama Camera are manufactured.
- ☆ 1972–New models of A8 Autograph and P31 and P32 terrestrial cameras introduced.
- ☆ 1976–The OR1 Avioplan begins production.
- ☆ 1980–Wild begins manufacturing of the AC1 Aviolyt and TA2 Aviotab.
- ☆ 1982–Wild BC1 begins delivery.

The historical developments discussed up until now involved terrestrial and aerial photogrammetric techniques. One of the unique challenges facing the scientific community was an understanding of life and features under water. Dimitri Rebikoff was born in 1921 in Paris, France. He is credited with a number of pioneering works in underwater photography including the electronic flash, film cameras and underwater photogrammetry. Collaborating with Prof. Ivanoff, Mr. Le Grad and Mr. Cuvier, he develops very accurate photogrammetric correction lenses for underwater applications. While photogrammetric instruments were important in the development of the mapping industry, photogrammetry was dependent upon survey control to perform orient the photography and perform the necessary intersection.

Control was an expensive and labor intensive operation, especially for projects encompassing large geographic areas. Numerous attempts to reduce these costs were made over the years. One such idea was the ABC (Airborne Control) Survey System invented by personnel at the U.S. Geological Survey (USGS), including Hugh "Red" Loving (1921- 1982) who was granted a patent for the Hoversight used in the helicopter. This survey control method was good for lower order (fourth-order horizontal and vertical control) surveys which were sufficient for USGS topographic surveys. A helicopter would hover over a control point. The pilot would maneuver over the point using the Hoversight were directly over the point, would radio to survey personnel who were set up over at least two control points whose positions were known. They would measure the distance as well as the horizontal and vertical angles to the helicopter. On board the helicopter, a plumb line was lowered to the control point to measure the height of the helicopter over the control point.

An important data collection tool, the camera is the main data acquisition tool. Its geometry needs to be recreated within the photogrammetric process to yield accurate spatial data. While many of the camera manufacturers tested and calibrated cameras, the U.S. Geological Survey began to calibrate aerial cameras in the late 1950's for contractors involved in USGS projects. Other federal agencies, both in and outside the military, also performed calibration services. In 1973, responsibility for camera calibration outside of the Department of Defense was transferred from the National Bureau of Standards to the USGS.

1.3 Analytical Photogrammetry

Much of the basic principles upon which modern analytical photogrammetry are based can be attributed to early photogrammetric pioneers. Most notable include the work by Finsterwalder, von Gruber, and Church. In 1947, Ralph O. Anderson, working for the Tennessee Valley Authority, developed a semi-analytical approach for analytical control. The orientation of the photos was done semigraphically. Once the pictures are fixed, control extension was completed analytically.

In the early 1950s, Everett Merritt published works on analytical photogrammetry. He worked for the Naval Photographic Interpretation Center in Washington, D.C. He developed a series of analytical solutions for camera calibration, space resection, interior and exterior orientation, relative and absolute orientation of stereo airs, and finally analytical control extension. While sharing similarities to Church's work, Merritt's approach was more complete, handling a number of different problems. Incidentally, Merritt later served as a member on the photography evaluation group for the Warren Commission on the Kennedy assassination.

The invention of the computer made significant advances to photogrammetric developments after 1950. It is responsible for the third cycle of photogrammetric development–analytical photogrammetry. In 1953, Dr. Hellmut Schmid, at the Ballistic Research Laboratory, Aberdeen, Maryland, developed the principles of modern multi-station analytical photogrammetry using matrix notation. The important features of his approach included a rigorously correct least squares solution, the simultaneous solution of any number of photographs, and a complete study of error propagation.

Dr. Paul Herget at the Ohio State University developed a new approach to analytical control extension using vector notation (similar to the approach used by Finsterwalder). A solution was obtained by minimizing the perpendicular distances

Figure 2.14: BC4 Satellite Triangulation Camera Station Operated by the U.S. Coast and Geodetic Survey.

between corresponding image light rays to points on the ground. The result is that the equation for computing the ground control points had the same form as the equation used in describing the relative orientation. Herget's method was the foundation for the Direct Geodetic Restraint Method used by the U.S. Geological Survey. From the National Research Council in Canada, G. H. Schut used the coplanarity concept to analytical triangulation. While an advocate for a simultaneous block adjustment, Schut recognized the limitations of computer technology at the time and, therefore, developed a cantilever strip adjustment. This was achieved by computing, first, the relative orientation of each photograph. Next, each photograph was oriented along the strip and finally the strip coordinates were adjusted to the ground control.

Duane Brown (1929-1994) is also responsible for continued work in analytical photogrammetry while working with Schmid and later in private industry. After receiving his undergraduate degree in Mathematics from Yale University in 1951, Brown joined Schmidt at the Ballistic Research Laboratory where he was involved in geodesy using the ballistic cameras to determine the orbital path of satellites. He moved to the RCA Missile Test Project in 1955 where he developed new approaches to camera calibration and the mathematical formulation of the bundle adjustment. This is significant because it involved a simultaneous solution of the exterior orientation parameters of the camera and the coordinates of the survey points along with the interior orientation and systematic radial lens distortion. Prior to this, the exterior orientation parameters were developed independently, and then the photogrammetric intersection of the ground points was determined. His work on first order partitioned regression pioneered this algorithm to many different areas of photogrammetry and geodesy (Figure 2.14).

In 1961, Duane Brown joined the Instrument Corporation of Florida and two years later purchased the Research and Analysis Division, where he was the Director, and formed DBA (Duane Brown and Associates). DBA soon established itself as a leader in close-range photogrammetry and analytical photogrammetry. Brown continued to refine the bundle adjustment for large photogrammetric blocks to include self-calibration.

The importance of self-calibration is that the accuracy and reliability of the photogrammetric adjustment improved. He recognized that the environment has an effect on the process and by calibrating the camera to extract the camera parameters in the environment in which the photography was acquired, the new camera parameters were a better representation than those derived from conventional calibration procedures. In 1962 he was able to apply his principles to the survey of the radio telescope at Greenbank, WV. Using a modified ballistic camera, Brown was able to achieve accuracies at around the 1:50,000 level, or about 2 mm.

During his stay at DBA, the company developed a number of high-accuracy, large-format, and close-range photogrammetric cameras. His work in photogrammetry also included adjusting for decentering distortion, principal point calibration, and film unflatness where he was able to show considerable deformation of glass plates. Some of the work undertaken by DBA involved mapping the moon for the Apollo program. In 1977 he founded Geodetic Services, Inc (GSI) through his purchase of the Photogrammetric and Geodetic Services Division from DBA Systems.

In 1979 he was given a patent for the reseau platen that he developed. Duane Brown was also involved in many geodetic activities and was sometimes referred to as "the father of the short-arc method of geodesy". While at GIS, Duane Brown began to prepare for the next stage that photogrammetry was moving towards. This project involved taking photogrammetry from a tool requiring highly-trained technicians to perform the measurement to a turn-key system that could be used directly by the client.

This transformation required a number of tasks for GIS. The first development was the introduction of retro-reflecting targets which offered significant improvement over conventional photogrammetric targeting. Next, he helped in the mathematical development required by programmers to move the bundle adjustment software that ran on large main-frame computers to personal computers. This new software was called STARS (Simultaneous Triangulation and Resection Software). Next he developed the CRC-1 (Close Range Camera-1) which utilized film instead of glass plates, an ultra-flat platen, back-projected reseau, and continuously focusable lens. Finally, he helped develop the AutoSet-1 monocomparator that was a faster, operator-independent, fully-automated measuring instrument. This completed the turn-key system which was also called

STARS (replace Software with System). This system provided Duane Brown with the tools to achieve a milestone he was working towards- breaking the micrometer level in photogrammetric accuracies. He was the first to achieve accuracy better than 1:1,000,000 on a project. Moreover, he helped see that use of industrial photogrammetry grow a thousand fold during the 1980s.

Conventional aerotriangulation requires a good metric camera with camera calibration parameters. This precludes the use of non-metric, off-the shelf cameras for precise work. Houssam Mahmoud (Sam) Karara (1928 –1992), along with Y.I. Abdel-Azis, developed the Direct Linear Transformation (DLT) in 1971.

Recognizing that there are a number of potential applications with non-metric cameras, they found a way to perform a direct transformation from comparator measurements to object-space coordinates that did not require camera calibration data, transformation of comparator to photo coordinates and initial approximations of unknowns. The DLT method used two polynomial equations as the basis of the transformation. These polynomials were derived from the collinearity condition and the affine coordinate transformation.

Sam Karara was a professor at the University of Illinois at Champaign-Urbana and was widely known for his work in close-range photogrammetry. He served as editor-in-chief and author for both editions of the "Handbook on Non-Topographic Photogrammetry" published by the American Society of Photogrammetry and Remote Sensing.

The father of the analytical plotter is Uuno (Uki) Vilho Helava (192-1994) (Figure 2.15). Born in Finland, Helava moved to Canada as a research fellow at the National Research Council (NRC). While at NRC, Helava developed the analytical plotter in 1957. This instrument used servocontrol instead of the optical or mechanical construction of previous instruments (Konecny, 1985). A Computer was used not only to drive the instrument around the stereomodel but also to digitally transform

Figure 2.15: Uuno (Uki) Helava.

coordinates between the image and the map. In 1965, Helava left the NRC and became a consultant to the Ottico Meccanica Italiana in Italy. He began working at Bendix Corporation in Southfield, Michigan in 1967 and became president of Helava and Associates, Inc. in 1979.

1.4 Digital Photogrammetry

One of the pioneers in digital photogrammetry was Gilbert Louis Hobrough (1918-2002). Born in Toronto, Ontario, he grew into a man of many interests. During his life he has been awarded at least 47 patents in such diverse areas as "phonograph turntable and pickup, high-fidelity loudspeaker design, radar and barometric altimetry, three-dimensional machine vision and laser interferometry. His photogrammetric career began with his employment at Photographic Survey Corporation Ltd. In 1951. His initial contributions to the science involved the development of an electronic dodging printer. He also built an airborne profile recorder. This instrument required a radar system to measure the range from the aircraft to the ground with about a one-foot accuracy and a reference barometer with the same comparable accuracy.

Figure 2.16: Raytheon-Wild B8 Stereomat.

In 1957, Hobrough demonstrated his concept of image correlation on a Kelsh plotter. Due to the technology at the time, the correlation process was an analog one where the hardware was used to compare the gray levels of the imagery. This development led to the creation of the Raytheon-Wild B8 Stereomat (Figure 2.16). While novel, the instrument was not successful because of the reluctance of the photogrammetric practitioners to adopt this new technology, probably due to its inflexibility and issues of reliability.

In 1961, Hobrough moved to California where he worked with George Wood on the Automatic Registration Electronic Stereoscope (ARES). The purpose of this instrument was to correlate high resolution reconnaissance photography with high-precision survey photography in order to enable more precise measurement of changeable ground conditions. This project was later sold to Itek Corporation, who Hobrough and Wood subsequently worked for.

In 1967, Hobrough moved to Vancouver, Canada, to establish Hobrough Ltd. While there, he developed the Gestalt Photo Mapper (GPM) (Figure 2.17). This was an automated orthophotographic system utilizing correlation of stereo imagery. It consisted of a printer that was generally located in a darkroom, a scanner, correlator and computer system, operator console, and an input/output device.

Five years later he left this company to work with his son. Uki Helava also played a central role in the development of digital photogrammetry, first as a research scientist at Bendix and later at Helava Associates, Inc. (1979). Helava Associates eventually became a subsidiary of General Dynamics in 1986 where he helped develop digital photogrammetric workstations for the Defense Mapping Agency (now called NIMA). When General Dynamicsdivested its Electronics Division in 1992, Helava Associates became GDE Systems. It formed a joint

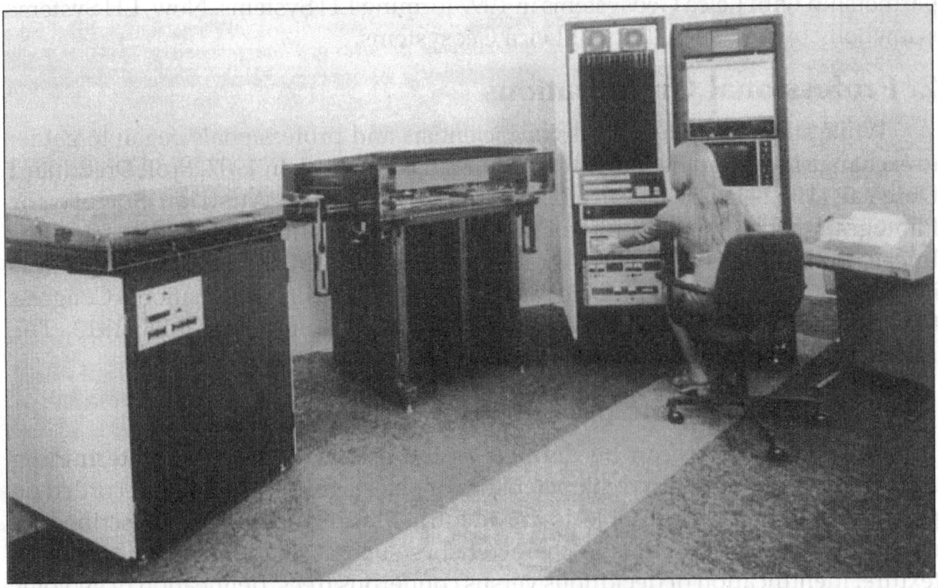

Figure 2.17: Gestault Photo Mapper.

Figure 2.18: Prof. Eduard DoleŸal.

partnership with Leica Geosystems in 1997 forming LH Systems. Now, LH Systems is a wholly owned subsidiary of Leica Geosystems.

2. Professional Organizations

While still a fledging profession, scientists and professionals began to gather to exchange ideas and developments within the industry. In 1907, Prof. Dr. Eduard DoleŸal (1862–1955) (Figure 2.18) helped form the Austrian Society for Photogrammetry, the first photogrammetric organization. Two years later, the International Society for Photogrammetry was established and DoleŸal was installed as its first president. DoleŸal alsoorganized the first International Congress for Photogrammetry held in Vienna, Austria on September 24-26, 1913. The American Society for Photogrammetry was formed in 1934.

Principles of Photogrammetry

Photogrammetry can be defined as the science and art of determining qualitative and quantitative characteristics of objects from the images recorded on photographic emulsions. Objects are identified and qualitatively described by observing photographic image characteristics such as shape, pattern, tone, and texture. Identification of deciduous versus coniferous trees, delineation of geologic landforms, and inventories of existing land use are examples of qualitative observations obtained from photography.

The quantitative characteristics of objects such as size, orientation, and position are determined from measured image positions in the image plane of the camera taking the photography. Tree heights, stockpile volumes, topographic maps, and horizontal and vertical coordinates of unknown points are examples of quantitative measurements obtained from photography.

1. Difference between Maps and Photographs

The main difference between maps and photographs lies in the method of projection (refer to Figure 2.19). A map is called an orthographic projection where each point on the earth's surface is projected onto the map surface in a manner that is perpendicular to the map surface. Thus, the mapped position represents the true horizontal position of the point in question, just as if the observed was viewing the scene directly above the point/feature. This allows the user to accurately scale between points to obtain distance, as an example.

The photograph uses a central perspective projection. Here, all of the light rays are brought to a bundle at the camera lens and are then projected onto the photograph. As one can see from Figure 2.19, the datum position of a point is not represented accurately on the photograph. The effect of this type of projection is that the distances on the photograph do not represent the true horizontal positions on the ground, but instead represent the projected position.

Photogrammetrists have the ability to project an image orthographically onto a map plane. This is called orthophotography. It requires additional processing of the image, either in an analog (map) form or digitally. This latter approach is the most common method used today.

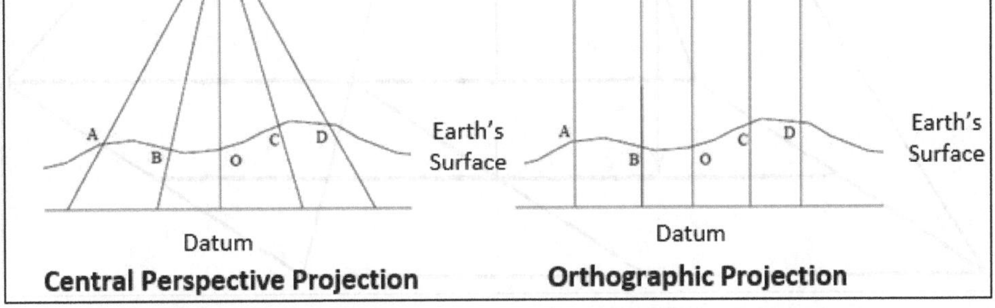

Figure 2.19: Difference between a Central and Orthographic Projection.

2. Geometry of Aerial Photography

The geometry of a single vertical photograph is shown in Figure 2.20 and 2.21. The photographic negative is shown for completeness, but in practice it is typical to work with the photographic positive printed on paper, film, or glass. The front nodal point of the camera lens is defined as the exposure station of the photograph. The nodal points are those points in the camera lens system such that any light ray entering the lens and passing through the front nodal point will emerge from the rear nodal point travelling parallel to the incident light ray. Therefore, the positive photograph can be shown on the object side of the camera lens, positioned such that the object point, the image point, and the exposure station all lie on the same straight line. The line through the lens nodal points and perpendicular to the image plane intersects the image plane at the principal point. The distance measured from the rear nodal point to the negative principal point or from the front nodal point to the positive principal point is equal to the focal length (*f*) of the camera lens.

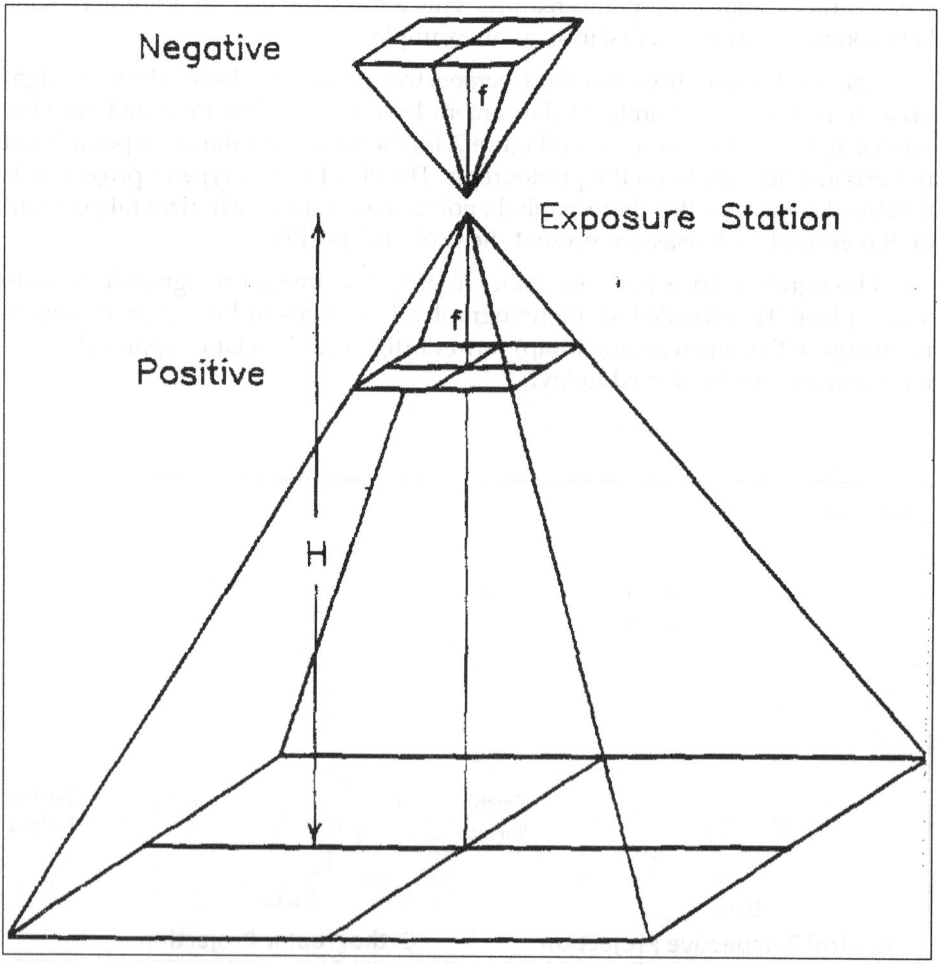

Figure 2.20: Single Vertical Photograph Geometry.

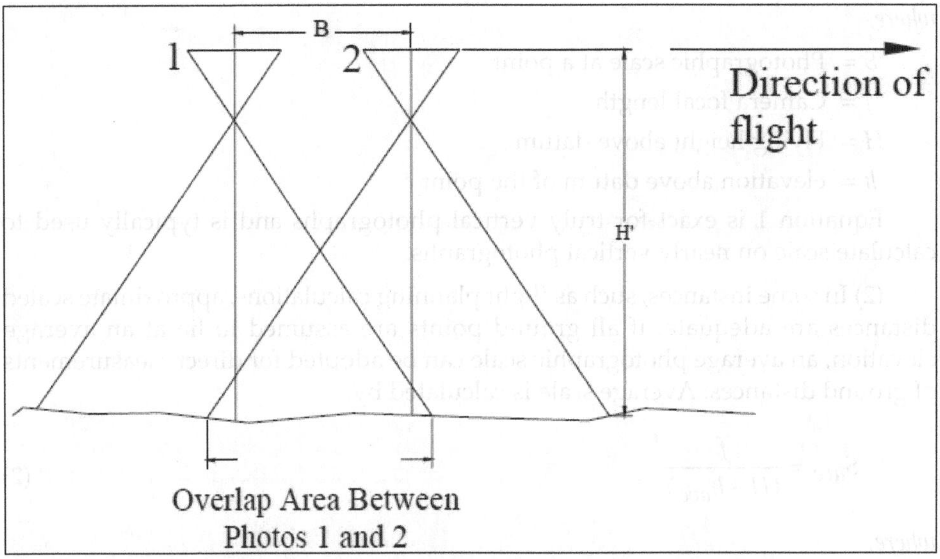

Figure 2.21: Geometry of Aerial Photography.

2.1 Single Vertical Aerial Photography

Vertical photographs, exposed with the optical axis vertical or as nearly vertical as possible, are the principal kind of photographs used for mapping. If the axis is perfectly vertical, the resulting photograph is termed a "truly vertical" photograph. In spite of the precautions taken to maintain the vertical camera axis, small tilts are invariably present; but these tilts are usually less than 1 degree and they rarely exceed 3 degrees. Photographs containing these small, unintentional tilts are called "near vertical" or "tilted" photographs. Many of the equations developed in this chapter are for truly vertical photographs, but for certain work, they may be applied to near vertical photos without serious error. Photogrammetric principles and practices have been developed to account for tilted photographs, and no accuracy whatsoever need be lost in using tilted photographs.

(a) Photographic Scale

The scale of an aerial photograph can be defined as the ratio between an image distance on the photograph and the corresponding horizontal ground distance. Note that if a correct photographic scale ratio is to be computed using this definition, the image distance and the ground distance must be measured in parallel horizontal planes. This condition rarely occurs in practice since the photograph is likely to be tilted and the ground surface is seldom a flat horizontal plane. Therefore, scale will vary throughout the format of a photograph, and photographic scale can be defined only at a point.

(1) The scale at a point on a truly vertical photograph is given by

$$S = \frac{f}{H - h} \tag{1}$$

where,

S = Photographic scale at a point

f = Camera focal length

H = Flying height above datum

h = elevation above datum of the point

Equation 1 is exact for truly vertical photographs and is typically used to calculate scale on nearly vertical photographs.

(2) In some instances, such as flight planning calculations, approximate scaled distances are adequate. If all ground points are assumed to lie at an average elevation, an average photographic scale can be adopted for direct measurements of ground distances. Average scale is calculated by:

$$S_{ave} = \frac{f}{(H - h_{ave})} \tag{2}$$

where,

h_{ave} is the average ground elevation in the photo. Then referring to the vertical photograph shown in Figure 2.22, the approximate horizontal length of the line AB is:

$$D = \frac{d(H - h_{ave})}{f} \tag{3}$$

where,

D = Horizontal ground distance

d = Photograph image distance

The flat terrain assumption, however, introduces scale variation errors. For accurate determinations of horizontal distances and angles, the scale variation caused by elevation differences between points must be accounted for in the photogrammetric solution.

(b) Horizontal Ground Coordinates

Horizontal ground distances and angles can be computed using coordinate geometry if the horizontal coordinates of the ground points are known. Figure 2.22 and 2.23 illustrates the photogrammetric solution to determine horizontal ground coordinates.

(1) Horizontal ground coordinates can be calculated by dividing each photocoordinate by the true photographic scale at the image point. In equation form, the horizontal ground coordinates of any point are given by

$$X_p = \frac{X_p(H - h_p)}{f} \tag{4}$$

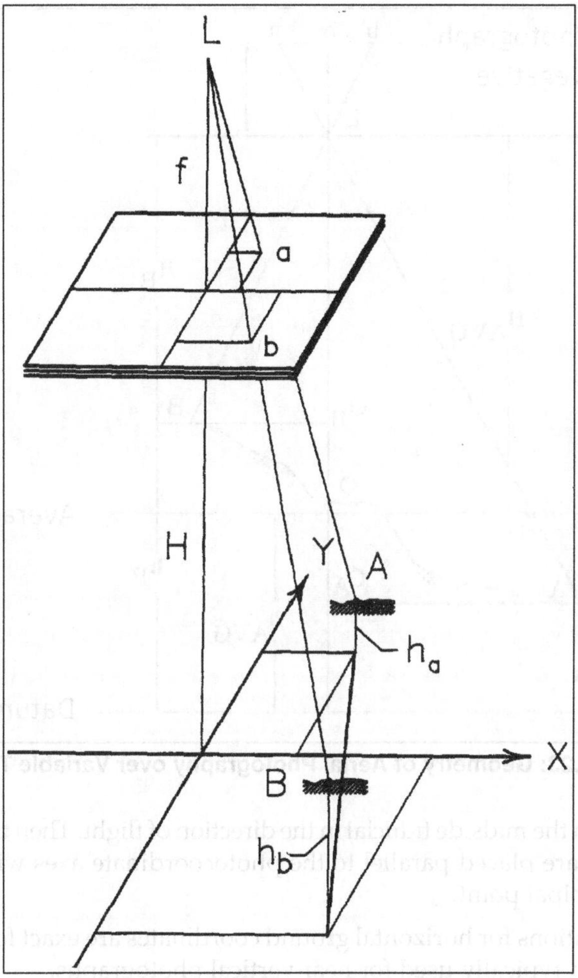

Figure 2.22: Horizontal Ground Coordinates from Single Vertical Photograph.

$$Y_p = \frac{y_p(H - h_p)}{f}$$

where,

X_p, Y_p = Ground coordinates of point p

x_p, y_p = Photocoordinates of point p

h_p = Ground elevation of point p

Note that these equations use a coordinate system defined by the photocoordinate axes having an origin at the photo principal point and the x-axis

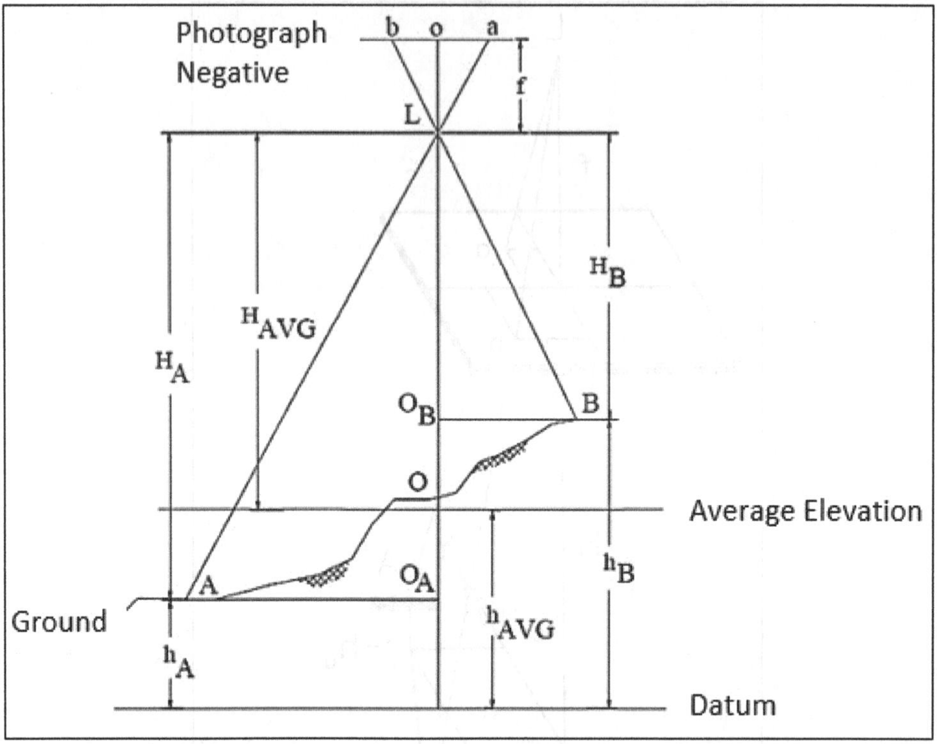

Figure 2.23: Geometry of Aerial Photography over Variable Terrain.

typically through the midside fiducial in the direction of flight. Then the local ground coordinate axes are placed parallel to the photocoordinate axes with an origin at the ground principal point.

(2) The equations for horizontal ground coordinates are exact for truly vertical photographs and typically used for near vertical photographs.

(3) After the horizontal ground coordinates of points A and B in Figure 2.23 are computed, the horizontal distance is given by

$$\sqrt{D_{AB} = (X_a - X_b)^2 + (Y_a - Y_b)^2} \qquad (5)$$

This solution is not an approximation because the effect of scale variation caused by unequal elevations is included in the computation of the ground coordinates. It is important to note, however, that the elevations h_A and h_B must be known before the horizontal ground coordinates can be computed. The need to know elevation can be overcome if a stereo solution is used.

(c) Relief Displacement

Relief displacement is another characteristic of the perspective geometry recorded by an aerial photograph. The displacement of an image point caused by changes in ground elevation is closely related to photographic scale variation. Relief displacement is evaluated when analyzing or planning mosaic or orthophoto projects. Relief displacement is also a tool that can be used in photo interpretation to determine heights of vertical objects.

(1) The displacement of photographic images caused by differences in elevation is illustrated in Figure 2.24. The image displacement is always along radial lines from the principal point of a truly vertical photograph or the nadir of a tilted photograph. The magnitude of relief displacement is given by the formula

$$d = \frac{rh}{H} \tag{6}$$

where,

D = Image displacement

r = Radial distance from the principal point to the image point

H = Flying height above ground

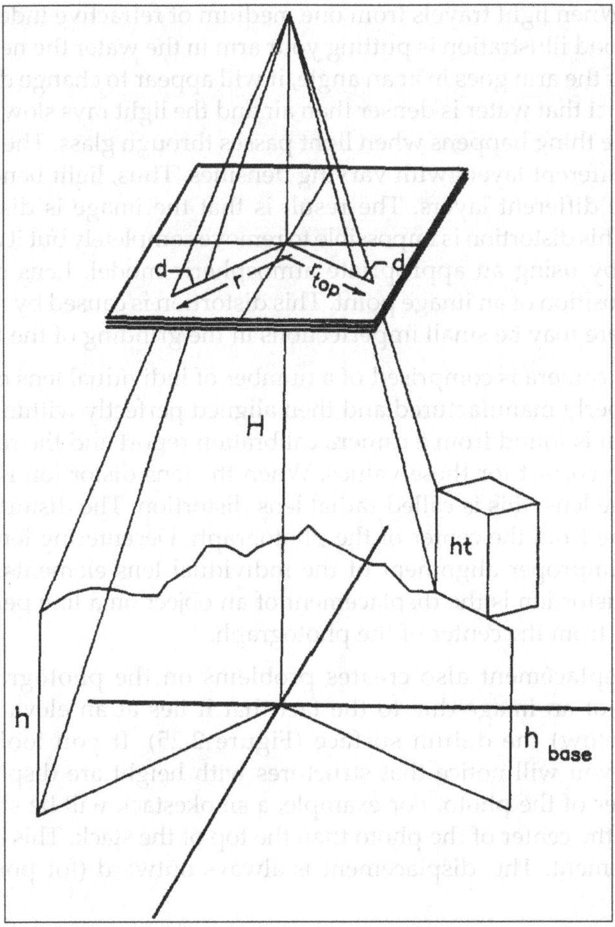

Figure 2.24: Relief Displacement on a Vertical Photograph.

(2) Since the image displacement of a vertical object can be measured on the photograph, Equation (6) can be solved for the height of the object to obtain

$$h_t = \frac{d(H - h_{base})}{r_{top}} \tag{7}$$

where,

h_t = Vertical height of the object

h_{base} = Elevation at the object base above datum

Remember, heights of vertical objects can be determined on a single photograph, but elevation above datum cannot be determined.

2.2 Distortion in Photography

There are a number of distortions associated with photography that need to be accounted for by the photogrammetrist. A large error is associated with the atmosphere. When light travels from one medium of refractive index to another it gets bent. A good illustration is putting your arm in the water the next time you go to the beach. If the arm goes in at an angle, it will appear to change directions. This is due to the fact that water is denser than air and the light rays slow down and get bent. The same thing happens when light passes through glass. The atmosphere is made up of different layers with varying densities. Thus, light bends as it passes through these different layers. The result is that the image is displaced on the photograph. This distortion is impossible to remove completely but it can be reduced significantly by using an appropriate atmospheric model. Lens distortion also changes the position of an image point. This distortion is caused by manufacturing problems. There may be small imperfections in the grinding of the lens' surfaces.

An aerial camera is comprised of a number of individual lens elements. Each has to be properly manufactured and then aligned perfectly within the lens cone. Lens distortion is found from a camera calibration report and the results are used in mapping to correct for these values. When the lens distortion is due to faulty grinding of the lens, this is called radial lens distortion. The distortion will occur on a radial line from the center of the photograph. Decentering lens distortion is caused from improper alignment of the individual lens elements. The effect of decentering distortion is the displacement of an object on a line perpendicular to the radial line from the center of the photograph.

Relief displacement also creates problems on the photograph. This is a displacement of an image due to the fact that it lies at an elevation above (or sometimes below) the datum surface (Figure 2.25). If you look at an aerial photography you will notice that structures with height are displaced outward from the center of the photo. For example, a smokestack will be shown with the base closer to the center of the photo than the top of the stack. This effect is due to relief displacement. This displacement is always outward (for points above the datum).

Figure 2.25: Aerial Photograph Showing the Effects of Relief Displacement.

Increasing the flying height decreases the effects of relief displacement whereas the effects get larger for points farther away from the center of the photo. The photogrammetrist accounts for this distortion by placing the photography on a stereoplotter, which handles these changes in relief excellently. Thus, in stereophotogrammetry, this problem is alleviated.

The principles that have been discussed up to this point have assumed that the photographs were perfectly vertical, a situation that is almost impossible to occur. When the optical axis of the camera is unintentionally tilted from the vertical by some small amount, usually less than 2°-3°, the photography is nearly vertical and is generally referred to as a vertical photograph. When the photograph is intentionally tilted, it is referred to as an oblique photograph. There are two different types of oblique photos. If the photograph contains the horizon then it is referred to as a high oblique while a low oblique does not show the horizon.

2.3 Exterior Orientation of Tilted Photographs

Unavoidable aircraft tilts cause aerial photographs to be exposed with the camera axis tilted slightly from vertical, and the resulting pictures are called tilted photographs. The equations given above are exact for truly vertical photographs, and they are used with near vertical photography for planning, estimating, and photo interpretation. However, an accurate photogrammetric solution using aerial photographs must account for the camera position and tilt at the instant of exposure.

(a) The exterior orientation of a photograph is its spatial position and angular orientation with respect to the ground coordinate system. Six independent parameters are required to define exterior orientation. The space position is normally given by threedimensional coordinates of the exposure station in a ground coordinate system. The vertical coordinate corresponds to the flying height above datum. Angular orientation is the amount and direction of tilt in the photo. Three angles are sufficient to define angular

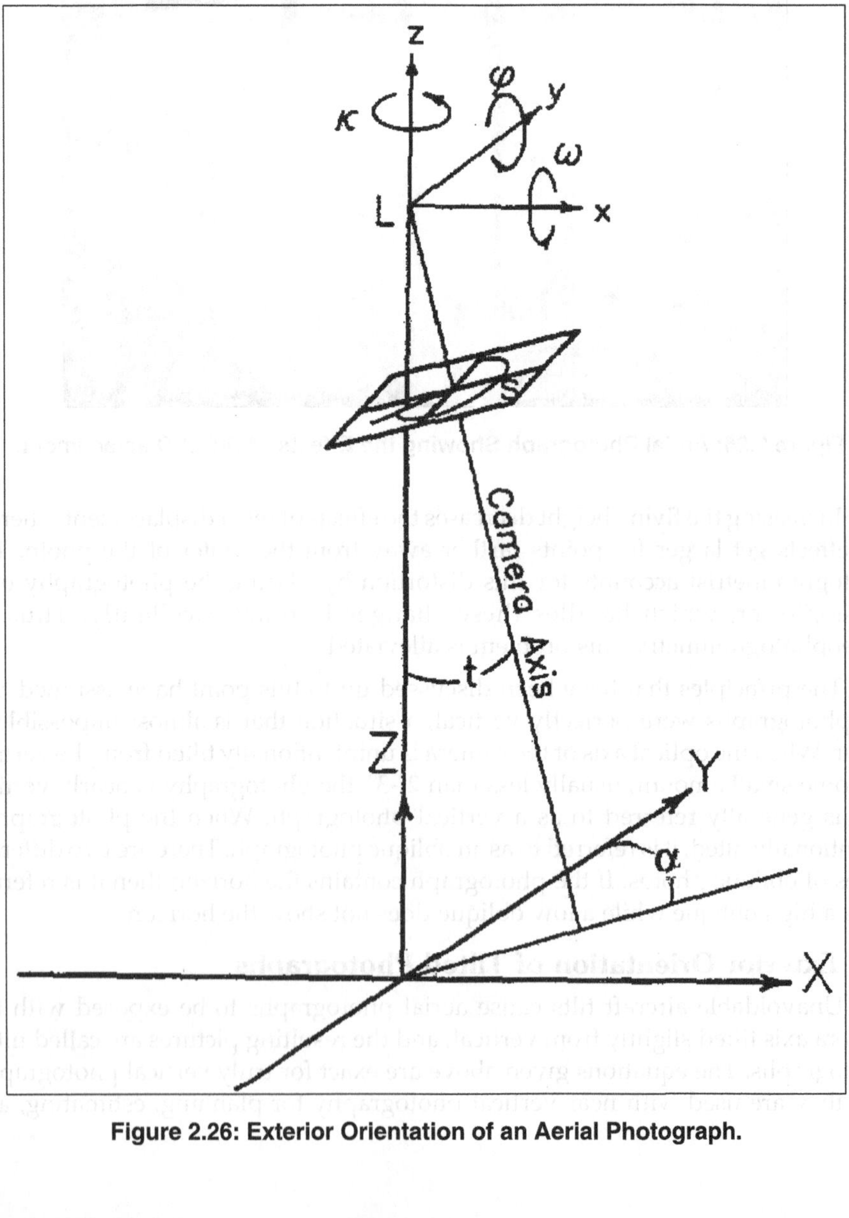

Figure 2.26: Exterior Orientation of an Aerial Photograph.

orientation, and two different systems are commonly used: the tilt-swing-azimuth system (t-s-α) and the omega-phi-kappa system (ω, ϕ, κ). The omega-phikappa system possesses certain computational advantages over the tilt-swing-azimuth system, but the tilt-swing-azimuth system is perhaps more easily understood. Figure 2.26 illustrates the parameters used to express exterior orientation of an aerial photograph.

(b) The tilt-swing-azimuth system is appropriate for hand calculations. An auxiliary photocoordinate system is defined with an origin at the photo nadir and y' axis along the direction of tilt. The expression for scale on a tilted photograph is:

$$S = \frac{\dfrac{f}{\cos t} - (y' \sin t)}{H - h} \qquad (8)$$

Where:

y' = auxiliary photocoordinate
t = photo tilt angle

(c) The tilt of aerial mapping photography is seldom large enough to require using tilted photograph equations for hand calculations when planning and estimating projects. However, Equation 8 does show that scale is a function of tilt, and scale variation occurs on a tilted photo even over flat terrain.

(d) Since the swing angle *S* of a truly vertical photograph is undefined, the omega-phi-kappa angular orientation system is preferred when expressing exterior orientation of any photograph. In the omega-phi-kappa system, the angular orientation of a tilted photograph is given in terms of three *sequential* rotation angles, omega, phi, and kappa. These angles, shown also in Figure 2.26, uniquely define the angular relationships between the three image coordinate system axes of a tilted photo and the three axes of the ground coordinate system. Omega is a rotation about the *x* photographic axis, phi is about the *y*-axis, and kappa is about the *z*-axis.

(e) The angular orientation of a truly vertical photograph taken with the flight line in the ground X or east direction is

$$\omega = 0$$

$$\phi = 0$$

$$\kappa = 0$$

The omega-phi-kappa angular orientation system is used in analogic and analytical solutions to express the exterior orientation of a photograph and produce accurate map information from aerial photographs.

2.4 Stereoscopic Vision

Stereoscopic vision determines the distance to an object by intersecting two lines of sight. In the human vision system, the brain senses the parallactic angle between the converging lines of sight and unconsciously associates the angle with a distance. Overlapping aerial photographs can be viewed stereoscopically with the aid of a stereoscope. The stereoscope forces the left eye to view the left photograph and the right eye to view the right photograph. Since the right photograph images the same terrain as the left photograph, but from a different exposure station, the brain perceives a parallactic angle when the two images are fused into one. As the viewer scans the entire overlap area of the two photographs, a continuous stereomodel of the ground surface can be seen. The stereomodel can be measured in three dimensions, yielding the elevation and horizontal position of unknown points. The limitation that elevation cannot be determined in a single photograph solution is overcome by the use of stereophotography.

(a) Lens Stereoscope

A lens or pocket stereoscope is a low-cost instrument that is very useful in the field as well as the office. It offers a fixed magnification, typically 2.5X. The lens stereoscope is useful for photo interpretation, control point design, and verification of mapped planimetric and topographic features.

(b) Mirror Stereoscope

A mirror stereoscope can be used for the same functions as a lens, but is not appropriate for field use. The mirror stereoscope has a wider field of view at the nominal magnification ratio. Since photographs can be held fixed for stereo viewing under a mirror stereoscope, the instrument is useful for simple stereoscopic measurements. Mirror stereoscopes can be equipped with binocular eyepieces that yield 6X and 9X magnification. The high magnification helps to identify, interpret, and measure photographed features.

(c) Floating Mark

Stereoscopic measurements are possible if a floating mark is introduced in the viewing system. Identical halfmarks, such as a small filled circle or a small cross, are put in the field of view of each eye. As the stereomodel surface is viewed, the two halfmarks are viewed against the photographed scene by each eye. If the halfmark positions are properly adjusted, the brain will fuse their images into a single floating mark that appears in three-dimensional space relative to the stereomodel. By moving the halfmarks parallel to the viewer's eye base, the floating mark can be adjusted until it is perceived to lie on the stereomodel surface. At this point, the two halfmarks are on the identical or conjugate image points of two different photographs. The horizontal position and elevation of the mark can be determined and plotted on a map. The importance of the floating mark is that all points in a stereomodel can be measured and mapped. Thus, indistinct points, such as hilltops, centers of road intersections, and contours can be mapped in three dimensions from the stereomodel.

2.5 Geometry of Aerial Stereophotographs

The basic unit of photogrammetric mapping is the stereomodel formed in the overlapping ground coverage of successive photographs along a flight line. Figure 2.27 illustrates the ground coverage along three contiguous flight lines in a block of aerial photographs. Along each flight line, the overlap of photographs, termed end lap, is typically designed to be 60 percent. End lap must be at least 55 percent to ensure continuous stereoscopic coverage and provide a minimum triple overlap area where stereomodels can be matched together. Between adjacent flight lines, the overlap of strips, termed side lap, is typically designed to be 30 percent. Side lap must be at least 20 percent to ensure continuous stereoscopic coverage.

(a) Stereomodel dimensions: For project planning and estimating, photograph and stereomodel ground dimensions are computed by assuming truly vertical photography and flat terrain at average ground elevation.

(1) Using average photographic scale, the ground coverage G of one side of a square format photograph is:

$$G = \frac{d}{S_{ave}} \tag{9}$$

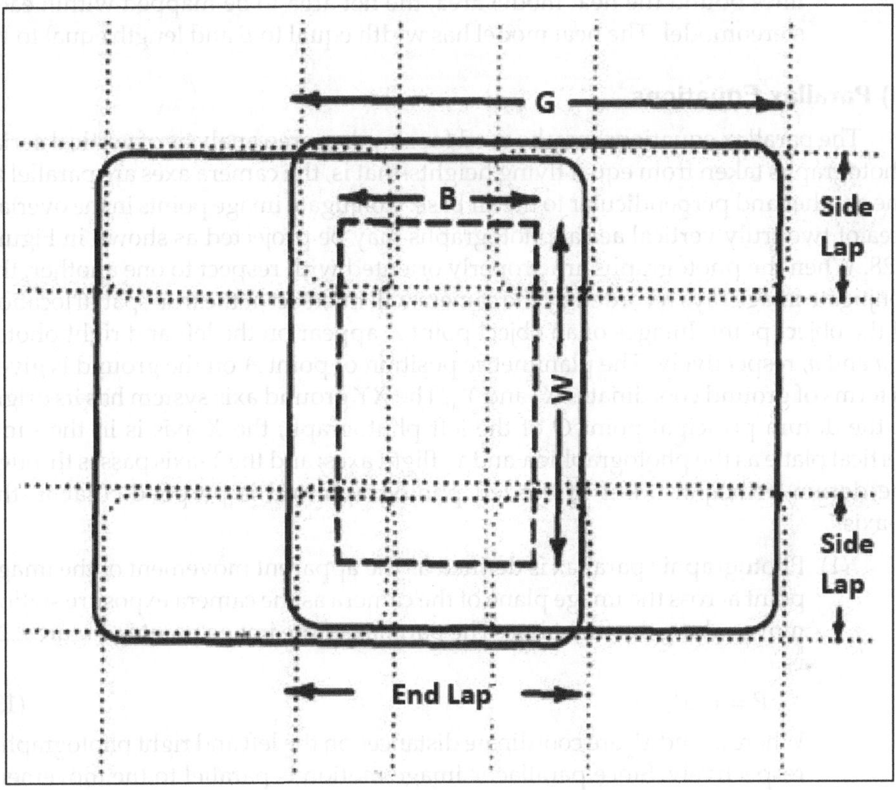

Figure 2.27: Aerial Photography Stereomodel and Neat Model.

Where *d* is the negative format dimension. The flying height above datum is also found using average scale and average ground elevation.

$$H = h_{ave} + \frac{f}{S_{ave}} \tag{10}$$

(2) Let *B* represent the air base between exposures in the strip. Then from the required photo end lap E_{lap}

$$B = G\left\{\frac{1 - E_{lap}}{100}\right\} \tag{11}$$

(3) Let *W* represent the distance between adjacent flight lines. Then from the required side lap S_{lap}

$$W = G\left\{\frac{1 - S_{lap}}{100}\right\} \tag{12}$$

(4) Match lines between contiguous stereomodels pass through the center of the triple overlap area and the center of the side lap area. These match lines bound the neat model area, the net area to be mapped within each stereomodel. The neat model has width equal to *B* and length equal to *W*.

(b) Parallax Equations

The parallax equations may be used for simple stereo analysis of vertical aerial photographs taken from equal flying heights-that is, the camera axes are parallel to one another and perpendicular to the air base. Conjugate image points in the overlap area of two truly vertical aerial photographs may be projected as shown in Figure 2.28. When the photographs are properly oriented with respect to one another, the conjugate image rays recorded by the camera will intersect at the true spatial location of the object point. Images of an object point *A* appear on the left and right photos at *a* and *á*, respectively. The planimetric position of point *A* on the ground is given in terms of ground coordinates X_A and Y_A. The *XY* ground axis system has its origin at the datum principal point *O* of the left photograph; the *X*-axis is in the same vertical plane as the photographic *x* and *x′* flight axes; and the *Y*-axis passes through the datum principal point of the left photograph and is perpendicular to the *X*-axis.

(1) Photographic parallax is defined as the apparent movement of the image point across the image plane of the camera as the camera exposure station moves along the flight line. The parallax of the image point in Figure 2.28 is:

$$P = x_{a-} x'_a \tag{13}$$

Where x_a and x'_a are coordinate distances on the left and right photographs, respectively. Since parallactic image motion is parallel to the movement of the camera, the parallax coordinate system must be parallel to the

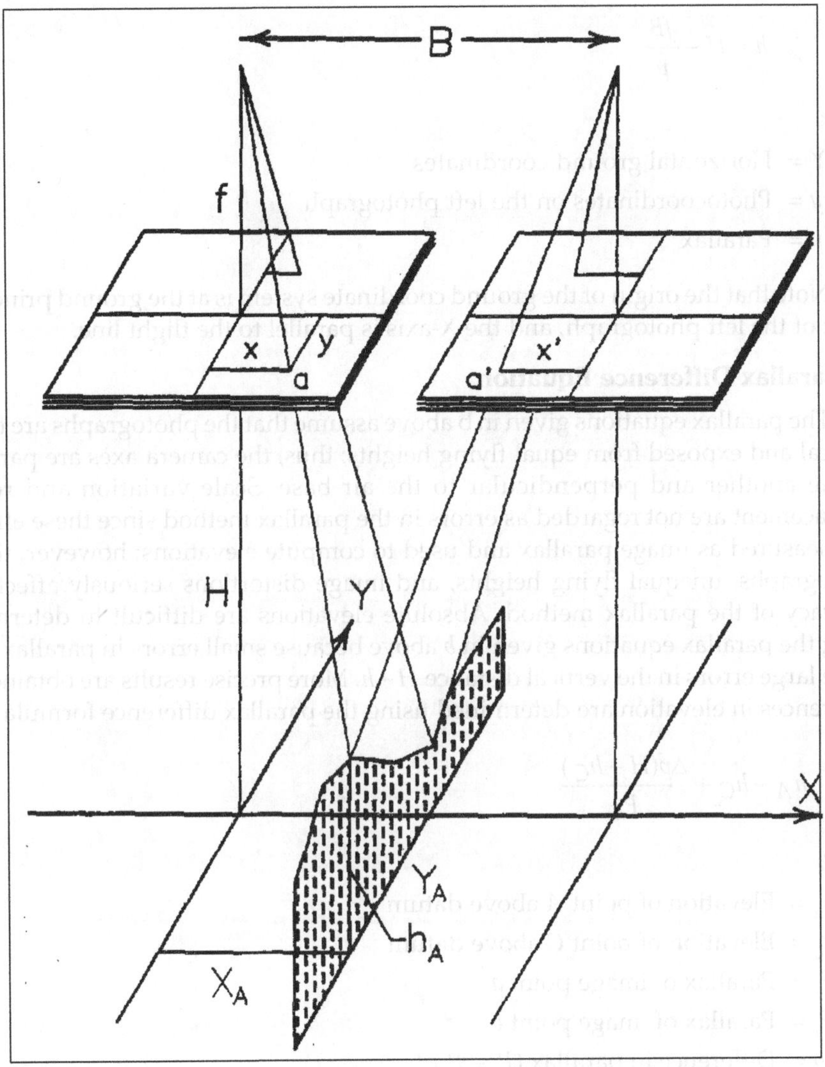

Figure 2.28: Stereopair of Vertical Aerial Photographs.

direction of flight. All parallax occurs along the x-axis in the axis of flight photocoordinate system. The y and y' coordinates are equal.

(2) Given truly vertical aerial photographs and photocoordinates measured in the axis of flight system, the following parallax equations can be derived:

$$X = \frac{xB}{p}$$

$$Y = \frac{yB}{p}$$

$$h = H - \frac{fB}{p} \tag{14}$$

where,

X, Y = Horizontal ground coordinates

x, y = Photocoordinates on the left photograph

p = Parallax

Note that the origin of the ground coordinate system is at the ground principal point of the left photograph, and the X-axis is parallel to the flight line.

(c) Parallax Difference Equation

The parallax equations given in b above assume that the photographs are truly vertical and exposed from equal flying heights; thus, the camera axes are parallel to one another and perpendicular to the air base. Scale variation and relief displacement are not regarded as errors in the parallax method since these effects are measured as image parallax and used to compute elevations; however, tilted photographs, unequal flying heights, and image distortions seriously affect the accuracy of the parallax method. Absolute elevations are difficult to determine using the parallax equations given in *b* above because small errors in parallax will cause large errors in the vertical distance $H - h$. More precise results are obtained if differences in elevation are determined using the parallax difference formula

$$h_A - h_C + \frac{\Delta p(H - h_C)}{P_a} \tag{15}$$

where,

h_A = Elevation of point A above datum

h_C = Elevation of point C above datum

P_a = Parallax of image point a

P_c = Parallax of image point c

Δp = Difference in parallax $(P_c - P_a)$

The formula should be applied to points that are close to one another on the photo format. The differencing technique cancels out the systematic errors affecting the parallax of each point. If C is a vertical control point, the absolute elevation of A can be determined by this method.

2.6 Fundamental Photogrammetric Problems

The fundamental photogrammetric problems are resection and intersection. All photogrammetric procedures are composed of these two basic problems.

(a) Resection

Resection is the process of recovering the exterior orientation of a single photograph from image measurements of ground control points. In a spatial resection, the image rays from total ground control points (horizontal position and elevation known) are made to resect through the lens nodal point (exposure station) to their image position on the photograph. The resection process forces the photograph to the same spatial position and angular orientation it had when the exposure was taken. The solution requires at least three total control points that do not lie in a straight line, and the interior orientation parameters, focal length, and principal point location. In aerial photogrammetric mapping, the exact camera position and orientation are generally unknown. The exterior orientation must be determined from known ground control points by the resection principle.

(b) Intersection

Intersection is the process of photogrammetrically determining the position of ground points by intersecting image rays from two or more photographs. If the interior and exterior orientation parameters of the photographs are known, then conjugate image rays can be projected from the photograph through the lens nodal point (exposure station) to the ground space. Two or more image rays intersecting at a common point will determine the horizontal position and elevation of the point. Map positions of points are determined by the intersection principle from correctly oriented photographs.

2.7 Photogrammetric Solution Methods

Correct and accurate photogrammetric solutions must include all interior and exterior orientation parameters. Each orientation parameter must be modeled if the recorded image ray is to be correctly projected and an accurate photogrammetric product obtained. Interior orientation parameters include the camera focal length and the position of the photo principal point. Typically the interior orientation is known from camera calibration. Exterior orientation parameters include the camera position coordinates and the three orientation angles.

Typically, the exterior orientation is determined by resection principles as part of the photogrammetric solution. The remaining parameters are the ground coordinates of the point to be mapped. Planimetric and topographic details are mapped by intersecting conjugate image rays from two correctly oriented photographs. Methods of solving the fundamental photogrammetric problems may be classified as analog or analytical solutions.

(a) Analog Solutions

Analog photogrammetric solutions use optical or mechanical instruments to form a scale model of the image rays recorded by the camera. An optical analog instrument projects a transparency of the image through a lens such that the camera bundle of image rays is accurately reproduced and the projected image is brought to focus at some finite distance from the lens. A mechanical analog instrument uses a straight metal rod, a space rod, to represent the image ray from the image point, through the lens perspective center, to the modelled ground point. Analog instruments are limited in function (focal length, model scale enlargement, flight geometry) by the physical constraints of the analog mechanism. They are limited in accuracy by the calibration of the analog mechanism and by unmodelled systematic errors. Analog instruments cannot effectively compensate for differential or nonlinear image and film deformation errors.

(b) Analytical Solutions

An analytical photogrammetric solution uses a mathematical model to represent the image rays recorded by the camera. The image ray is assumed to be a straight line through the image point, the exposure station, and the ground point. The following collinearity equation expresses this condition:

$$x = x_o - f\frac{[m_{11}(X - X_L) + m_{12}(Y - Y_L) + m_{13}(Z - Z_L)]}{[m_{31}(X - X_L) + m_{32}(Y - Y_L) + m_{33}(Z - Z_L)]}$$

$$y = y_o - f\frac{[m_{21}(X - X_L) + m_{22}(Y - Y_L) + m_{32}(Z - Z_L)]}{[m_{31}(X - X_L) + m_{32}(Y - Y_L) + m_{33}(Z - Z_L)]} \quad (16)$$

where,

x, y = Measured photocoordinates

x_o, y_o = Principal point photocoordinates

m_{ij} = Nine direction cosines expressing the angular orientation

X, Y, Z = Ground point coordinates

X_L, Y_L, Z_L = Exposure station coordinates

The collinearity condition equations include all interior and exterior orientation parameters required to solve the resection and intersection problems accurately. Analytical solutions consist of systems of collinearity equations relating measured image photocoordinates to known and unknown parameters or the photogrammetric problem. The equations are solved simultaneously to determine the unknown parameters. However, since there are usually redundant measurements producing more equations than there are unknowns in the problem, a least squares adjustment is used to estimate the unknown parameters. The least squares adjustment algorithm includes residuals v_x and v_y on the measured photocoordinates that estimate random measurement error.

3. Terrestrial Photogrammetry

(a) Definition

Terrestrial photogrammetry refers to applications where the camera is supported on the surface of the earth. Typically the camera is mounted on a tripod and the camera axis is oriented in or near the horizontal plane. When the distance to the object is less than approximately 300 m, the method is often referred to as close-range photogrammetry.

(b) Applications

Terrestrial photogrammetry techniques can be applied to measuring and mapping large vertical surfaces. These applications include topographic mapping of rugged terrain and vertical cliffs, deformation measurement of structures, and architectural mapping of buildings. Close-range applications include measuring and calibrating irregular surfaces such as antenna reflectors and industrial assembly jigs, and three-dimensional mapping of industrial installations and construction sites.

(c) Instrumentation

(i) Cameras

Cameras for engineering applications of terrestrial photogrammetry should be stable, calibrated metric cameras similar to aerial cameras. For highest accuracy, some cameras expose the image directly on glass planes to ensure image plane flatness.

Film type cameras should have a film-flattening device. Terrestrial cameras are available in a range of focal lengths and format size. If the camera is combined with angle measuring capability for orienting the camera axis in a known direction, the instrument is referred to as aphototheodolite. Stereometric cameras are two cameras rigidly fixed to a base bar. Since the base is usually short, stereometric cameras fall into the close-range category.

(ii) Stereoplotters

Some specialized stereoplotters have been developed specifically for terrestrial applications; however, these instruments are not typically available in the United States photogrammetric industry.

Some analog stereoplotters can accommodate some terrestrial and close-range cameras; but often incompatible focal lengths, format sizes, and depth of field constraints make analog stereoplotters unusable for this work. Analytical stereoplotters can accommodate terrestrial and close-range cameras because no mechanical constraints are placed on the mathematical projection. Of course, monocomparators and analytical software work equally well for precise applications of point mapping.

(d) Methods

The photogrammetric principles discussed in this chapter apply also to the terrestrial solutions. Terrestrial photogrammetry differs from aerial photogrammetry in the fact that the camera is accessible. Thus, the camera position and orientation can be measured. Theoretically this eliminates the need for control points and resection to orient the camera. When considering a terrestrial or close-range application, additional consideration must be given to optimum camera location geometry, focusing requirements of the camera, and depth of field requirements to capture the object to be measured.

3.1 Aerotriangulation

Aerotriangulation is the simultaneous space resection and space intersection of image rays recorded by an aerial mapping camera. Conjugate image rays projected from two or more overlapping photographs intersect at the common ground points to define the three-dimensional space coordinates of each point. The entire assembly of image rays is fit to known ground control points in an adjustment process. Thus, when the adjustment is complete, ground coordinates of unknown ground points are determined by the intersection of adjusted image rays.

3.1.1 Geometric Principle

Photogrammetric control extension requires that a group of photographs be oriented with respect to one another in a continuous strip or block configuration. The exterior orientation of any photograph that does not contain ground control is determined entirely by the orientation of the adjacent photographs. In a given aerotriangulation configuration, each photograph contributes to the exterior orientation of the adjacent photographs through the pass points located in the triple overlap areas.

The term triple overlap refers to the ground area shared by two adjacent stereomodels along the strip. The triple overlap area is imaged on three consecutive photographs. Thus, when end lap is specified to be 60 percent, the alternate photographs will overlap each other by 20 percent at average terrain elevation.

3.1.2 Methods

A small number of ground control points are, however, essential in the execution of these methods. We distinguish two methods:

(a) Radial Triangulation

These methods are based on angular measurements on the photographs resulting in only planimetric coordinates X and Y. The control, obtained by these methods, has proved to be highly satisfactory substitute for ground control. With only a few ground control points around the periphery of the area of survey, a network of supplemental control can be established at the desired locations, which is suitable for many types of mapping.

This can be done by simple methods wherein no calculation or measurement is involved. Photo-interpreter often requires the extension of control for planimetric survey. These simple methods are, therefore, very suitable for him in the making of his base maps. The methods may be:

(i) Graphical method

(ii) Slotted template method, if better accuracy is required.

(iii) Analytical method

In case a natural resource service needs 3 coordinates, X, Y and Z, of all control points, it will be better to ask for assistance of a photogrammetric service.

(b) Aerial Triangulation in Space

Aerial Triangulation in Space resulting in X, Y and Z coordinates of each point. This triangulation is carried out in precision stereo plotting instruments (or in analytical plotters) wherefrom coordinates of points fixed are obtained in machine coordinate system. These are transferred into ground system with the help of ground control points. There are three types of aerial triangulation in space:-

(i) Bridging Method

(ii) (I)M.T. (Independent Model Triangulation)

(iii) Bundle Block Adjustment (Plate Coordinates)

(i) Bridging Method

The type of triangulation (Spatial Aerial Triangulation) is carried out on Precise Photogrammetric Stereo plotting machines. This involves, reconstruction of exact stereo model, on which measurement of model coordinates of control points is carried out along the three axes. Once first stereo model in a strip is absolutely oriented, and measured, the second stereo model is constructed by orienting the third projector (photograph) with respect to the second projector, (which is already in absolute orientation), and thus the second stereo model, so generated is also in Absolute Orientation. After measurement of second stereo model, the third stereo model is generated in the same way and so on. This method is knows as Bridging, and is adopted on instruments, which have the provision of constructing successive stereo models with the help of projectors. This type of instruments includes the most basic ones, *e.g.*, Multiplex, having a large number of projectors (one for each photograph) as well as more recent ones, *e.g.*, WILD Autograph A7, with only two camera projectors, used for first model with first photo on left and second photo on right projector. The second model is created by second photo on right and third photo on left projector, and third model is created with third photo on left and forth photo on right projector, and so on. The first, third, fifth models etc. are observed with some changes in observation and measuring system (known as base out condition).

(ii) IMT Method

Another technique of spatial aerial triangulation is known as Independent Model Triangulation, in which each stereo model, after Relative Orientation on stereo plotter, is measured independently for all the control points for X, Y and Z values. These models coordinates of all of the stereo models, are later on used to connect the stereo models to form a full strip, which is operated upon for Absolute Orientation with the help of ground control points by mathematical solution. As in this method part of work is done on stereo plotter and rest by computation, it is also known as Semi Analytical Method.

(iii) Bundle Block Adjustment (Plate Coordinates)

In case of purely Analytical Method, the precise image coordinates are measured on each of the photograph, by using either a stereo comparator or a mono comparator, for all the points, and entire rest of operation is done by mathematical computations, finally to yield the ground coordinates of all the control points. This involves huge amount of computations, for which a computer is indispensable.

4. Ground Control Datums and Coordinate Reference Systems

Photogrammetric ground control consists of any points whose positions are known in a ground reference coordinate system, and whose images can be positively identified in the photographs. Ground control provides the means for orienting or relating aerial photographs and stereomodels to the ground. Almost every phase of photogrammetric work requires some ground control.

Photogrammetric control is generally classified as either *horizontal control* (the position of the point is known with respect to a horizontal datum), or *vertical control* (the elevation of the point is known with respect to a vertical datum). Sometimes both horizontal and vertical object space positions of points are known so that these points serve a dual control purpose. Such points may be referred to as *total control.*

4.1 Coordinate Datums and Reference Systems

A datum is a reference surface on which a coordinate system is defined. In surveying and mapping, many datums and many coordinate systems can be defined that will express the position of points on the earth's surface. It is of utmost importance to know the coordinate system used to reference any control survey or map. Control survey or map data sets referenced to different coordinate systems cannot be combined unless a transformation of one coordinate system into the other is performed. A detailed discussion of coordinate datums and transformations is given by Soler and Hothem (1989).

(a) Horizontal Geodetic Datum

Datums are often separated into horizontal and vertical datums. A horizontal geodetic datum uses an ellipsoid as the reference surface. The size and shape of the ellipsoid are chosen to give the best fit to the earth on a global basis or on a regional or continental basis. Many different ellipsoids have been proposed. The datum definition is completed by orienting a specific ellipsoid to the earth and defining a coordinate system.

(i) *Curvilinear coordinates*: Geodetic longitude and latitude are curvilinear coordinates that define the position of a point on the curved ellipsoid surface. The coordinates are expressed in angular units. (Longitudeis the angle measured in the plane of the equator from the reference meridian through Greenwich, England, to the meridian through the point. Latitude is the angle in the plane of the meridian through the point from the equator to the normal through the point.) A geodetic height can also be defined as the distance along the normal from the surface of the ellipsoid to the point.

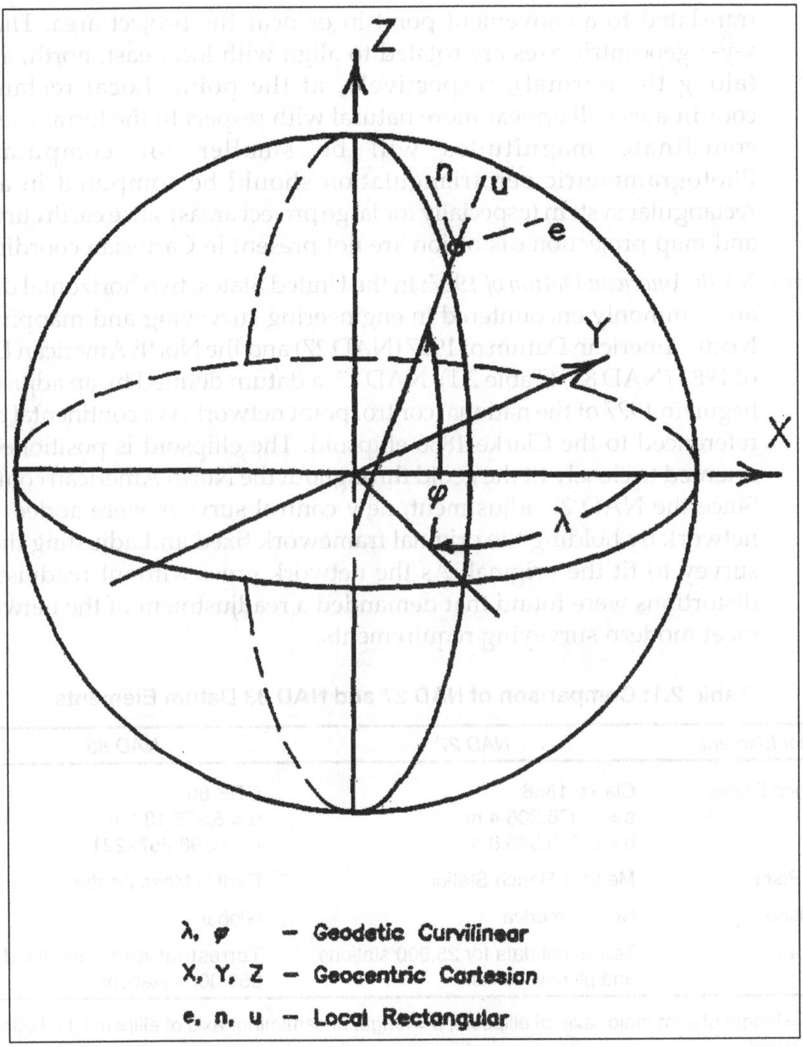

Figure 2.29: Geocentric Coordinates.

(ii) *Geocentric coordinates*: Geocentric coordinates are defined for a given datum by placing the origin of a right-handed spatial rectangular coordinate system at the center of the reference ellipsoid. The z-axis coincides with the semiminor axis of the ellipsoid. The x- and y-axes are in the plane of the equator with the x-axis through 0-degree longitude. Simple Cartesian coordinate computations can be performed in the geocentric system. The ellipsoid parameters and curved reference surface are eliminated from the computations (Figure 2.29).

(iii) *Local rectangular coordinates*: A local rectangular coordinate system can be defined for a given project area. The geocentric coordinate axes are translated to a convenient point in or near the project area. Then the x-y-z geocentric axes are rotated to align with local east, north, and up (along the normal), respectively, at the point. Local rectangular coordinates will appear more natural with respect to the terrain, and the coordinate magnitudes will be smaller for computations. Photogrammetric aerotriangulation should be computed in a local rectangular system (especially for large project areas) since earth curvature and map projection distortion are not present in Cartesian coordinates.

(iv) *North American Datum of 1927*: In the United States, two horizontal datums are commonly encountered in engineering surveying and mapping: the North American Datum of 1927 (NAD 27) and the North American Datum of 1983 (NAD 83) (Table 2.1). NAD 27, a datum defined by an adjustment begun in 1927 of the national control point network, is a continental datum referenced to the Clarke 1866 ellipsoid. The ellipsoid is positioned and oriented to closely fit the geoid throughout the North American continent. Since the NAD 27 adjustment, new control surveys were added to the network by holding the original framework fixed and adjusting the new survey to fit the original. As the network grew without readjustment, distortions were found that demanded a readjustment of the network to meet modern surveying requirements.

Table 2.1: Comparison of NAD 27 and NAD 83 Datum Elements

Datum Element	*NAD 27*	*NAD 83*
Reference Ellipse	Clarke 1866 a = 6,378,206.4 m b = 6,356,583.8 m	GRS 80 a = 6,378,137 m f = 1/298.2572221
Datum Point	Meades Ranch Station	Earth's Mass Center
Best Fitting	North America	Global
Adjustment	Terrestrial data for 25,000 stations and piecewise fits	Terrestrial and satellite data for 250,000+ stations

Note: a = length of semimajor axis of ellipsoid; b = length of semiminor axis of ellipsoid; f = flattening of ellipsoid.

(v) *North American Datum of 1983*: NAD 83 is a rigorous least squares adjustment of the entire national control network incorporating modern high-precision traverse and satellite measurements. NAD 83 is a global datum referenced to the Geodetic Reference System of 1980 ellipsoid. The ellipsoid is earthcentered and designed to be a best fit to the global geoid. Thus, the NAD 83 datum will have a larger geoid-to-ellipsoid separation than the NAD 27 datum in the United States. Since NAD 83 is a readjustment of the original observations in the control network, it is not possible to directly transform coordinates from one datum to the other. Interpolation methods using control points adjusted in both datums have been programmed by the NGS for the purpose of estimating the NAD 83 coordinates of points surveyed in the NAD 27 datum. Although the NAD 27 datum is still used, it should gradually be replaced by the NAD 83 datum.

(b) Vertical Geodetic Datum

A vertical geodetic datum is a reference system for measuring vertical position or elevation. The reference surface is the geoid, which is an equipotential gravity surface that best approximates mean sea level. Elevation above mean sea level (the geoid) is measured along the direction of gravity, perpendicular to the equipotential surfaces. In the United States, the vertical geodetic datum is the National Geodetic Vertical Datum of 1929. The elevation of a point can be combined with the horizontal position on the ellipsoid to define a three-dimensional coordinate, latitude, longitude, and height (ϕ, λ, h), by the relationship:

$$h = H + N \tag{17}$$

where,

h = Height above the ellipsoid

H = Elevation above the geoid

N = Geoid separation above the ellipsoid

(c) Map Projections

A map projection is a mathematical function that defines a relationship of coordinates between the ellipsoid and a plane. The purpose of a map projection is to display the curved earth surface on a flat plane that can be printed on a sheet of paper. If field survey observations are reduced to the plane, the computation and adjustment of horizontal control positions can be carried out in two-dimensional Cartesian coordinates. The need for more involved geodetic survey computations can be eliminated.

(i) The final map reference system is a plane developed from a regular mathematical surface. Developable surfaces that can be rolled out into a plane include a plane, a cone, and a cylinder. The ideal map projection would display all distances and directions correctly, all areas would retain their correct shape and relative size, meridians and parallels would intersect at right angles, and great circles and rhumb lines would be represented as straight lines. Obviously, the earth's surface, being spherical, cannot be developed upon a plane without distortion of some of these quantities.

(ii) In engineering surveying and mapping, correct depiction of shapes is important. Thus, the characteristic of conformality is enforced in projections used for large-scale mapping. This is accomplished by mathematically constraining the projection scale factor at a point, whatever it may be, such that it is the same in all directions from that point. This characteristic of a projection preserves angles between infinitesimal lines. That is, all lines on the grid cut each other at the same angles as do the corresponding lines on the ellipsoid for very short lines. Hence, for a small area, there is no local distortion of shape. But since the scale must change from point to point, distortion of shape can exist over large areas. Remember that the distortion present in any map projection is not an "error" since it is a mathematically defined function. The distortion must simply be accounted for in transforming data from map to ground.

(iii) The most commonly encountered map projections in engineering surveying and mapping are the State Plane Coordinate Systems (SPCS). NGS has developed SPCS's for each state, and many states have formally adopted the systems by legislative action. The SPCS's are based on the Lambert Conformal Conic Projection or the transverse Mercator projection.

(iv) State Plane Coordinate Systems are defined for both the NAD 27 and NAD 83 datums. For the NAD 27 SPCS definition, the unit of length is the US Survey Foot. For the NAD 83 SPCS definition, the unit of length is variable among the states. Care must be exercised when using NAD 83 SPCS values in feet since either the US Survey Foot or the International Foot may be used in a specific state or locality.

(v) Another commonly used plane coordinate projection system is the UTM projection. Since this is the standard military reference system, operational support mapping work may need to be compiled using this system. The UTM system uses a Mercator cylindrical projection with the axis of the cylinder oriented perpendicular to the earth's polar axis. The UTM coordinate zones are uniformly spaced around the equator at 6-degree intervals and extend north and south from the equator.

5. Map Substitutes-Mosaics, Photo Maps, and Orthophotographs

Map substitute refers to those products where the compiled planimetric line map is replaced or augmented by the aerial photographic images. Since the photographic image will have scale variation caused by the perspective view of the camera, photo tilt, and unequal flying heights, each product must be evaluated according to how the scale variation problem is treated. Scale variation may be ignored, as in an index mosaic, and an approximate map substitute is obtained; or the scale variation can be compensated, as in an orthophoto process, and a nearly true map substitute can be obtained.

5.1 Aerial Mosaics

Aerial mosaics are constructed from sets of individual adjoining aerial photographs. Typically, the outer edge of the photo coverage of each print is trimmed back to a selected match line, and the photos assembled by carefully matching ground detail along the match line. A photo index mosaic is a rough composite of a number of individual photographs of a flight line or set of flight lines overlaid one on top of the other without trimming the photo prints. This composite may itself be photographed onto a single piece of film. Mosaic and index assemblies form a continuous representation of the terrain covered by the photography. Photo maps are maps using a photograph (or mosaic) as the base to which limited cartographic detail such as names, route numbers, etc., are added. Generally, a photo map is constructed from an orthophoto, or from several orthophotos mosaicked together.

5.1.1 Uses of Mosaics

The increased need for presenting a pictorial view of the earth's surface has led to the aerial mosaic as a means of showing a complete view of large areas. Because a single photo is limited in area, groups of photos are combined into mosaics to provide the aerial picture. Mosaics are of principal use for presenting synoptic views of a relatively large area, and as indexes of individual photographs. The features are usually labeled or "annotated" to facilitate the recognition of critical areas, or overlays may be prepared to show planning developments or contemplated changes in an area. For example, a mosaic can show the location of a proposed highway. Using this technique, laypersons, and particularly property owners adjacent to the proposed highway, can visualize its effect on their holdings. Photo indexes are used as a visual record of the photographs of a project and as a reference of availability of photographs over a particular area of terrain. Photo maps are particularly useful for land use, land cover delineation, land planning, zoning, tax maps, and preliminary engineering design.

5.1.2 Methods of Mosaic Construction

There are three principal construction methods for the making of mosaics:

(a) From Paper Prints

This is the simplest method. The paper prints that are to be joined are laid one on top of the other so that commonly imaged terrain is matched. Then the paper prints are cut, along a feature if possible, so that the edge joins will be as inconspicuous as possible. The photographs are matched and pasted to a base. As necessary, the photographs are stretched to assure a good match of features across the join line. The completed mosaic may be photographed if a reduced scale mosaic, such as an index mosaic, is desired.

(b) From Negatives or Diapositives

This method is called photomechanical mosaicking. The negatives to be used for the construction are overlaid and stud registered. After the join lines are selected, a mask is made that is opaque on one side of the join line and clear on the other. A negative of the mask is made for the other photo of the pair to be joined. These masks are also stud registered. After all negatives and masks have been stud registered, a piece of unexposed film is put down (in the darkroom) over the studs and the first photo negative and its corresponding mask are registered and an exposure made. The mask permits only that portion of the unexposed film under the negative to be exposed. Successive portions of the film are exposed as this process is repeated for each negative and mask until the whole film is exposed as a single mosaic.

(c) From Digital Imagery

Digital (or digitized) imagery can be used to create a mosaic. This method has been used most commonly with small-scale satellite photographs. The method is very computer intensive. Either by human or automated correlation, common points on the two images to be joined are selected. The computer can then match each pixel of the one image with the corresponding pixel of the second image. A join line is decided upon and the two images digitally "sheared" along the join line. The two sheared images are then brought together into a single file having a common pixel coordinate system.

5.1.3 Types of Mosaics and Air Photo Plan Maps

Mosaics or air photo plans may be uncontrolled, semiuncontrolled, or controlled. They may be constructed from unrectified, rectified, or differentially rectified photographs. A controlled mosaic or air photo plan is prepared using photographs that have been rectified to an equal scale, while an uncontrolled mosaic is prepared by a "best fit" match of a series of individual photographs.

5.2 Orthophotographs

Orthophotographs are photographs constructed from vertical or near-vertical aerial photographs, such that the effects of central perspective, relief displacement, and tilt are (practically) removed. The resulting orthophotograph is an orthographic project.

(a) Orthophotomaps

Orthophoto maps are orthophotographs with overlaid line map data. The common line data overlays include grids, property lines, political boundaries, geographic names, and other selected cultural features.

(b) Limitation of Orthophotographs

The original aerial negative from which the orthophotograph is made is a central projection and, as such, displays relief displacement and obscuration of features. For example, a building will obscure the terrain that lies behind it. This obscuration results in gaps of information that can be gotten only from other sources of information, such as field survey or separate photograph. Also, optical orthorectifying devices have a finite width and length slot width that can "rectify" only to an average elevation within that slot. Thus, small objects, such as trees, buildings, etc., are not rectified at all. Normally, only a "smoothed" approximation to the actual terrain is rectified. Digital rectifiers with spot diameters as small as 5 to 25 micrometers can, to a much greater degree, overcome this latter shortcoming.

5.2.1 Uses of Orthophotographs

Orthophotos and orthophoto maps are widely used for general planning purposes. The complete and accurate display of all features in a project area is an ideal medium for demonstrating terrain features to laypersons. Proposed designs of engineering projects can be superimposed on the orthophoto map for a vivid understanding of work to be accomplished. Many factors must be considered before deciding on a photomap product instead of a topographical map product. The intended use of the product needs to be classified exactly, together with the accuracy required and whether height information is required or not. For many USACE engineering, operations, and maintenance activities, rectified airphoto plan sheets may be used in lieu of fully rectified orthophotographs.

5.2.2 Methods of Orthophotograph Preparation

Orthophotographs are prepared from pairs of overlapping aerial photographs using specially designed orthoplotting instruments. The photographs are oriented in the instrument in the conventional manner for standard stereo photogrammetric mapping. The requirements for ground control or control to be established by aerotriangulation are the same as for photogrammetric mapping. The instrument provides the means of scanning the stereomodel to effect corrections for the varying scales caused by topographic relief. The tilt and other distortions are corrected in the orientation of the stereomodel.

Elements of Analytical Photogrammetry

1. Concept of Image and Object Space

Photogrammetry is the science of obtaining reliable information about objects and of measuring and interpreting this information. The task of obtaining information is called data acquisition, a process. Figure 2.30 depicts the data acquisition process. Light rays reflected from points on the object, say from point *A*, form a divergent bundle which is transformed to convergent bundle by the lens. The principal rays of each bundle of all object points pass through the center of the entrance and exit pupil, unchanged in direction. The front and rear nodal points are good approximations for the pupil centers.

Another major task of photogrammetry is concerned with reconstructing the object space from images. This entails two problems: geometric reconstruction (*e.g.*, the position of objects) and radiometric reconstruction (*e.g.*, the gray shades of a surface). The latter problem is relevant when photographic products are generated, such as orthophotos.

Photogrammetry is mainly concerned with the geometric reconstruction. The object space is only partially reconstructed, however. With partial reconstruction we mean that only a fraction of the information recorded from the object space is used for its representation. Take a map, for example. It may only show the perimeter of buildings, not all the intricate details which make up real buildings.

Obviously, the success of reconstruction in terms of geometrical accuracy depends largely on the similarity of the image bundle compared to the bundle of principal rays that entered the lens during the instance of exposure. The purpose of camera calibration is to define an image space so that the similarity becomes as close as possible.

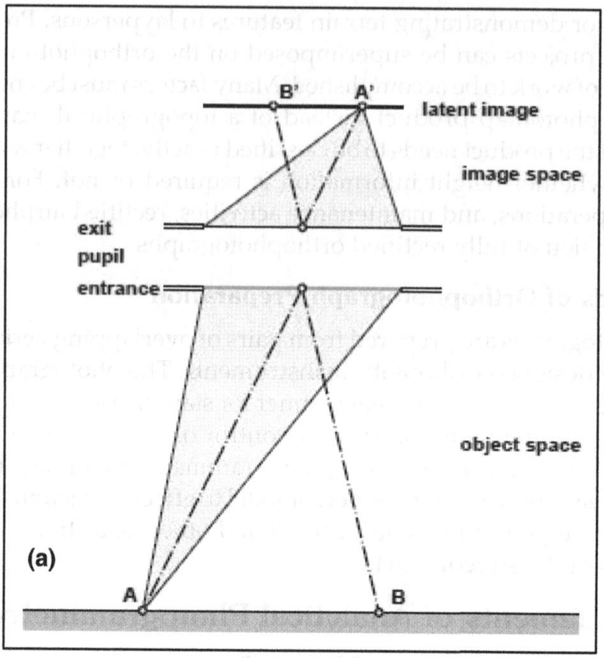

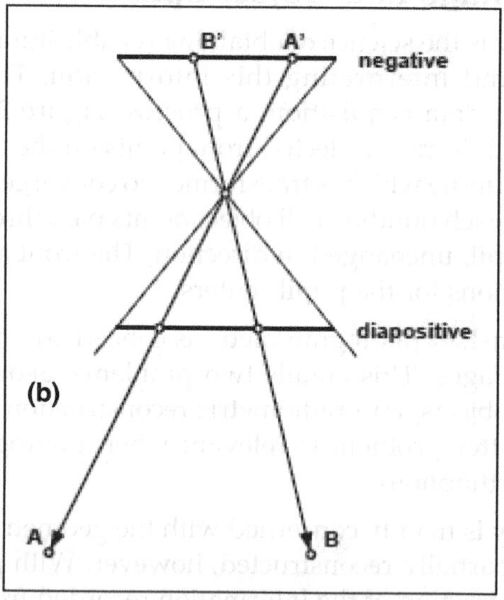

Figure 2.30: (a) The Data Acquisition Process is Depicted. (b) Illustrate the Reconstruction Process.

The geometrical relationship between image and object space can best be established by introducing suitable coordinate systems for referencing both spaces. We describe the coordinate systems in the next section. Various relationships exist between images and object space. In Table 2.2 the most common relationships are summarized, together with the associated photogrammetric procedures and the underlying mathematical models. In this chapter we describe these procedures and the mathematical models, except aerotriangulation (block adjustment). For one and the same procedure, several mathematical models may exist. They differ mainly in the degree of complexity, that is, how closely they describe physical processes. For example, a similarity transformation is a good approximation to describe the process of converting measured coordinates to photo-coordinates. This simple model can be extended to describe more closely the underlying measuring process. With a few exceptions, we will not address the refinement of the mathematical model.

Table 2.2: Summary of the Most Important Relationships between Image and Object Space.

Relationship between	*Procedure*	*Mathematical Model*
Measuring system and photo-coordinate system	Interior orientation	2-D transformation
Photo-coordinate system and object coordinate system	Exterior orientation	Collinearity eq.
Photo-coordinate systems of a stereopair	Relative orientation	Collinearity eq. Coplanarity condition
Model coordinate system and object coordinate system	Absolute orientation	7-Parametertransformation
Several photo-coordinate systems and object coordinate system	Bundle block adjustment	Collinearity equation
Several model coordinate systems and object coordinate system	Independent model block adjustment	7-Parameter transformation

2. Coordinate Systems

2.1 Photo-Coordinate System

The photo-coordinate system serves as the reference for expressing spatial positions and relations of the image space. It is a 3-D Cartesian system with the origin at the perspective center. Figure 2.31 depicts a diapositive with fiducial marks that define the fiducial center *FC*. During the calibration procedure, the offset between fiducial center and principal point of autocollimation, *PP*, is determined, as well as the origin of the radial distortion, *PS*. The x, y coordinate plane is parallel to the photograph and the positive x"axis points toward the flight direction.

Positions in the image space are expressed by point vectors. For example, point vector **p** defines the position of point P on the diapositive (see Figure 2.31). Point vectors of positions on the diapositive (or negative) are also called *image vectors*. We have for point P

$$p = \begin{bmatrix} x_p \\ y_p \\ -c \end{bmatrix} \tag{1}$$

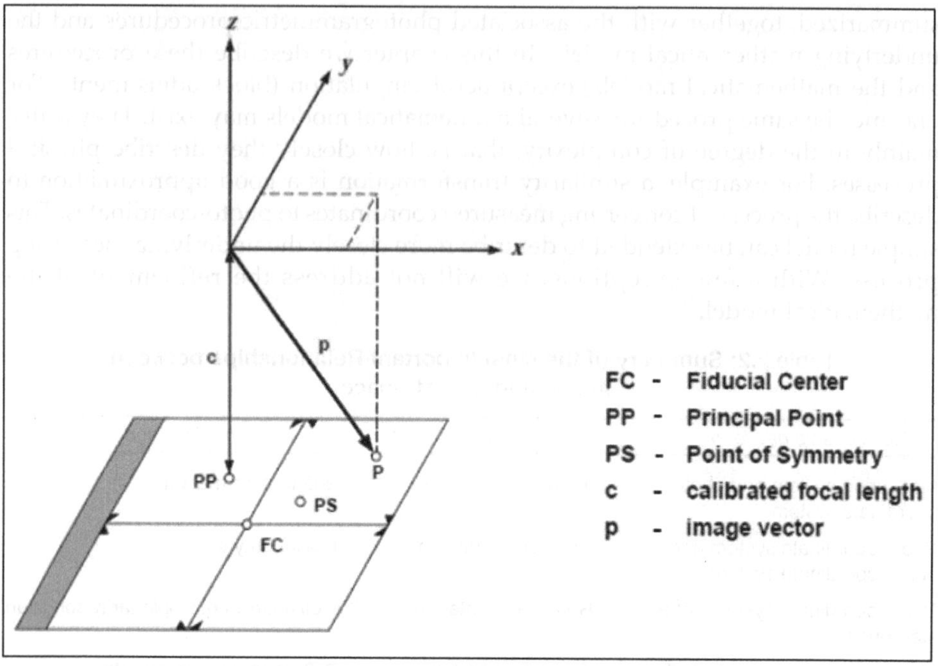

Figure 2.31: Definition of the Photo-coordinate System.

Note that for a diapositive the third component is negative. This changes to a positive value in the rare case a negative is used instead of a diapositive.

2.2 Object Space Coordinate Systems

In order to keep the mathematical development of relating image and object space simple, both spaces use 3-D Cartesian coordinate systems. Positions of control points in object space are likely available in another coordinate systems, *e.g.*, State Plane coordinates. It is important to convert any given coordinate system to a Cartesian system before photogrammetric procedures, such as orientations or aerotriangulation, are performed.

3. Interior Orientation

We have already introduced the term *interior orientation* in the discussion about camera calibration, to define the metric characteristics of aerial cameras. Here we use the same term for a slightly different purpose. From Table 2.2 we conclude that the purpose of interior orientation is to establish the relationship between a measuring system and the photo-coordinate system. This is necessary because it is not possible to measure photo-coordinates directly. One reason is that the origin of the photo-coordinate system is only mathematically defined; since it is not visible it cannot coincide with the origin of the measuring system.

Figure 2.32 illustrates the case where the diapositive to be measured is inserted in the measuring system whose coordinate axis are *xm, ym*. The task is to determine

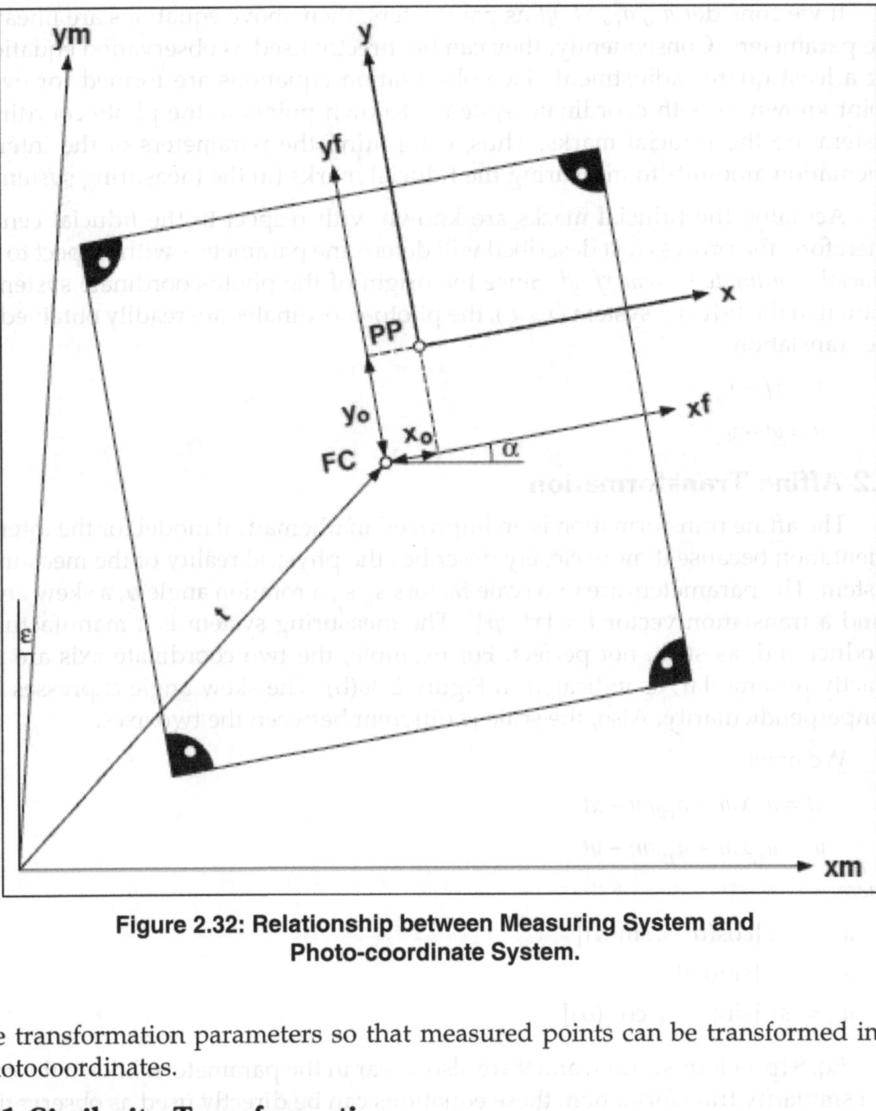

Figure 2.32: Relationship between Measuring System and Photo-coordinate System.

the transformation parameters so that measured points can be transformed into photocoordinates.

3.1 Similarity Transformation

The simplest mathematical model for interior orientation is a similarity transformation with the four parameters: translation vector **t**, scale factor *s*, and rotation angle α.

$$xf = s(xm\cos(\alpha) - ym\sin(\alpha)) - xt \tag{2}$$

$$yf = s(xm\sin(\alpha) + ym\cos(\alpha)) - yt \tag{3}$$

These equations can also be written in the following form:

$$xf = a_{11}xm - a_{12}ym - xt \tag{4}$$

$$yf = a_{12}xm + a_{11}ym - yt \tag{5}$$

If we consider a_{11}, a_{12}, xt, yt as parameters, then above equations are linear in the parameters. Consequently, they can be directly used as observation equations for a leastsquares adjustment. Two observation equations are formed for every point known in both coordinate systems. Known points in the photo-coordinate system are the fiducial marks. Thus, computing the parameters of the interior orientation amounts to measuring the fiducial marks (in the measuring system).

Actually, the fiducial marks are known with respect to the fiducial center. Therefore, the process just described will determine parameters with respect to the *fiducial coordinate system xf, yf*. Since the origin of the photo-coordinate system is known in the fiducial system (x_0, y_0), the photo-coordinates are readily obtained by the translation

$$x = xf - x_0 \tag{6}$$

$$y = yf - y_0 \tag{7}$$

3.2 Affine Transformation

The affine transformation is an improved mathematical model for the interior orientation because it more closely describes the physical reality of the measuring system. The parameters are two scale factors s_x, s_y, a rotation angle α, a skew angle , and a translation vector $\mathbf{t} = [xt, yt]^T$. The measuring system is a manufactured product and, as such, not perfect. For example, the two coordinate axis are not exactly rectangular, as indicated in Figure 2.30(b). The skew angle expresses the nonperpendicularity. Also, the scale is different between the two axes.

We have

$$xf = a_{11}xm + a_{12}ym - xt \tag{8}$$

$$yf = a_{21}xm + a_{22}ym - yt \tag{9}$$

where,

$$a_{11} = sx\,[\cos(\alpha - \varepsilon \sin(\alpha)]$$

$$a_{12} = sy\,[\sin(\alpha)]$$

$$a_{21} = sx\,[\sin(\alpha + \varepsilon \cos(\alpha)]$$

Eq. 8 (previous section) and 9 are also linear in the parameters. Like in the case of a similarity transformation, these equations can be directly used as observation equations. With four fiducial marks we obtain eight equations leaving a redundancy of two.

3.3 Correction for Radial Distortion

As discussed in previous section, radial distortion causes off-axial points to be radially displaced. A positive distortion increases the lateral magnification while a negative distortion reduces it. Distortion values are determined during the process of camera calibration. They are usually listed in tabular form, either as a function of the radius or the angle at the perspective center. For aerial cameras the distortion values are very small. Hence, it suffices to linearly interpolate the distortion. Suppose we want to determine the distortion for image point x_p, y_p. The radius is $r_p = (x^2_p +$

$y^2_p)^{1/2}$. From the table we obtain the distortion dr_i for $r_i < r_p$ and dr_j for $r_j > r_p$. The distortion for r_p is interpolated

$$dr_p = \frac{(dr_j - dr_i)r_p}{(r_j - r_i)} \tag{10}$$

As indicated in Figure 2.33 the corrections in x- and y-direction are

$$dr_x = \frac{x_p}{r_p} dr_p \tag{11}$$

$$dr_y = \frac{y_p}{r_p} dr_p \tag{12}$$

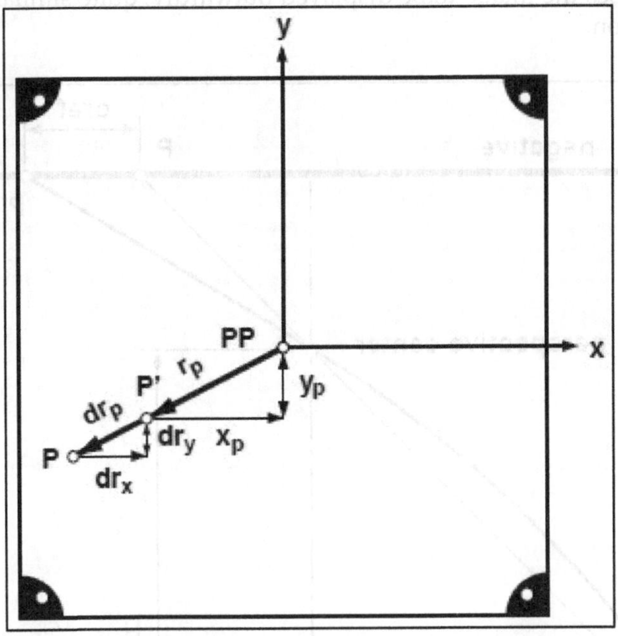

Figure 2.33: Correction for Radial Distortion.

Finally, the photo-coordinates must be corrected as follows:

$$x_p = x_p - dr_x = x_p\left(1 - \frac{dr_p}{r_p}\right) \tag{13}$$

$$y_p = y_p - dr_y = y_p\left(1 - \frac{dr_p}{r_p}\right) \tag{14}$$

The radial distortion can also be represented by an odd-power polynomial of the Form

$$dr = p_0 r + p_1 r^3 + p_2 r^5 + \ldots\ldots\ldots\ldots \qquad (15)$$

The coefficients *pi* are found by fitting the polynomial curve to the distortion values. Equation (15) is a linear observation equation. For every distortion value, an observation equation is obtained.

In order to avoid numerical problems (ill-conditioned normal equation system), the degree of the polynom should not exceed nine.

3.4 Correction for Refraction

Figure 2.34 shows how an oblique light ray is refracted by the atmosphere. According to Snell's law, a light ray is refracted at the interface of two different media. The density differences in the atmosphere are in fact different media. The refraction causes the image to be displayed outwardly, quite similar to a positive radial distortion.

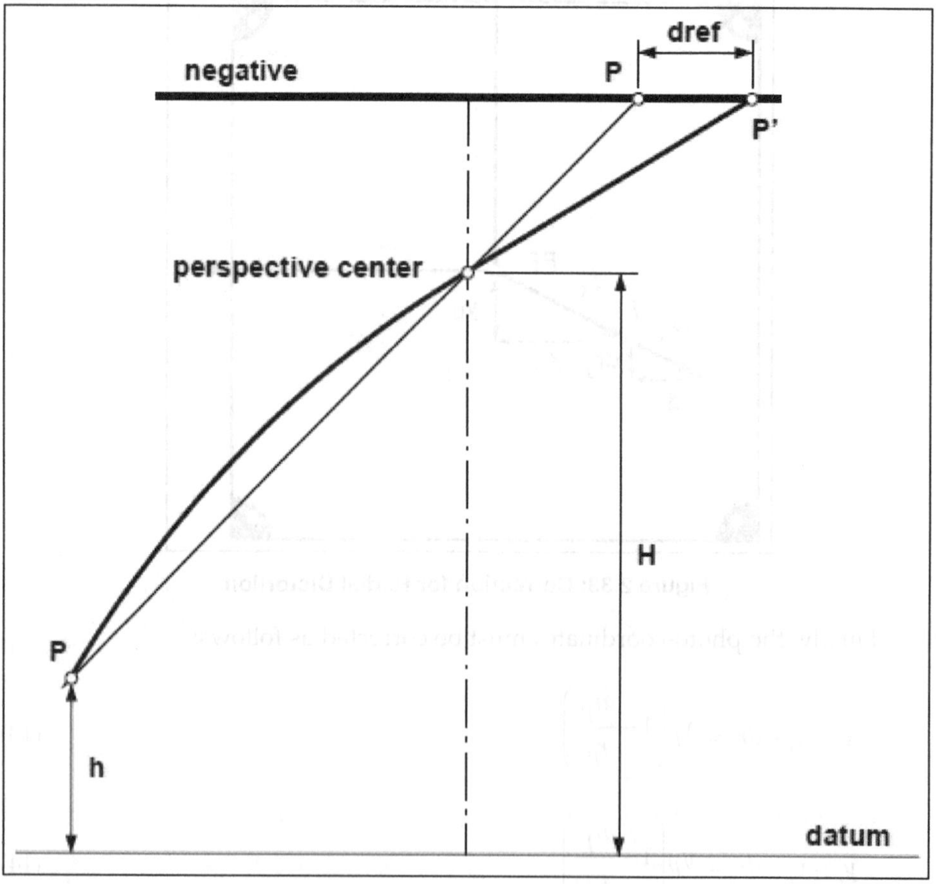

Figure 2.34: Correction for Refraction.

The radial displacement caused by refraction can be computed by:

$$d_{ref} = K\left(r + \frac{r^3}{c^2}\right) \tag{16}$$

$$K = \left[\frac{2410H}{H^2 - 6H + 250} - \frac{2410H^2}{(h^2 - 6H + 250)H}\right]10^{-6} \tag{17}$$

These equations are based on a model atmosphere defined by the USAir Force. The flying height H and the ground elevation h must be in units of kilometers.

3.5 Correction for Earth Curvature

As mentioned in the beginning of this Chapter, the mathematical derivation of the relationships between image and object space are based on the assumption that for both spaces, 3-D Cartesian coordinate systems are employed. Since ground control points may not directly be available in such a system, they must first be transformed, say from a State Plane coordinate system to a Cartesian system.

The X and Y coordinates of a State Plane system are Cartesian, but not the elevations. Figure 2.35 shows the relationship between elevations above a datum and elevations in the 3-D Cartesian system. If we approximate the datum by a sphere, radius $R = 6372.2$ km, then the radial displacement can be computed by

$$d_{earth} = \frac{r^3(H - ZP)}{2c^2R} \tag{18}$$

Like radial distortion and refraction, the corrections in x- and y-direction is readily determined by Equation (14). Strictly speaking, the correction of photo-coordinates due to earth curvature is not a refinement of the mathematical model. It is much better to eliminate the influence of earth curvature by transforming the object space into a 3-D Cartesian system before establishing relationships with the ground system. This is always possible, except when compiling a map. A map, generated on an analytical plotter, for example, is most likely plotted in a State Plane coordinate system. That is, the elevations refer to the datum and not to the XY plane of the Cartesian coordinate system. It would be quite awkward to produce the map in the Cartesian system and then transform it to the target system. Therefore, during map compilation, the photocoordinates are "corrected" so that conjugate bundle rays intersect in object space at positions related to reference sphere.

3.6 Summary of Computing Photo-Coordinates

We summarize the main steps necessary to determine photo-coordinates. The process to correct them for systematic errors, such as radial distortion, refraction and earth curvature is also known as *image refinement*. Figure 2.36 depicts the coordinate systems involved, an imaged point P, and the correction vectors dr, d_{ref}, d_{earth}.

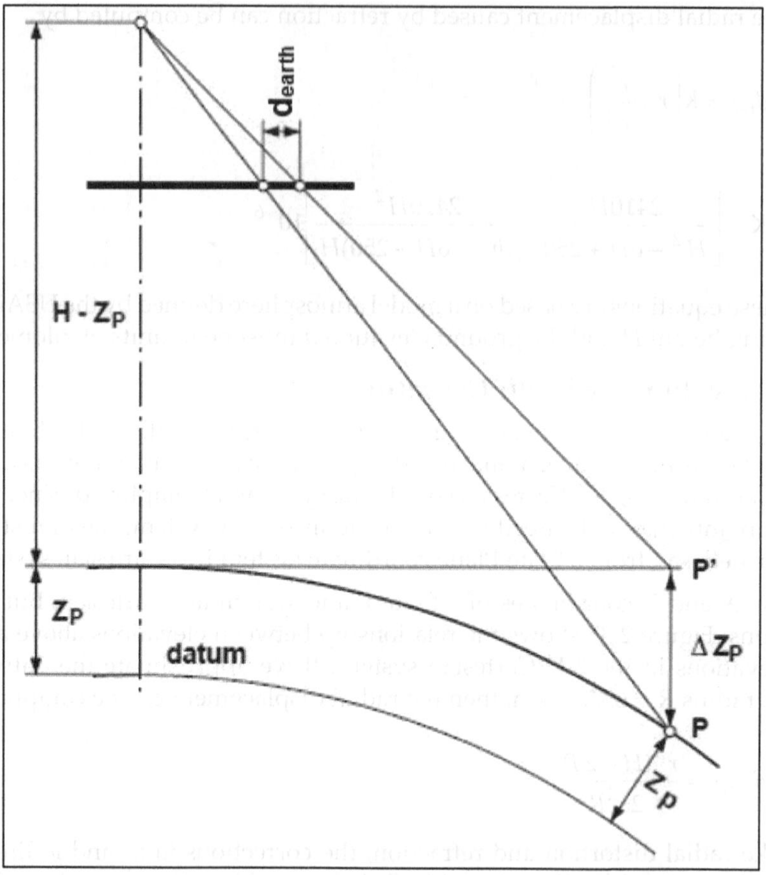

Figure 2.35: Correction of Photo-coordinates Due to Earth Curvature.

- ☆ Insert the diapositive into the measuring system (*e.g.*, comparator, analytical plotter) and measure the fiducial marks in the machine coordinate system x_m, y_m. Compute the transformation parameters with a similarity or affine transformation. The transformation establishes a relationship between the measuring system and the fiducial coordinate system.
- ☆ Translate the fiducial system to the photo-coordinate system [Equation (7)].
- ☆ Correct photo-coordinates for radial distortion. The radial distortion dr_p for point P is found by linearly interpolating the values given in the calibration protocol [Equation (10)].
- ☆ Correct the photo-coordinates for refraction, according to Equation (17). This correction is negative. The displacement caused by refraction is a functional relationship of $d_{ref} = f(H, h, r, c)$. With a flying height $H = 2,000$ m, elevation above ground $h = 500$ m we obtain for a wide angle camera

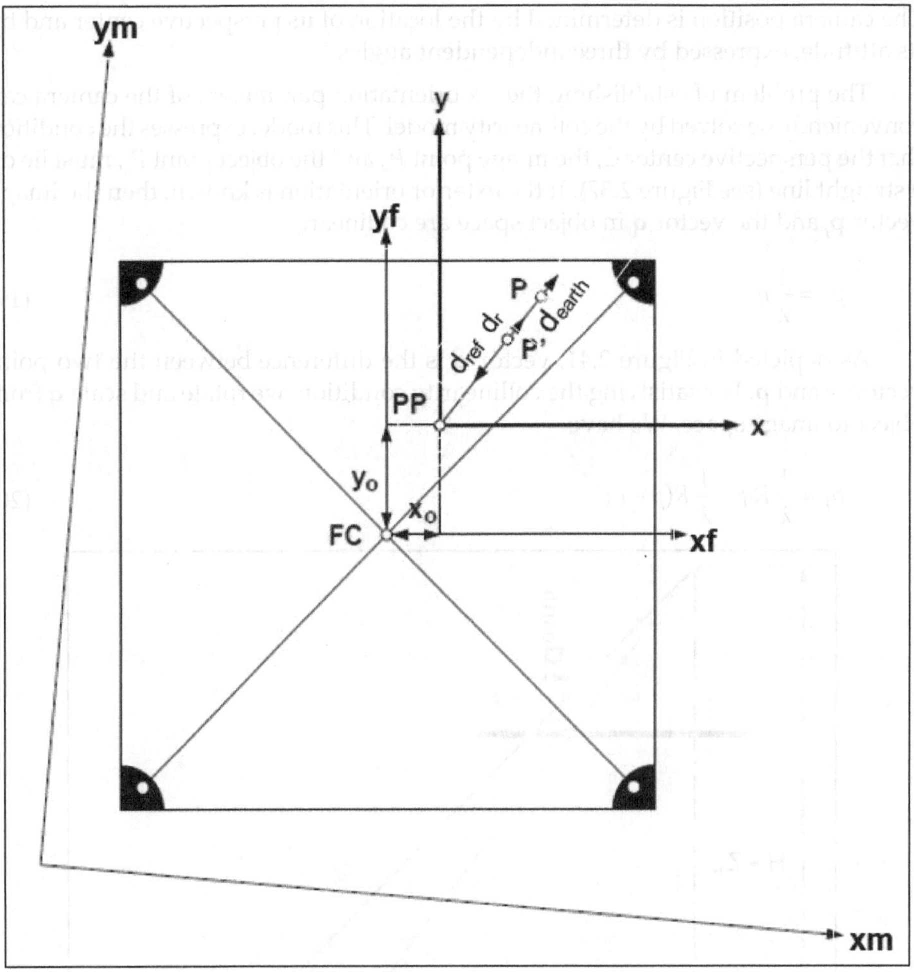

Figure 2.36: Interior Orientation and Image Refinement.

($c \approx 0.15$ m) a correction of "4µm for r = 130 mm. An extreme example is a superwide angle camera, $H = 9,000$ m, $h = 500$ m, where d_{ref} = "34 µm for the same point.

☆ Correct for earth curvature only if the control points (elevations) are not in a cartesian coordinate system or if a map is compiled. Using the extreme example as above, we obtain $d_{earth} = 65\ \mu m$. Since this correction has the opposite sign of the refraction, the combined correction for refraction and earth curvature would be $d_{comb} = 31\ \mu m$. The correction due to earth curvature is larger than the correction for refraction.

4. Exterior Orientation

Exterior orientation is the relationship between image and object space. This is accomplished by determining the camera position in the object coordinate system.

The camera position is determined by the location of its perspective center and by its attitude, expressed by three independent angles.

The problem of establishing the six orientation parameters of the camera can conveniently be solved by the collinearity model. This model expresses the condition that the perspective center C, the image point P_i, and the object point P_o, must lie on a straight line (see Figure 2.37). If the exterior orientation is known, then the image vector $\mathbf{p}_i$ and the vector $\mathbf{q}$ in object space are collinear:

$$p_i = \frac{1}{\lambda} q \qquad (19)$$

As depicted in Figure 2.41, vector **q** is the difference between the two point vectors **c** and **p**. For satisfying the collinearity condition, we rotate and scale **q** from object to image space. We have

$$p_i = \frac{1}{\lambda} Rq = \frac{1}{\lambda} R(p - c) \qquad (20)$$

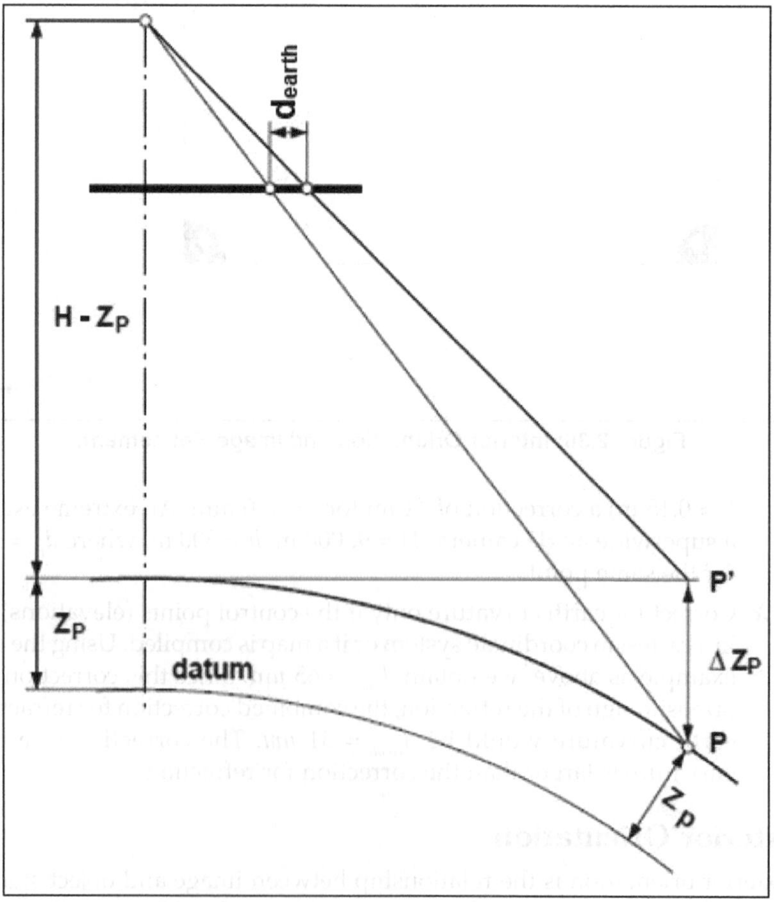

Figure 2.37: Exterior Orientation.

With **R** an orthogonal rotation matrix with the three angles ω, φ and κ:

$$\mathbf{R} = \begin{vmatrix} \cos\varphi\cos\kappa & -\cos\varphi\sin\kappa & \sin\varphi \\ \cos\omega\sin\kappa + \sin\omega\sin\varphi\cos\kappa & \cos\omega\cos\kappa - \sin\omega\sin\varphi\sin\kappa & -\sin\omega\cos\varphi \\ \sin\omega\sin\kappa - \cos\omega\sin\varphi\cos\kappa & \sin\omega\cos\kappa + \cos\omega\sin\varphi\sin\kappa & \cos\omega\cos\varphi \end{vmatrix} \quad (21)$$

Eq. 20 renders the following three coordinate equations.

$$x = \frac{1}{\lambda}\,(X_P - X_C)r_{11} + (Y_P - Y_C)r_{12} + (Z_P - Z_C)r_{13} \quad (22)$$

$$1\,y = \frac{1}{\lambda}\,(X_P - X_C)r_{21} + (Y_P - Y_C)r_{22} + (Z_P - Z_C)r_{23} \quad (23)$$

$$-c = \frac{1}{\lambda}\,(X_P - X_C)r_{31} + (Y_P - Y_C)r_{32} + (Z_P - Z_C)r_{33} \quad (24)$$

By dividing the first by the third and the second by the third equation, the scale factor $1/\lambda$ is eliminated leading to the following two collinearity equations:

$$x = -c\,\frac{(X_P - X_C)r_{11} + (Y_P - Y_C)r_{12} + (Z_P - Z_C)r_{13}}{(X_P - X_C)r_{31} + (Y_P - Y_C)r_{32} + (Z_P - Z_C)r_{33}} \quad (25)$$

$$y = -c\,\frac{(X_P - X_C)r_{21} + (Y_P - Y_C)r_{22} + (Z_P - Z_C)r_{23}}{(X_P - X_C)r_{31} + (Y_P - Y_C)r_{32} + (Z_P - Z_C)r_{33}} \quad (26)$$

with:

$$p_i = \begin{bmatrix} x \\ y \\ -f \end{bmatrix} \quad p = \begin{bmatrix} X_P \\ Y_P \\ Z_P \end{bmatrix} \quad c = \begin{bmatrix} X_C \\ Y_C \\ Z_C \end{bmatrix}$$

The six parameters: X_C, Y_C, Z_C, ω, φ, κ are the unknown elements of exterior orientation. The image coordinates x, y are normally known (measured) and the calibrated focal length c is a constant. Every measured point leads to two equations, but also adds three other unknowns, namely the coordinates of the object point (X_P, Y_P, Z_P). Unless the object points are known (control points), the problem cannot be solved with only one photograph.

The collinearity model as presented here can be expanded to include parameters of the interior orientation. The number of unknowns will be increased by three. This combined approach lets us determine simultaneously the parameters of interior and exterior orientation of the cameras.

There are only limited applications for single photographs. We briefly discuss the computation of the exterior orientation parameters, also known as single photograph resection, and the computation of photo-coordinates with known orientation parameters. Single photographs cannot be used for the main task of photogrammetry, the reconstruction of object space. Suppose we know the exterior orientation of a photograph. Points in object space are not defined, unless we also know the scale factor $1/\lambda$ for every bundle ray.

4.1 Single Photo Resection

The position and attitude of the camera with respect to the object coordinate system (exterior orientation of camera) can be determined with help of the collinearity equations. Euation (26) express measured quantities as a function of the exterior orientation parameters. Thus, the collinearity equations can be directly used as observation equations, as the following functional representation illustrates.

$$x, y = f(\underbrace{X_C, Y_C, Z_C, \omega, \varphi, \kappa,}_{\text{Exterior orientation}} \underbrace{X_P, Y_P, Z_P}_{\text{Object point}})$$

For every measured point two equations are obtained. If three control points are measured, a total of 6 equations is formed to solve for the 6 parameters of exterior orientation.

The collinearity equations are not linear in the parameters. Therefore, Eduation (25) (previous section) and 26 must be linearized with respect to the parameters. This also requires approximate values with which the iterative process will start.

4.2 Computing Photo Coordinates

With known exterior orientation elements photo-coordinates can be easily computed from Eqs. 26. This is useful for simulation studies where synthetic photocoordinates are computed. Another application for the direct use of the collinearity equations is the real-time loop of analytical plotters where photo-coordinates of ground points or model points are computed after relative or absolute orientation.

5. Orientation of a Stereopair

5.1 Model Space, Model Coordinate System

The application of single photographs in photogrammetry is limited because they cannot be used for reconstructing the object space. Even though the exterior orientation elements may be known it will not be possible to determine ground points unless the scale factor of every bundle ray is known. This problem is solved by exploiting stereopsis that is by using a second photograph of the same scene, taken from a different position.

Two photographs with different camera positions that show the same area, at least in part, is called a *stereopair*. Suppose the two photographs are oriented such that *conjugate points* (corresponding points) intersect. We call this intersection space *model space*. In order for expressing relationships of this model space we introduce

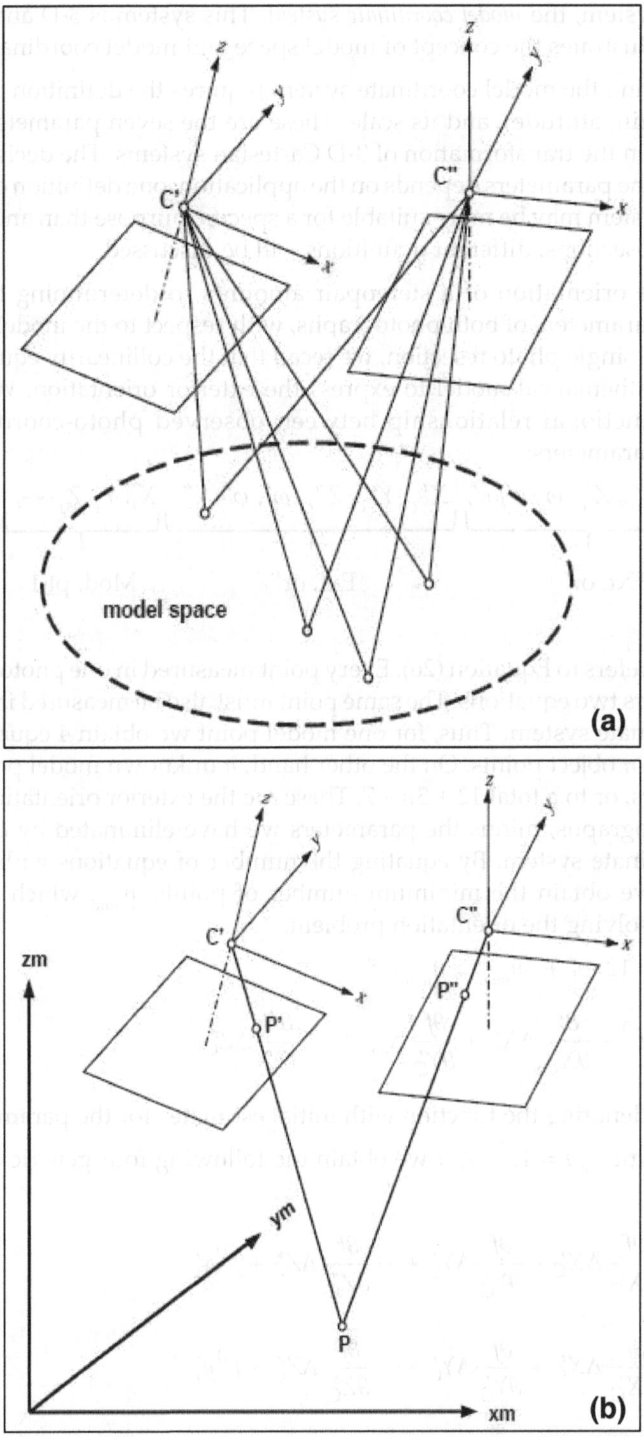

Figure 2.38: The Concept of Model Space (a) and Model Coordinate System (b).

a reference system, the *model coordinate system*. This system is 3-D and Cartesian. Figure 2.38 illustrates the concept of model space and model coordinate system.

Introducing the model coordinate system requires the definition of its spatial position (origin, attitude), and its scale. These are the seven parameters we have encountered in the transformation of 3-D Cartesian systems. The decision on how to introduce the parameters depends on the application; one definition of the model coordinate system may be more suitable for a specific purpose than another. In the following subsections, different definitions will be discussed.

Now the orientation of a stereopair amounts to determining the exterior orientation parameters of both photographs, with respect to the model coordinate system. From single photo resection, we recall that the collinearity equations form a suitable mathematical model to express the exterior orientation. We have the following functional relationship between observed photo-coordinates and orientation parameters:

$$x, y = f\,(\underbrace{X'_C, Y'_C, Z'_C, \omega', \varphi', \kappa'}_{\text{Ext. or}'},\ \underbrace{X''_C, Y''_C, Z''_C, \omega'', \varphi'', \kappa''}_{\text{Ext. or}''},\ \underbrace{X_1, Y_1, Z_1}_{\text{Mod. pt 1}}\text{---},\ \underbrace{Xn, Yn, Zn}_{\text{Mod. pt n}}) \tag{28}$$

Where f refers to Equation (26). Every point measured in one photo-coordinate system renders two equations. The same point must also be measured in the second photo-coordinate system. Thus, for one model point we obtain 4 equations, or $4n$ equations for n object points. On the other hand, n unknown model points lead to $3n$ parameters, or to a total $12 + 3n - 7$. These are the exterior orientation elements of both photographs, minus the parameters we have eliminated by defining the model coordinate system. By equating the number of equations with number of parameters we obtain the minimum number of points, n_{min}, which we need to measure for solving the orientation problem.

$$4\,n_{min} = 12 - 7 + 3n_{min} \Rightarrow n_{min} = 5 \tag{29}$$

$$x, y \approx f^0 + \frac{\vartheta f}{\vartheta X'_C}\Delta X'_C + \frac{\vartheta f}{\vartheta Y'_C}\Delta Y'_C + \cdots \frac{\vartheta f}{\vartheta Z''_C}\Delta Z''_C \tag{30}$$

With f° denoting the function with initial estimates for the parameters.

For a point P_i, $i = 1, ---, n$ we obtain the following four generic observation equations

$$r'_{xi} = \frac{\vartheta f}{\vartheta X'_C}\Delta X'_C + \frac{\vartheta f}{\vartheta Y'_C}\Delta Y'_C + \cdots \frac{\vartheta f}{\vartheta Z''_C}\Delta Z''_C + f^0 x'_i$$

$$r'_{yi} = \frac{\vartheta f}{\vartheta X'_C}\Delta X'_C + \frac{\vartheta f}{\vartheta Y'_C}\Delta Y'_C + \cdots \frac{\vartheta f}{\vartheta Z''_C}\Delta Z''_C + f^0 y'_i \tag{31}$$

$$r''_{xi} = \frac{\vartheta f}{\vartheta X'_C}\Delta X'_C + \frac{\vartheta f}{\vartheta Y'_C}\Delta Y'_C + \cdots \frac{\vartheta f}{\vartheta Z''_C}\Delta Z''_C + f^0 x''_i$$

$$r''_{yi} = \frac{\vartheta f}{\vartheta X'_C}\Delta X'_C + \frac{\vartheta f}{\vartheta Y'_C}\Delta Y'_C + \cdots \frac{\vartheta f}{\vartheta Z''_C}\Delta Z''_C + f^0 y''_i$$

As mentioned earlier, the definition of the model coordinate system reduces the number of parameters by seven. Several techniques exist to consider this in the least squares approach.

(i) The simplest approach is to eliminate the parameters from the parameter list. We will use this approach for discussing the dependent and independent relative orientation.

(ii) The knowledge about the 7 parameters can be introduced in the mathematical model as seven independent pseudo observations (*e.g.*, $\Delta X_C = 0$), or as condition equations which are added to the normal equations. This second technique is more flexible and it is particularly suited for computer implementation.

5.2 Dependent Relative Orientation

The definition of the model coordinate system in the case of a dependent relative orientation is depicted in Figure 2.39. The position and the orientation is identical to one of the two photo-coordinate systems, say the primed system. This step amounts to introducing the exterior orientation of the photo-coordinate system as known. That is, we can eliminate it from the parameter list. Next, we define the scale of the model coordinate system. This is accomplished by defining the distance between the two perspective centers (base), or more precisely, by defining the X-component.

With this definition of the model coordinate system we are left with the following functional model

$$x, y = f(\underbrace{ym''_c, zm''_c, \omega'', \phi'', \kappa''}_{\text{ext. or}''}, \underbrace{xm_1, ym_1, zm_1}_{\text{model pt 1}}, \cdots, \underbrace{xm_n, ym_n, zm_n}_{\text{model pt n}}) \quad (32)$$

With 5 points we obtain 20 observation equations. On the other hand, there are 5 exterior orientation parameters and 5×3 model coordinates. Usually more than 5 points are measured. The redundancy is $r = n - 5$. The typical case of relative orientation on a stereoplotter with the 6 von Gruber points leads only to a redundancy of one. It is highly recommended to measure more, say 12 points, in which case we find $r = 7$.

With a non linear mathematical model we need be concerned with suitable approximations to ensure that the iterative least squares solution converges. In the case of the dependent relative orientation we have:

$$f^0 = f(yc^0_c, zm^0_c, \omega^0, \phi^0, \kappa^0, xm^0_1, ym^0_1, zm^0_1, \cdots, xm^0_n, ym^0_n, zm^0_n) \quad (33)$$

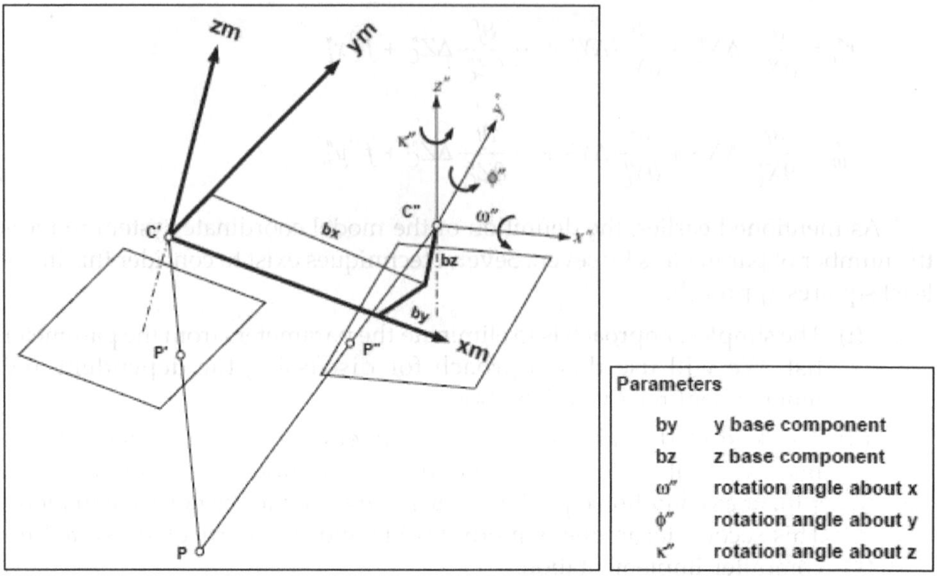

Figure 2.39: Definition of the Model Coordinate System and Orientation Parameters in the Dependent Relative Orientation.

The initial estimates for the five exterior orientation parameters are set to zero for aerial applications, because the orientation angles are smaller than five degrees, and $xm_c >> ym_c, xm_c >> zm_c =! ym^o_c = zm^o_c = 0$. Initial positions for the model points can be estimated from the corresponding measured photo-coordinates. If the scale of the model coordinate system approximates the scale of the photo-coordinate system, we estimate initial model points by

$$\begin{aligned} xm^o_i &\approx x'_i \\ ym^o_i &\approx y'_I \\ zm^o_i &\approx z'_i \end{aligned} \tag{34}$$

The dependent relative orientation leaves one of the photographs unchanged; the other one is oriented with respect to the unchanged system. This is of advantage for the conjunction of successive photographs in a strip. In this fashion, all photographs of a strip can be joined into the coordinate system of the first photograph.

5.3 Independent Relative Orientation

Figure 2.40 illustrates the definition of the model coordinate system in the independent relative orientation. The origin is identical to one of the photo-coordinate systems, *e.g.*, in Figure 2.39 it is the primed system. The orientation is chosen such that the positive *xm*-axis passes through the perspective center of the other photo-coordinate system. This requires determining two rotation angles in the primed photo-coordinate system. Moreover, it eliminates the base components *by*, *bz*. The rotation about the *x*-axis (ω) is set to zero. This means that the *ym*-axis is

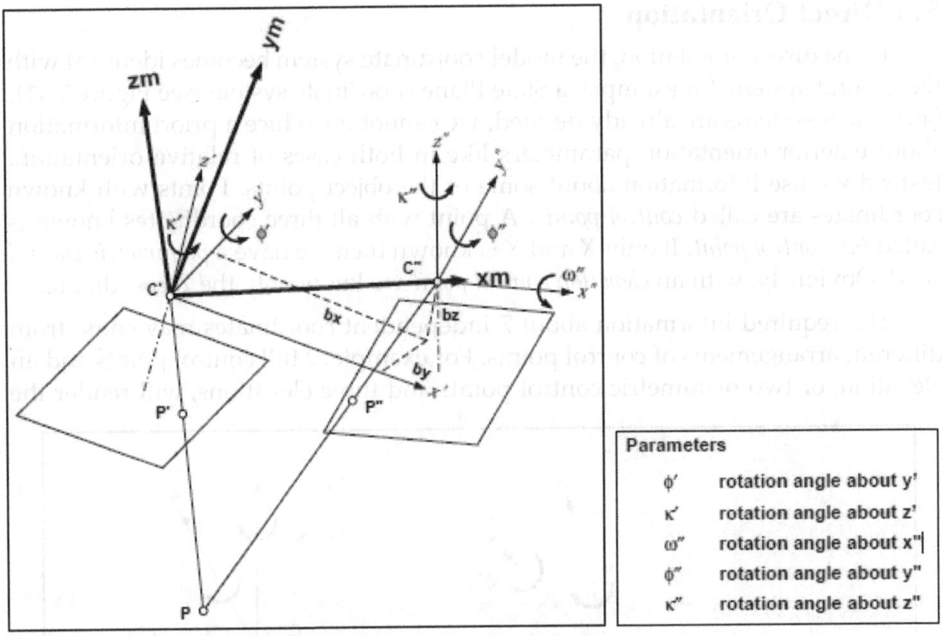

Figure 2.40: Definition of the Model Coordinate System and Orientation Parameters in the Independent Relative Orientation.

in the $x - y$ plane of the photo-coordinate system. The scale is chosen by defining $xm''_c = bx$.

With this definition of the model coordinate system we have eliminated the position of both perspective centers and one rotation angle. The following functional model applies

$$x, y = f\ (\underbrace{\varphi', \kappa'}_{\text{Ext. or}'},\ \underbrace{\omega'', \varphi'', \kappa''}_{\text{Ext. or}''},\ \underbrace{xm_1, ym_1, zm_1}_{\text{Mod. pt 1}},\ ---,\ \underbrace{xm_n, ym_n, zm_n}_{\text{Mod. pt n}}) \tag{35}$$

The number of equations, number of parameters and the redundancy are the same as in the dependent relative orientation. Also, the same considerations regarding initial estimates of parameters apply.

Note that the exterior orientation parameters of both types of relative orientation are related. For example, the rotation angles φ', κ' can be computed from the spatial direction of the base in the dependent relative orientation.

$$\phi' = \arctan\left(\frac{Zm''c}{bx}\right) \tag{36}$$

$$\kappa' = \arctan\left(\frac{Ym''c}{(bx^2 + Zm^2c)^{1/2}}\right) \tag{37}$$

5.4 Direct Orientation

In the direct orientation, the model coordinate system becomes identical with the ground system, for example, a State Plane coordinate system (see Figure 2.41). Since such systems are already defined, we cannot introduce a priori information about exterior orientation parameters like in both cases of relative orientation. Instead we use information about some of the object points. Points with known coordinates are called *control points*. A point with all three coordinates known is called *full control point*. If only X and Y is known then we have a *planimetric control point*. Obviously, with an *elevation control point* we know only the Z coordinate.

The required information about 7 independent coordinates may come from different arrangements of control points. For example, 2 full control points and an elevation, or two planimetric control points and three elevations, will render the

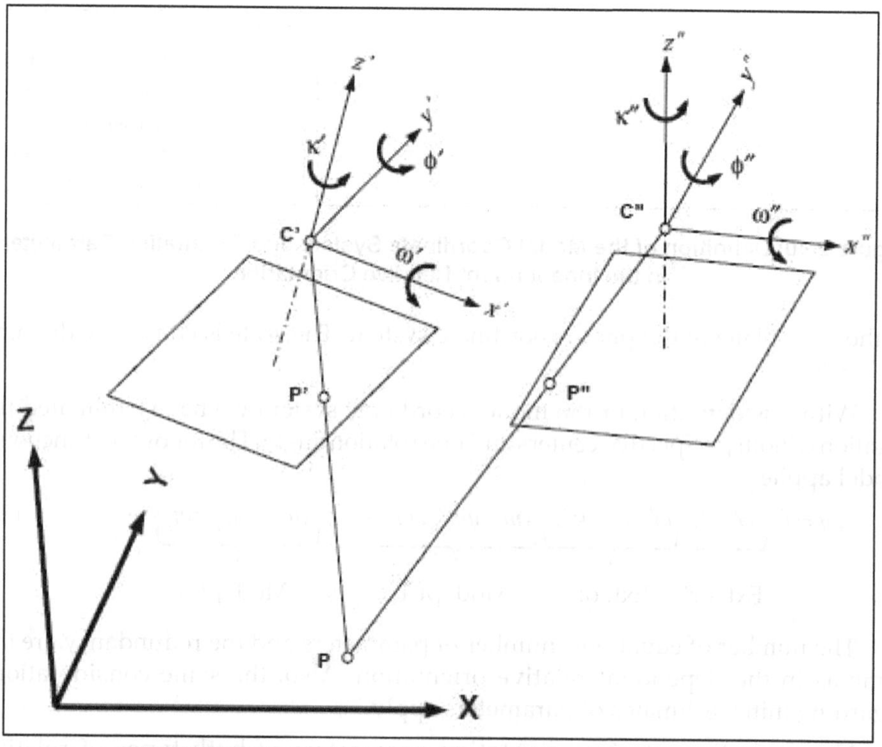

Parameters	
X'_C, Y'_C, Z'_C	position of perspective center left
ω', ϕ', κ'	rotation angles left
X''_C, Y''_C, Z''_C	position of perspective center right
$\omega'', \phi'', \kappa''$	rotation angles right

Figure 2.41: Direct Orientation of a Stereopair with Respect to a Ground Control Coordinate System.

necessary information. The functional model describing the latter case is given below:

$$x, y = f\,(\underbrace{X'_{C'}, Y'_{C'}, Z'_{C'}, \omega', \varphi', \kappa'}_{\text{Ext. or}'},\ \underbrace{X''_{C'}, Y''_{C'}, Z''_{C'}, \omega'', \varphi'', \kappa''}_{\text{Ext. or}''},\ \underbrace{Z_{1'}, Z_{2'}, X_{3'}, Y_{3'}, X_{4'}, Y_{4'}, X_{5'}, Y_{5}}_{\text{Unknown coord. of ctr. pts}}) \quad (38)$$

The Z-coordinates of the planimetric control points 1 and 2 are not known and thus remain in the parameter list. Likewise, $X-Y$ -coordinates of elevation control points 3, 4, 5 are parameters to be determined. Let us check the number of observation equations for this particular case. Since we measure the five partial control points on both photographs we obtain 20 observation equations. The number of parameters amounts to 12 exterior orientation elements and 8 coordinates. So we have just enough equations to solve the problem. For every additional point 4 more equations and 3 parameters are added. Thus, the redundancy increases linearly with the number of points measured.

Additional control points increase the redundancy more, *e.g.*, full control points by 4, an elevation by 2. Like in the case of relative orientation, the mathematical model of the direct orientation is also based on the collinearity equations. Since it is non-linear in the parameters we need good approximations to assure convergence. The estimation of initial values for the exterior orientation parameters may be accomplished in different ways. To estimate $X^0_{C'}, Y^0_C$ for example, one could perform a 2-D transformation of the photo coordinates to planimetric control points. This would also result in a good estimation of κ^o and of the photo scale which in turn can be used to estimate Z^0_C = scale c. For aerial applications we set $\omega^0 = \varphi^0 = 0$. With these initial values of the exterior orientation one can compute approximations $X^0_{i'}$, Y^0_i of object points where $Z_i = h_{aver}$.

Note that the minimum number of points to be measured in the relative orientation is 5. With the direct orientation, we need only three points assuming that two are full control points. For orienting stereopairs with respect to a ground system, there is no need to first perform a relative orientation followed by an absolute orientation. This traditional approach stems from analog instruments where it is not possible to perform a direct orientation by mechanical means.

5.5 Absolute Orientation

With absolute orientation we refer to the process of orienting a stereomodel to the ground control system. Figure 2.42 illustrates the concept. This is actually a very straightforward task which we discussed earlier under 7-parameter transformation. Note that the 7-parameter transformation establishes the relationship between two 3-D Cartesian coordinate systems. The model coordinate system is Cartesian, but the ground control system is usually not Cartesian because the elevations refer to a separate datum. In that case, the ground control system must first be transformed into an orthogonal system.

The transformation can only be solved if a priori information about some of the parameters is introduced. This is most likely done by control points. The same

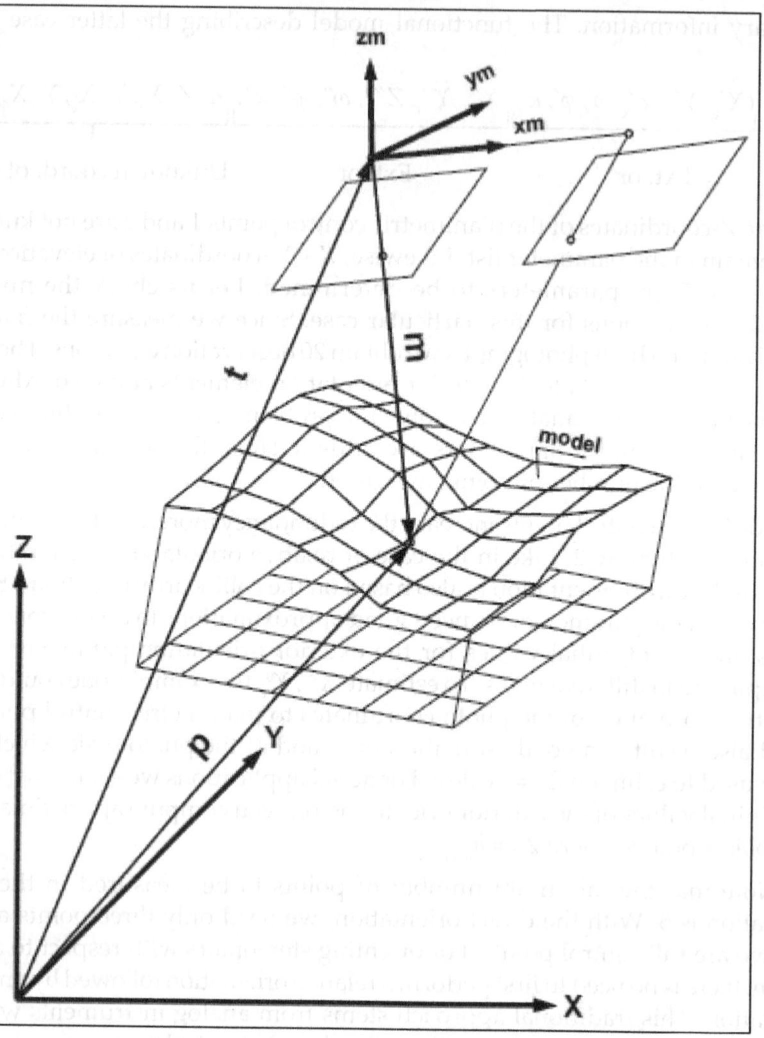

Figure 2.42: Absolute Orientation Entails the Computation of the Transformation Parameters between Model and Ground Coordinate System.

considerations apply as just discussed for the direct orientation. From Figure 2.42 we read the following vector equation which relates the model to the ground control coordinate system:

$$\mathbf{p} = s\mathbf{R}\mathbf{pm} - \mathbf{t} \tag{39}$$

Where $\mathbf{pm} = (xm, ym, zm)^T$ is the point vector in the model coordinate system, $\mathbf{p} = (X, Y, Z)^T$ the vector in the ground control system pointing to the object point P and $\mathbf{t} = (X_t, Y_t, Z_t)T$ the translation vector between the origins of the 2 coordinate systems. The rotation matrix $\mathbf{R}$ rotates vector $\mathbf{pm}$ into the ground control system and s, the scale factor, scales it accordingly. The 7 parameters to be determined comprise 3 rotation angles of the orthogonal rotation matrix **R**, 3 translation parameters and one scale factor.

The following functional model applies:

$$x, y, z = f\ (\underbrace{X_t, Y_t, Z_t,}_{\text{Translation}}\ \underbrace{\omega, \_\varphi, \kappa,}_{\text{Orientation}}\ \underbrace{s)}_{\text{Scale}} \tag{40}$$

In order to solve for the 7 parameters at least seven equations must be available. For example, 2 full control points and one elevation control point would render a solution. If more equations (that is, more control points) are available then the problem of determining the parameters can be cast as a least-squares adjustment. Here, the idea is to minimize the discrepancies between the transformed and the available control points. An observation equation for control point P_i in vector form can be written as

$$\mathbf{ri} = s\mathbf{Rpmi} - \mathbf{t} - \mathbf{pi} \tag{41}$$

With **r** the residual vector $(r_x, r_y, r_z)^T$. Obviously, the model is not linear in the parameters. As usual, linearized observation equations are obtained by taking the partial derivatives with respect to the parameters. The linearized component equations are the approximations may be obtained by first performing a 2-D transformation with *x*, *y*-coordinates only.

Photogrammetric/Aerial Cameras

The photogrammetric or aerial cameras are a very specialized piece of equipment. The acquisition of good quality imagery is central to the accuracy of photogrammetric mapping. Three issues impact on image quality: camera lens, photographic film, and optimum exposure of the film along with the subsequent processing. Unlike conventional amateur cameras, photogrammetric cameras need to maintain a rigid distance between the focal plane (where the film lies during exposure) and the lens. This distance is called the focal length of the camera. In addition, the focal plane needs to be flat. If the film is not flat then parts of the resulting image can become displaced on the photograph. It will also lead to image blur at that location on the photograph making the interpretation of any images in that area difficult.

The aerial camera is an important a critical part of the photogrammetric process. The camera records the existence of phenomena at an instant in time. For precision mapping, it is important that a number of requirements be met (Bormann, 1969). The aerial camera should:

- ☆ The photography should provide a faithful image of even the minute's detail,
- ☆ The definition of photography should be clear,
- ☆ The tilt and crabs are within tolerable limits.
- ☆ Be of high optical lens quality to ensure that imagery is discernable,
- ☆ Be capable of performing under extreme conditions found in the aircraft, such as temperature and humidity,

- ☆ Be simple to use during the photo mission,
- ☆ Be equipped with safeguards to protect against operator blunders during the operation,
- ☆ Be as automatic as possible,
- ☆ Be able to preserve the elements of interior orientation so as to preserve the internal geometric relationships existing within the camera, and
- ☆ Be able to take a series of exposures or single photos with an instantaneous release of the shutter.

1. Classification of Sensing Devices

Figure 2.43 shows a classification of the different types of sensing devices. An example of an active sensing device is radar. An operational system sometimes

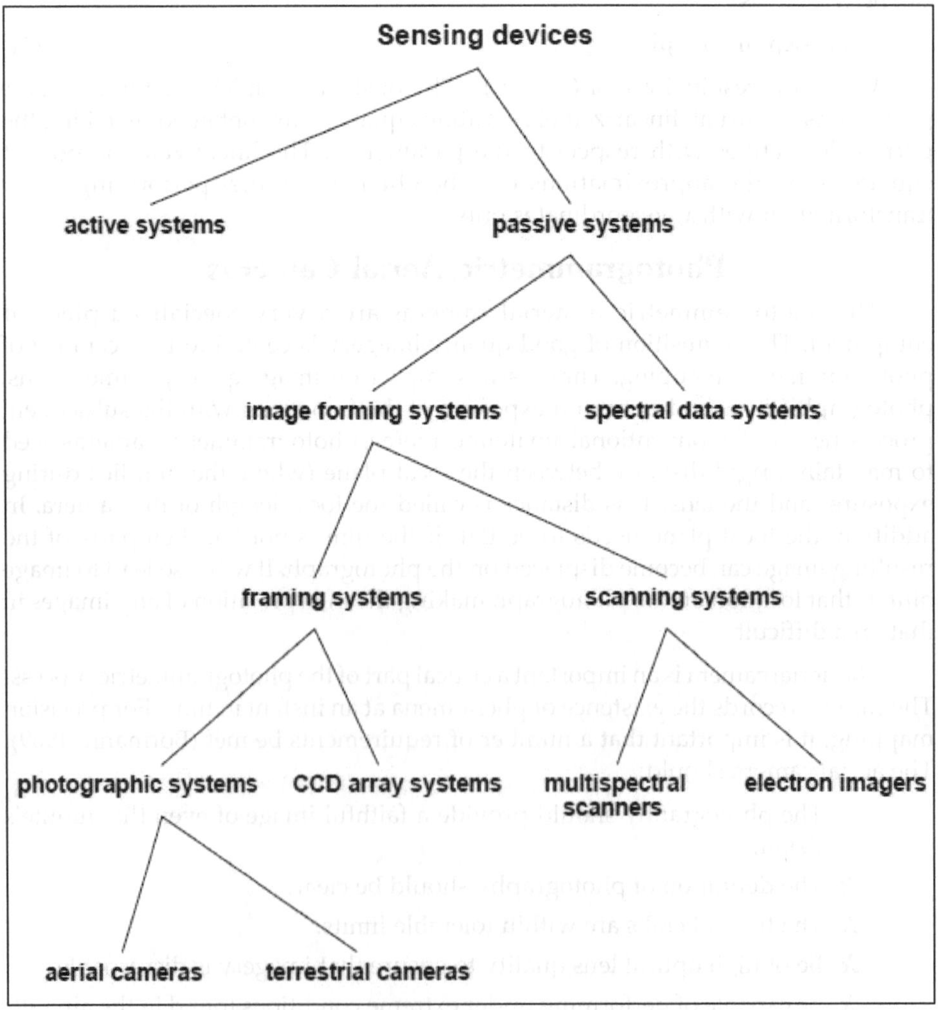

Figure 2.43: Classification of Sensing Devices.

used for photogrammetric applications is the side looking airborne radar (SLAR). Its chief advantage is the fact that radar waves penetrate clouds and haze. An antenna, attached to the belly of an aircraft directs microwave energy to the side, rectangular to the direction of flight. The incident energy on the ground is scattered and partially reflected. A portion of the reflected energy is received at the same antenna. The time elapsed between energy transmitted and received can be used to determine the distance between antenna and ground.

Passive systems fall into two categories: image forming systems and spectral data systems. We are mainly interested in image forming systems which are further subdivided into framing systems and scanning systems. In a framing system, data are acquired all at one instant, whereas a scanning system obtains the same information sequentially, for example scanline by scanline. Image forming systems record radiant energy at different portions of the spectrum. The spatial position of recorded radiation refers to a specific location on the ground. The imaging process establishes a geometric and radiometric relationship between spatial positions of object and image space.

Of all the sensing devices used to record data for photogrammetric applications, the photographic systems with metric properties are the most frequently employed. They are grouped into *aerial* cameras and *terrestrial* cameras. Aerial cameras are also called *cartographic* cameras. In this section we are only concerned with aerial cameras. *Panoramic* cameras are examples of non-metric aerial cameras.

In an ideal world, the airplane would travel in a level straight line with its heading direction and actual direction of travel the same. Unfortunately, the aircraft is susceptible to variations within the flight line. Pitch and yaw of the aircraft introduce tilt into the photography.

If there is a side wind, the aircraft must head into the wind in order to maintain the proper flight direction. The angle between the heading direction and the actual direction of the aircraft is called the crab angle. Crab is caused by drift of the aircraft. The effect of crab is a rotation of the aerial photo with respect to the flight direction thereby decreasing the amount of overlap within the photography (see Figure 2.44). If left uncorrected, this would require acquisition of additional photography and additional processing.

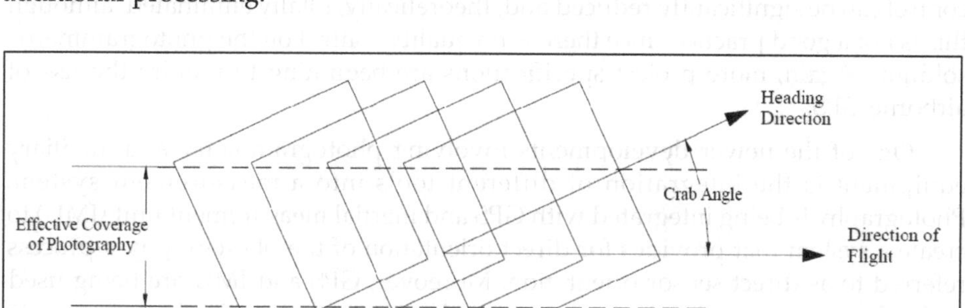

Figure 2.44: Aircraft Crab Angle.

To correct for these natural effects, aerial cameras are usually fixed to the aircraft with a stabilized camera mount. The mount will allow for the correction of crab along with the neutralization of the pitch and yaw of the aircraft. This is done by rotating the camera about the vertical axis so that it follows the direction of flight. Additionally, the camera is leveled just before the exposure is made thus fixing it in the desired vertical direction. It should be emphasized that residual crab and tilt will still exist, but the amount is small and within the specifications for the aerial photography. Newer stabilizing platforms use advanced sensors to determine the amount of tilt and these values are recorded for subsequent processing of the photography. Camera mounts are also equipped with vibration isolators. The airframe of a plane experiences a lot of vibration. If this were not dampened then this vibration would be transferred to the camera thereby creating a blurred image.

Modern aerial cameras are also equipped with very special peripheral instrumentation. For example, almost all use a technique called forward motion compensation (FMC). In fact, most project specifications require photography acquired with FMC. Remember that the aerial camera is moving during the exposure. As we all know from experience with our hand-held cameras, when the camera moves during the exposure the image becomes blurred. Since the distance between the aircraft and the ground is relatively large, the amount of blur is pretty small. But, flying closer to the ground increases this blur. As an example, when we drive on a freeway we see that the features close to the road go by very fast, like a blur, while features in the background almost appear to stand still. This same phenomenon works with aerial photography. To eliminate this blur, FMC is used to advance the film a slight amount during the exposure. This will reduce the amount of blur in the image. Blur is not completely compensated for because the camera may be tilted slightly or the ground is uneven. Therefore, some minor blurriness may still be present in the image but its effects are generally negligible.

A second item that more photogrammetrists are employing is GPS (global positioning system). In the early days, GPS was used to provide the ground control necessary to perform the mapping. Today, it is being used in conjunction with the aerial camera to record the location of the exposure station at the instant the photograph was taken. The advantage in using airborne-GPS is that the ground control can be significantly reduced and, theoretically, totally eliminated, although this is not a good practice since there is no quality control on the photogrammetric solution. Again, more project specifications are beginning to require the use of airborne GPS.

One of the newer developments involving photogrammetry and auxiliary equipment is the integration of different tools into a measurement system. Photography is being integrated with GPS and inertial measurement unit (IMU) to create a system that provides for direct orientation of the photography-a process referred to as direct sensor orientation. Moreover, GPS and IMU are being used with other sensors, such as laser scanners and radar, to provide timely and accurate digital data on the terrain. These tools are being integrated into a new development in geographic information systems (GIS) called telegeoinformatics.

The design of aerial cameras has significantly increased their accuracy and performance.

Improvement in lens design has been influenced by better manufacture of the lens elements, advances in applying the anti-reflective coatings and advanced computer programs that optimize the design of the lenses. Coupled with these items are better construction practices and quality control procedures like testing.

Modern lenses are practically free of any lens distortion. It is now possible to find distortion value at the 1μ to 2μ level or lower. In addition, the enhanced design has lead to improvements in the resolution characteristics. Tests show that corner values of up to 40 lp mm^{-1} for a standard 230 mm × 230 mm format camera are possible. Finally, many of the camera lenses have extended their spectral ranges to the infrared region thereby making the camera more universal in that both panchromatic and infrared film can be used with the same camera system.

A further development that has contributed to improved imagery is forward motion compensation (FMC). This feature advances the film during the exposure to compensate for the forward motion of the aircraft during the exposure. This is done by progressing the film-feed in magazine. Theatrically, the detail will be corrected for but this is only valid for truly vertical photography over flat terrain. Image blurring will still occur due to (Farrow, 1986):

- ☆ Geometric problems caused by the camera not being truly vertical,
- ☆ Residual vibrations exist with camera mounting, and
- ☆ Lateral movement of the air craft due to pitch and roll.

Different manufacturers accommodate fiducial mark imagery in diverse manners with FMC. Zeiss Jena, as an example, captures an image of the midpoint of the exposure cycle. Zeiss Oberkochan, on the other hand, exposes the fiducial during the whole exposure resulting in an elliptical image. The minor axis of this image is 100mm while the major axis is approximately 125 mm.

While photogrammetrists have used imagery for years without FMC capabilities, the compensation of forward motion has important economic advantages. First, because the imagery is clear, it should be possible to use smaller scale mapping, provided that the aerial film can resolve the detail that needs to be imaged on the film. Smaller scale, particularly for larger projects, saves money in the mapping. Even if the same scale is used for mapping, there are still economic advantages of FMC, most notably in the reduction of revising detail on the map that were hard to identify on the imagery.

2. Stability of Aerial Camera

Because of large capital investment required for purchasing an aerial camera, it would be beneficial to see if the interior orientation parameters change over time. Hakkarainen (1984) reported on by the tests performed ISPRS WG I/21. The results are listed as follows:

- ✰ The calibrated focal length (CFL) remained stable over time. Changes at the micrometer level were noticed but these were about the same as the standard deviation from the calibration of the CFL.
- ✰ The fiducial marks were found to be very stable.
- ✰ The mean radial lens distortion was also found to be stable. Moreover, camera age did not appear to be a factor.
- ✰ Changes in decentering distortion were noticed but their effect in a photogrammetry solution was deemed to be insignificant. The asymmetric distortions can be caused by decentering distortion within the individual lens elements, tension between the lens elements and camera cone, and some small residual effects from not having the focal plane perpendicular to the optical axis.
- ✰ It was found that dismounting the camera or service did affect the CFL by over 15 μm.

The recommendations for calibrations from this working group are:

- ✰ After each accident or service, calibrate the camera.
- ✰ For normal photogrammetric applications, a three-year calibration period is acceptable.

3. Film-based Cameras

3.1 Components of Aerial Cameras

A typical aerial camera consists of lens assembly, inner cone, focal plane, outer cone, drive mechanism, and magazine. These principal parts are shown in the schematic diagram of Figure 2.45 and 2.46.

Lens Assembly

The lens assembly, also called *lens cone,* consists of the camera lens (objective), the diaphragm, the shutter and the filter. The diaphragm and the shutter control the exposure. The camera is focused for infinity; that is, the image is formed in the

Figure 2.45: Aerial Camera Aviophot RC20 from Leica.

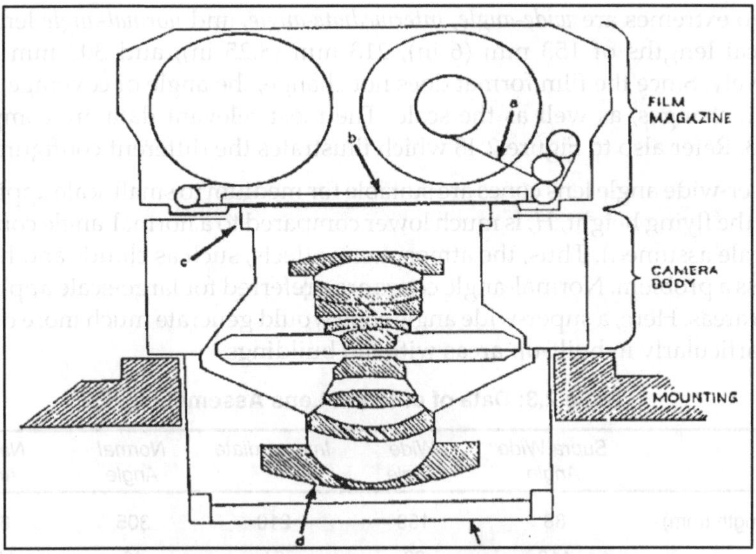

Figure 2.46: Schematic Diagram of Aerial Camera. (a: Film, b: Pressure plate, c: Focal plane frame, d: Lens, e: Filter).

focal plane. Figure 2.47 shows cross sections of lens cones with different focal lengths. *Superwide-angle* lens cones have a focal length of 88 mm (3.5 in). The other extreme are *narrow-angle* cones with a focal length of 610 mm (24 in). Between

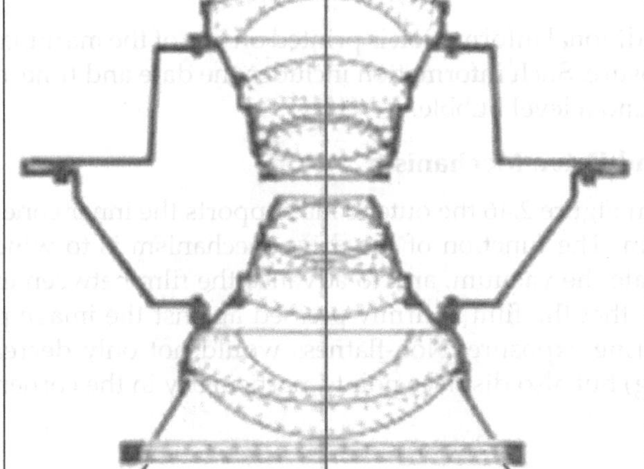

Figure 2.47: Cross-Sectional Views of Aerial Camera Lenses.

these two extremes are *wide-angle*, *intermediate-angle*, and *normal-angle* lens cones, with focal lengths of 153 mm (6 in), 213 mm (8.25 in), and 303 mm (12 in), respectively. Since the film format does not change, the angle of coverage, or field for short, changes, as well as the scale. The most relevant data are compiled in Table 2.3. Refer also to Figure 2.48 which illustrates the different configurations.

Super-wide angle lens cones are suitable for medium to small scale applications because the flying height, H, is much lower compared to a normal-angle cone (same photo scale assumed). Thus, the atmospheric effects, such as clouds and haze, are much less a problem. Normal-angle cones are preferred for large-scale applications of urban areas. Here, a super-wide angle cone would generate much more occluded areas, particularly in built-up areas with tall buildings.

Table 2.3: Data of different Lens Assemblies

	Supra Wide Angle	*Wide Angle*	*Intermediate*	*Normal Angle*	*Narrow Angle*
Focal length (mm)	88	153	210	305	610
Field (o)	119	82	64	46	24
Photo scale	7.2	4.0	2.9	2.0	1.0
Ground coverage	50.4	15.5	8.3	3.9	1.0

Inner Cone and Focal Plane

For metric cameras it is very important to keep the lens assembly fixed with respect to the focal plane. This is accomplished by the inner cone. It consists of a metal with low coefficient of thermal expansion so that the lens and the focal plane do not change their relative position. The focal plane contains *fiducial marks*, which define the fiducial coordinate system that serves as a reference system for metric photographs. The fiducial marks are either located at the corners or in the middle of the four sides.

Usually, additional information is printed on one of the marginal strips during the time of exposure. Such information includes the date and time, altimeter data, photo number, and a level bubble.

Outer Cone and Drive Mechanism

As shown in Figure 2.46 the outer cone supports the inner cone and holds the drive mechanism. The function of the drive mechanism is to wind and trip the shutter, to operate the vacuum, and to advance the film between exposures. The vacuum assures that the film is firmly pressed against the image plane where it remains flat during exposure. Non-flatness would not only decrease the image quality (blurring) but also displace points, particularly in the corners.

Magazine

Obviously, the magazine holds the film, both, exposed and unexposed. A film roll is 120 m long and provides 475 exposures. The magazine is also called film cassette. It is detachable, allowing to interchange magazines during a flight mission.

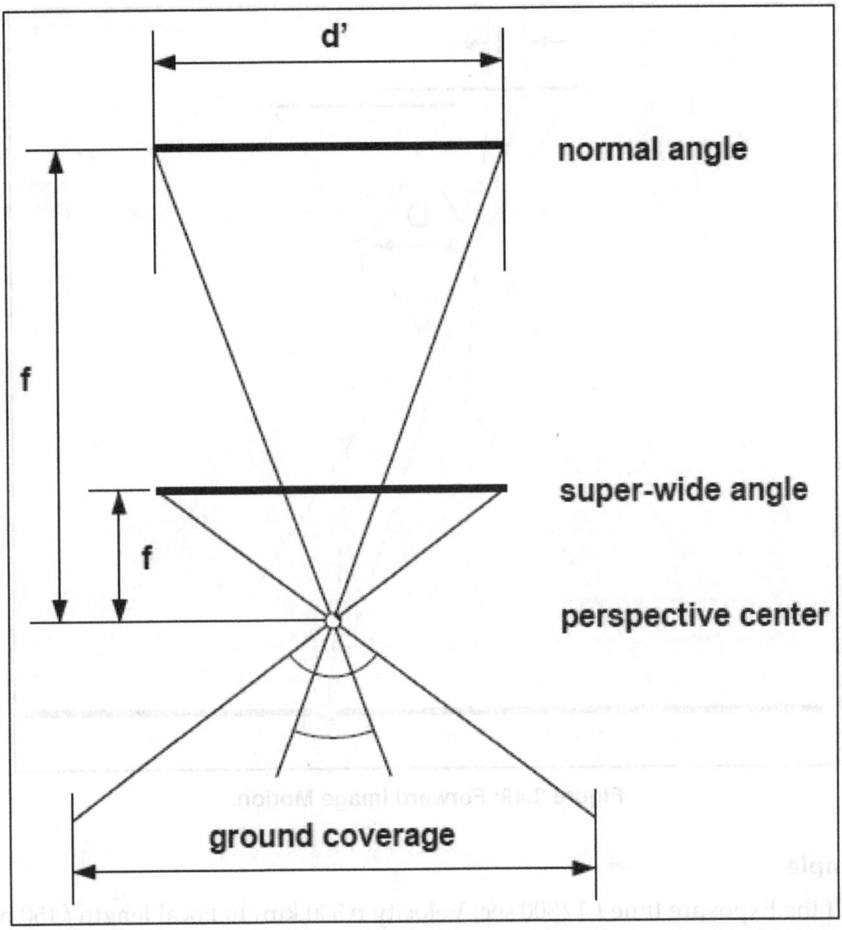

Figure 2.48: Angular Coverage, Photo Scale and Ground Coverage of Cameras with different Focal Lengths.

3.2 Image Motion

During the instance of exposure, the aircraft moves and with it the camera, including the image plane. Thus, a stationary object is imaged at different image locations, and the image appears to move. Image motion results not only from the forward movement of the aircraft but also from vibrations. Figure 2.49 depicts the situation for forward motion. An airplane flying with velocity v advances by a distance $D = v\ t$ during the exposure time t. Since the object on the ground is stationary, its image moves by a distance $d = D/m$ where m is the photo scale. We have:

$$d = \frac{vt}{m} = \frac{vtf}{H} \tag{1}$$

With f the focal length and H the flying height.

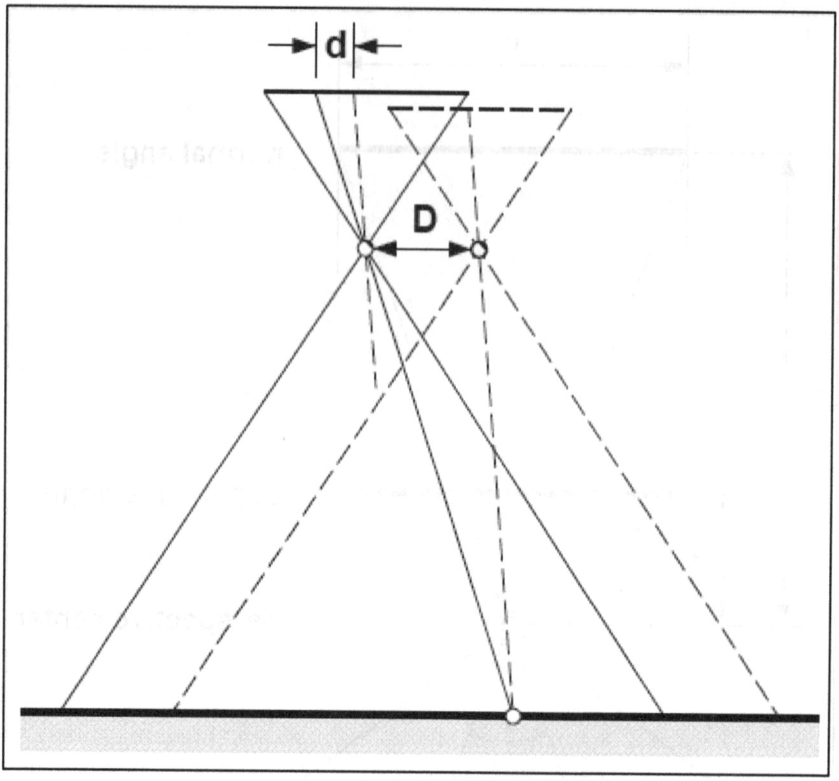

Figure 2.49: Forward Image Motion.

Example

If the Exposure time t 1/300 sec; Velocity v 300 km/h; Focal length f 150 mm; Flying height H 1500 m, thEn **image motion** *d is* 28 μm.

Image motion caused by vibrations in the airplane can also be computed using Equation (1). For that case, vibrations are expressed as a time rate of change of the camera axis (angle/sec). Suppose the camera axis vibrates by 2°/*sec*. This corresponds to a distance $D_v = 2H/\tilde{n}$ = 52.3 m. Since this "displacement" occurs in one second, it can be considered a velocity. In our example, this velocity is 188.4 km/sec, corresponding to an image motion of 18μm. Note that in this case, the direction of image motion is random.

As the example demonstrates, image motion may considerably decrease the image quality. For this reason, modern aerial cameras try to eliminate image motion. There are different mechanical/optical solutions, known as *image motion compensation*. The forward image motion can be reduced by moving the film during exposure such that the image of an object does not move with respect to the emulsion. Since the direction of image motion caused by vibration is random, it cannot be compensated by moving the film. The only measure is a shock absorbing camera mount.

3.3 Camera Calibration

During the process of camera calibration, the *interior orientation* of the camera is determined. The interior orientation data describe the metric characteristics of the camera needed for photogrammetric processes. The elements of interior orientation are:

1. The position of the perspective center with respect to the fiducial marks.
2. The coordinates of the fiducial marks or distances between them so that coordinates can be determined.
3. The calibrated focal length of the camera.
4. The radial and discentering distortion of the lens assembly, including the origin of radial distortion with respect to the fiducial system.
5. Image quality measures such as resolution.

There are several ways to calibrate the camera. After assembling the camera, the manufacturer performs the calibration under laboratory conditions. Cameras should be calibrated once in a while because stress, caused by temperature and pressure differences of an airborn camera, may change some of the interior orientation elements. Laboratory calibrations are also performed by specialized government agencies.

In *in-flight* calibration, a testfield with targets of known positions is photographed. The photo coordinates of the targets are then precisely measured and compared with the control points. The interior orientation is found by a least-square adjustment. We will describe one laboratory method, known as *goniometer calibration*. This will further the understanding of the metric properties of an aerial camera.

Figure 2.50 depicts a goniometer with a camera ready for calibration. The goniometer resembles a theodolite. In fact, the goniometer shown is a modified T4 high precision theodolite used for astronomical observations. To the far right of Figure 2.50(a) is a collimator. If the movable telescope is aimed at the collimator, the line of sight represents the optical axis. The camera is placed into the goniometer such that its vertical axis passes through the entrance pupil. Additionally, the focal plane is aligned perpendicular to the line of sight. This is accomplished by autoreflection of the collimator. Figure 2.50(b) depicts this situation; the fixed collimator points to the center of the grid plate which is placed in the camera's focal plane. This center is referred to as *principal point of autocollimation*, PPA.

Now, the measurement part of the calibration procedure begins. The telescope is aimed at the grid intersections of the grid plate, viewing through the camera. The angles subtended at the rear nodal point between the camera axis and the grid intersections are obtained by subtracting from the circle readings the zero position (reading to the collimator before the camera is installed). This is repeated for all grid intersections along the four semi diagonals.

Having determined the angles αi permits to compute the distances di from the center of the grid plate (PPA) to the corresponding grid intersections i by Equation (2)

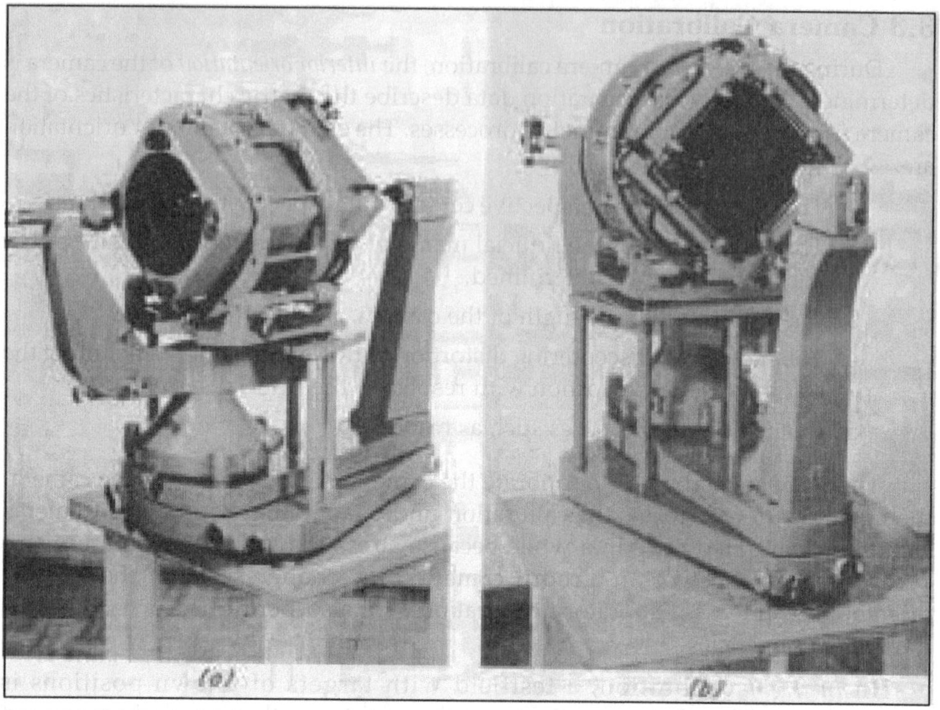

Figure 2.50: Two Views of a Goniometer with Installed Camera, Ready for Calibration.

$$d_i = f \tan(\alpha_i) \tag{2}$$

$$dr''_i = dg_i - d_i \tag{3}$$

The computed distances d_i are compared with the known distances dg_i of the grid plate. The differences dr''_i result from the radial distortion of the lens assembly. Radial distortion arises from a change of lateral magnification as a function of the distance from the center.

The differences dr''_i are plotted against the distances d_i. Figure 2.51(a) shows the result. The curves for the four semi diagonals are quite different and it is desirable to make them as symmetrical as possible to avoid working with four sets of distortion values.

This is accomplished by changing the origin from the PPA to a different point, called the *principal point of symmetry* (PPS). The effect of this change of the origin is shown in Figure 2.51(b). The four curves are now similar enough and the average curve represents the direction-independent distortion. The distortion values for this average curve are denoted by dr''_i.

The average curve is not yet well balanced with respect to the horizontal axis. The next step involves a rotation of the distortion curve such that $|dr_{min}| = |dr_{max}|$. A change of the focal length will rotate the average curve. The focal length with this desirable property is called *calibrated focal length, c*. Through the remainder of the text, we will be using *c* instead of *f*, that is, we use the calibrated focal length

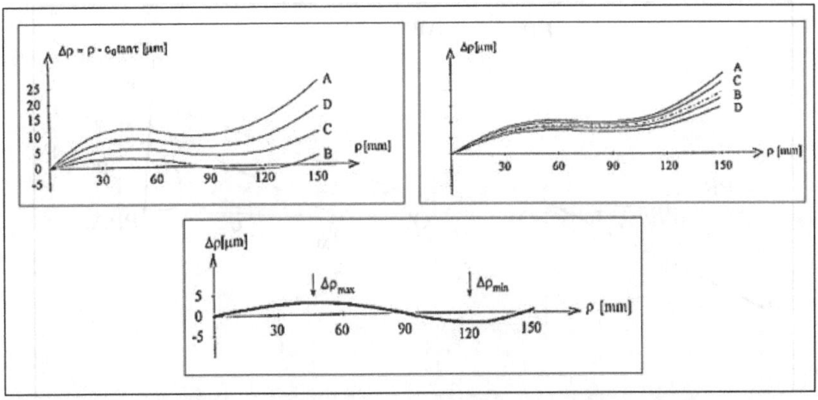

Figure 2.51: Radial Distortion Curves for the Four Semi-Diagonals (a). In (b) the curves are made symmetrical by shifting the origin to PPS. The final radial distortion curve in (c) is obtained by changing the focal length from *f* to *c*.

and not the optical focal length. After completion of all measurements, the grid plate is replaced by a photosensitive plate. The telescope is rotated to the zero position and the reticule is projected through the lens onto the plate where it marks the PPA. At the same time the fiducial marks are exposed. The processed plate is measured and the position of the PPA is determined with respect to the fiducial marks.

3.4 Camera Mounting

It is advisable to keep the maximum relative motions between image and film, arising from angular vibrations during the longest exposure, below a value of 0.002 mm. For a good camera mounting, the centre of support should be near the centre of gravity, the mount should be near the centre of gravity, the mount should feel soft and yield easily to hand pressure, with its natural frequency not higher than 5 cycles per second, and damping should be somewhat under damped.

3.5 Intervalometer

The use of an intervalometer, which controls the automatic exposure of the camera at, specified distance intervals, along the flight line result in correct forward overlap. The determined exposure interval is set on the intervalometer, which is then regulated by electric or mechanical impulses with varying flying speeds and flying heights.

3.6 Crab Compensation

Another important requirement is that the camera must be able to be turned into its mount to compensate for crab. The crab is determined through simple sighting devices and is eliminated by turning the camera through the 'angle of crab' and thus uniform overlap over the entire breadth of the photograph is ensured.

3.7 Summary of Interior Orientation

We summarize the most important elements of the interior orientation of an aerial camera by referring to Figure 2.52. The main purpose of interior orientation

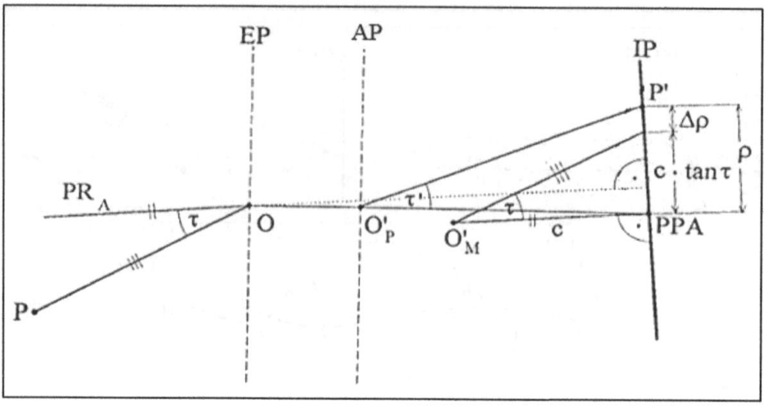

Figure 2.52: Illustration of Interior Orientation.

EP and AP are entrance and exit pupils. They intersect the optical axis at the perspective centers *O* and O_p. The mathematical perspective center O_m is determined such that angles at *O* and O_m become as similar as possible. Point *Ha*, also known as principal point of autocollimation, PPA, is the vertical drop of O_m to the image plane *B*. The distance O_m, H_a, c, is the calibrated focal length.

is to define the position of the perspective center and the radial distortion curve. A camera with known interior orientation is called *metric* if the orientation elements do not change.

An amateur camera, for example, is non-metric because the interior orientation changes every time the camera is focused. Also, it lacks a reference system for determining the PPA.

- ☆ The position of the perspective center is given by the PPA and the calibrated focal length *c*. The bundle rays through projection center and image points resemble most closely the bundle in object space, defined by the front nodal point and points on the ground.
- ☆ The radial distortion curve contains the information necessary for correcting image points that are displaced by the lens due to differences in lateral magnification. The origin of the symmetrical distortion curve is at the principal point of symmetry PPS. The distortion curve is closely related to the calibrated focal length.
- ☆ The position of the PPA and PPS is fixed with reference to the fiducial system. The intersection of opposite fiducial marks indicates the fiducial center FC. The three centers lie within a few microns. The fiducial marks are determined by distances measured along the side and diagonally.

Modern aerial cameras are virtually distortion free. A good approximation for the interior orientation is to assume that the perspective center is at a distance *c* from the fiducial center.

4. Photographic Processes

The most widely used detector system for photogrammetric applications is based on photographic material. It is analog system with some unique properties

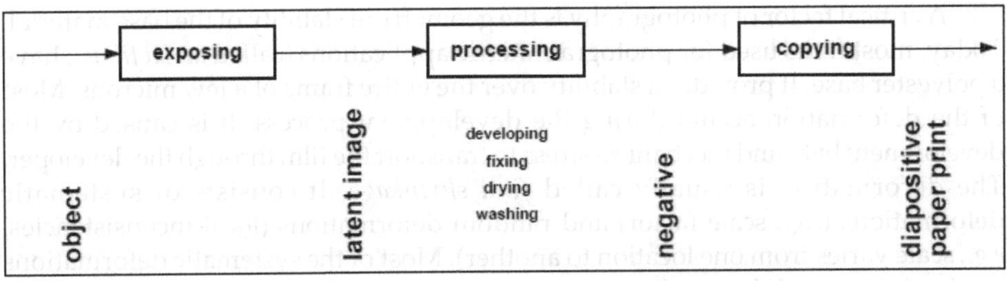

Figure 2.53: Overview of Photographic Processes.

which makes it superior to digital detectors such as CCD arrays. An aerial photograph contains on the order of one Gigabyte of data; the most advanced semiconductor chips have a resolution of $2K \times 2K$, or 4 MB of data.

In this section we provide an overview of photographic processes and properties of photographic material. Figure 2.53 provides an overview of photographic processes and introduces the terms *latent image, negative, (dia)positive* and *paper print*.

4.1 Photographic Material

Figure 2.54 depicts a cross-sectional view of color photography. It consists of three sensitized emulsions which are coated on the base material. To prevent transmitted light from being reflected at the base, back to the emulsion, an antihalation layer is added between emulsion and base.

The light sensitive emulsion consists of three thin layers of gelatine in which are suspended crystals of silver halide. Silver halide is inherently sensitive to near ultra violet and blue. In order for the silver halide to absorb energy at longer wavelengths, optical sensitizers, called *dyes*, are added. They have the property to transfer electromagnetic energy from yellow to near infrared to the silver halide.

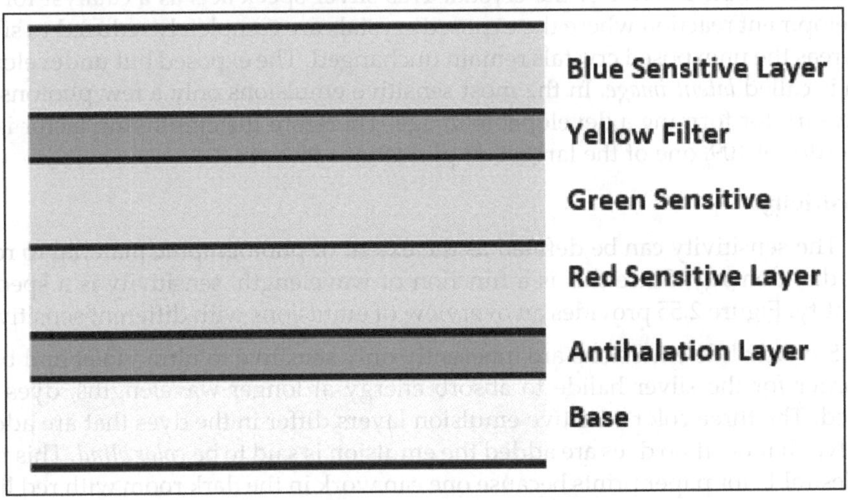

Figure 2.54: Cross Section of Film for Color Photography.

A critical factor of photography is the geometrical stability of the base material. Today, most films used for photogrammetric applications (called *aerial films*), have a polyester base. It provides a stability over the entire frame of a few microns. Most of the deformation occurs during the development process. It is caused by the development bath and mechanical stress to transport the film through the developer. The deformation is usually called *film shrinkage*. It consists of systematic deformations (*e.g.*, scale factor) and random deformations (local inconsistencies, *e.g.*, scale varies from one location to another). Most of the systematic deformations can be determined during the interior orientation and subsequently be corrected.

4.2 Photographic Processes

Exposure

Exposure H is defined as the quantity of radiant energy collected by the emulsion.

$$H = E\,t \tag{4}$$

Where E is the irradiance, and t the exposure time. H is determined by the exposure time and the aperture stop of the lens system. For fast moving platforms (or objects), the exposure time should be kept short to prevent blurring. In that case, a small f-number must be chosen so that enough energy interacts with the emulsion. The disadvantage with this setting is an increased influence of aberrations.

The sensitive elements of the photographic emulsion are microscopic crystals with diameters from 0.3 µm to 3.0 µm. One crystal is made up of 1010 silver halide ions. When radiant energy is incident upon the emulsion it is either reflected, refracted or absorbed. If the energy of the photons is sufficient to liberate an electron from a bound state to a mobile state then it is absorbed, resulting in a free electron which combines quickly with a silver halide ion to a silver atom.

The active product of exposure is a small aggregate of silver atoms on the surface or in the interior of the crystal. This silver speck acts as a catalyst for the development reaction where the exposed crystals are completely reduced to silver whereas the unexposed crystals remain unchanged. The exposed but undeveloped film is called *latent image*. In the most sensitive emulsions only a few photons are necessary for forming a developable image. Therefore the amplifying factor is on the order of 109, one of the largest amplifications known.

Sensitivity

The sensitivity can be defined as the extent of photographic material to react to radiant energy. Since this is a function of wavelength, sensitivity is a spectral quantity. Figure 2.55 provides an overview of emulsions with different sensitivity.

Silver halide emulsions are inherently only sensitive to ultra violet and blue. In order for the silver halide to absorb energy at longer wavelengths, dyes are added. The three color sensitive emulsion layers differ in the dyes that are added to silver halide. If no dyes are added the emulsion is said to be *color blind*. This may be desirable for paper prints because one can work in the dark room with red light

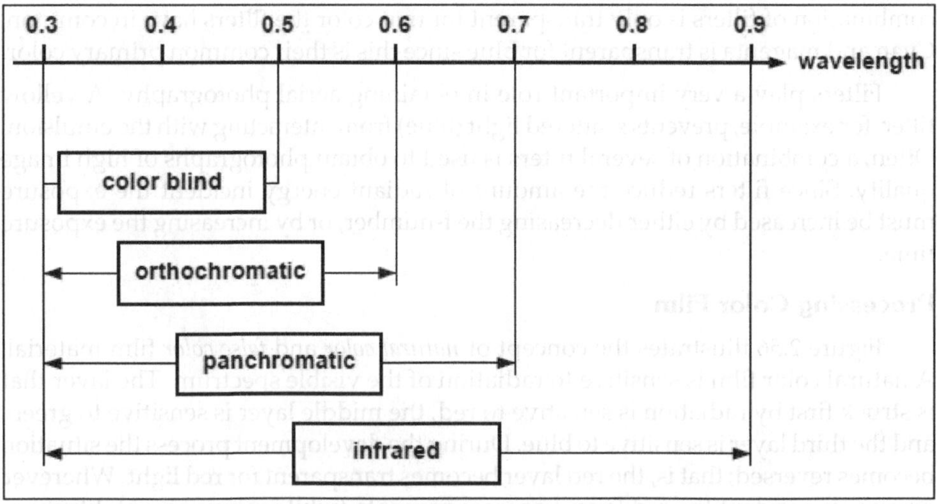

Figure 2.55: Overview of Photographic Material with different Sensitivity Range.

without affecting the latent image. Of course, color blind emulsions are useless for aerial film because they would only react to blue light which is scattered most causing a diffuse image without contrast.

In *orthochromatic* emulsions the sensitivity is extended to include the green portion of the visible spectrum. *Panchromatic* emulsions are sensitive to the entire visible spectrum; *infrared* film includes the near infrared.

Colors and Filters

The visible spectrum is divided into three categories: 0.4 to 0.5 µm, 0.5 to 0.6 µm, and 0.6 to 0.7 µm. These three categories are associated to the *primary* colors of blue, green and red. All other colors, approximately 10 million, can be obtained by an additive mixture of the primary colors. For example, white is a mixture of equal portions of primary colors. If two primary colors are mixed the three *additive colors* cyan, yellow and magenta are obtained. As indicated in Table 2.4, these additive colors also result from subtracting the primary colors from white light.

Table 2.4: Primary Colors and Additive Primary Colors

Additive Color Primary	*Additive Mixture of 2 Color Primaries*	*Subtraction from White Light*
Cyan	b + g	w - r
Yellow	g + r	w - b
Magenta	r + b	w - g

Subtraction can be achieved by using filters. Filter with a subtractive color primary is transparent for the additive primary colors. For example, a yellow filter is transparent for green and red. Such a filter is also called *minus blue* filter. A combination of filters is only transparent for that color the filters have in common. Cyan and magenta is transparent for blue since this is their common primary color.

Filters play a very important role in obtaining aerial photography. A yellow filter, for example, prevents scattered light (blue) from interacting with the emulsion. Often, a combination of several filters is used to obtain photographs of high image quality. Since filters reduce the amount of radiant energy incident the exposure must be increased by either decreasing the f-number, or by increasing the exposure time.

Processing Color Film

Figure 2.56 illustrates the concept of *natural color* and *false color* film material. A natural color film is sensitive to radiation of the visible spectrum. The layer that is struck first by radiation is sensitive to red, the middle layer is sensitive to green, and the third layer is sensitive to blue. During the development process the situation becomes reversed; that is, the red layer becomes transparent for red light. Wherever green was incident the red layer becomes magenta (white minus green); likewise, blue changes to yellow.

If this developed film is viewed under white light, the original colors are perceived. A closer examination of the right side of Figure 2.56 reveals that the sensitivity of the film is shifted towards longer wavelengths. A yellow filter prevents

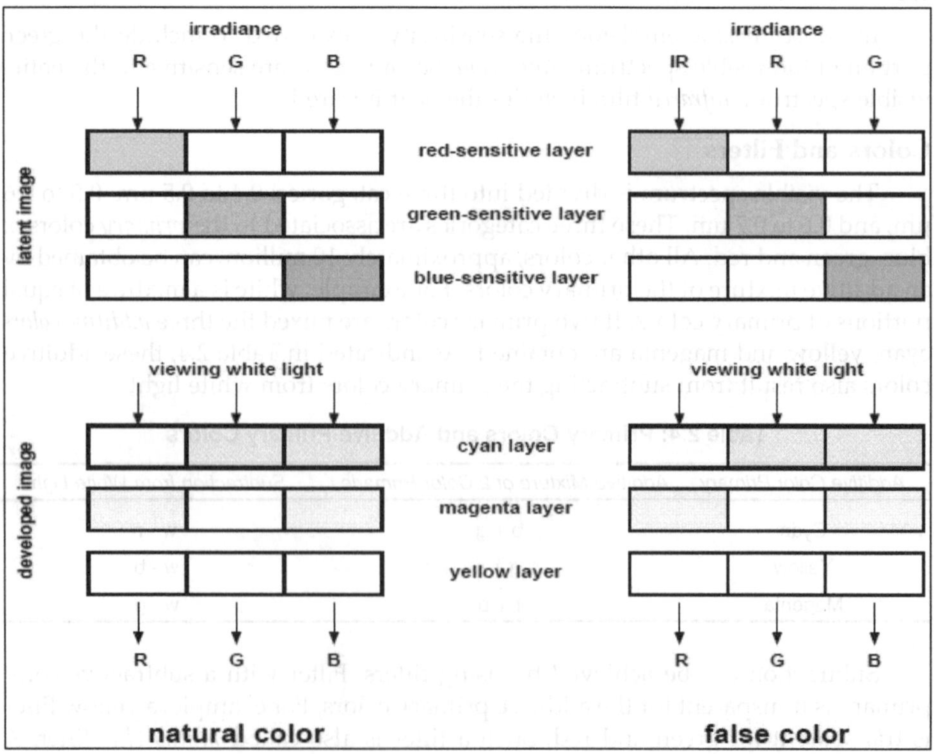

Figure 2.56: Concept of Processing Natural Color (left) and False Color Film (right).

blue light from interacting with the emulsion. The top most layer is now sensitive to near infrared, the middle layer to red and the third layer is sensitive to green. After developing the film, red corresponds to infrared, green to red, and blue to green. This explains the name false color film: vegetation reflects infrared most. Hence, forest, trees and meadows appear red.

4.3 Sensitometry

Sensitometry deals with the measurement of sensitivity and other characteristics of photographic material. The *density* (amount of exposure) can be measured by a *densitometer*. The density *D* is defined as the degree of blackening of an exposed film.

$$D = \log(O) \tag{5}$$

$$O = \frac{E_i}{E_t} \tag{6}$$

$$T = \frac{E_t}{E_i} = \frac{1}{O} \tag{7}$$

$$H = E_t \tag{8}$$

where,

O = Opacity, degree of blackening
E_i = Irradiance
E_t = Transmitted irradiance
T = Transmittance
H = Exposure

The density is a function of exposure *H*. It also depends on the development process. For example, the density increases with increasing development time. An underexposed latent image can be "corrected" to a certain degree by increasing the development time. Figure 2.57 illustrates the relationship between density and exposure. The *characteristic curve* is also called *D-log(H)* curve. Increasing exposure results in more crystals with silver specks that are reduced to black silver: a bright spot in the scene appears dark in the negative.

The characteristic curve begins at a threshold value, called *fog*. An unexposed film should be totally transparent when reduced during the development process. This is not the case because the base of the film has a transmittance smaller than unity. Additionally, the transmittance of the emulsion with unexposed material is smaller than unity. Both factor contribute to fog. The lower part of the curve, between point 1 and 2, is called *toe region*. Here, the exposure is not enough to cause a readable image. The next region, corresponding to correct exposure, is characterized by a straight line (between point 2 and 3). That is, the density increases linearly with the logarithm of exposure. The slope of the straight line is called *gamma* or *contrast*. A film with a slope of 450 is perceived as truly presenting the contrast in the scene. A

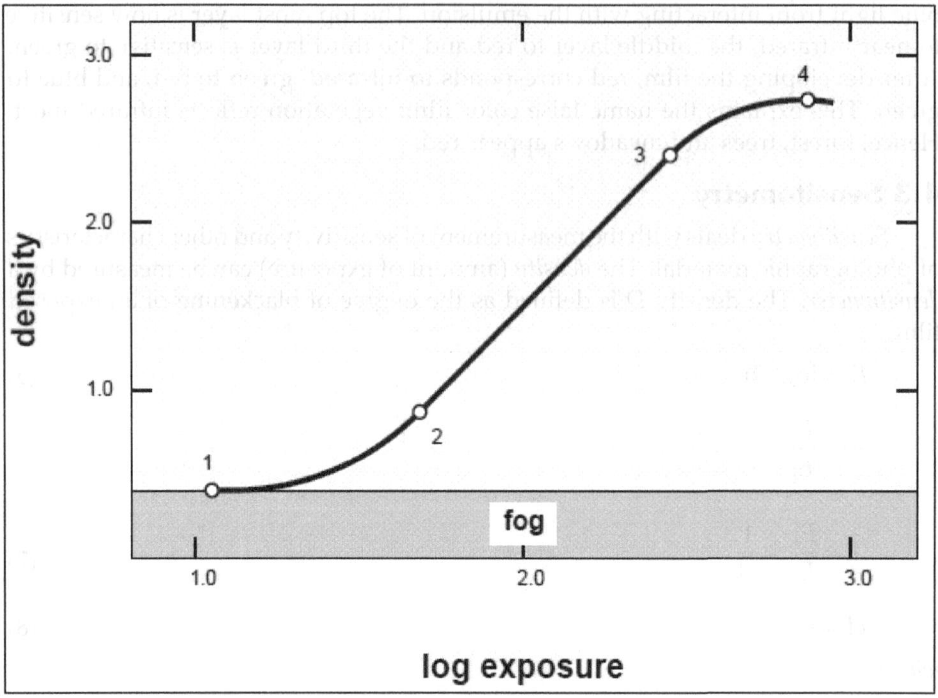

Figure 2.57: Characteristic Curve of a Photographic Emulsion.

film with a higher gamma exaggerates the scene contrast. The contrast is not only dependent on the emulsion but also on the development time. If the same latent image is kept longer in the development process, its characteristic curve becomes flatter.

The straight portion of the characteristic curve ends in the *shoulder region* where the density no longer increases linearly. In fact, there is a turning point, *solarization*, where D decreases with increasing exposure (point 4 in Figure 2.57). Clearly, this region is associated with over exposure.

4.4 Speed

The size and the density of the silver halide crystals suspended in the gelatine of the emulsion vary. The larger the crystal size the higher the probability that is struck by photons during the exposure time. Fewer photons are necessary to cause a latent image. Such a film would be called faster because the latent image is obtained in a shorter time period compared to an emulsion with smaller crystal size. In other words, a faster film requires less exposure. Unfortunately, there is no universally accepted definition of speed. There is, however, a standard for determining the speed of aerial films, known as Aerial film Speed (AFS).

The exposure used to determine AFS is the point on the characteristic curve at which the density is 0.3 units above fog. The exposure H needed to produce this density is used in the following definition

$$AFS = \frac{3}{2H} \tag{9}$$

Note that aerial film speed differs from speed as defined by ASA. Here, the exposure is specified which is necessary to produce a density 0.1 units above fog. Figure 2.58 shows two emulsions with different speed and different gamma. Since emulsion A requires less exposure to produce the required density at 0.3 above fog, it is faster than emulsion B ($HA < HB$).

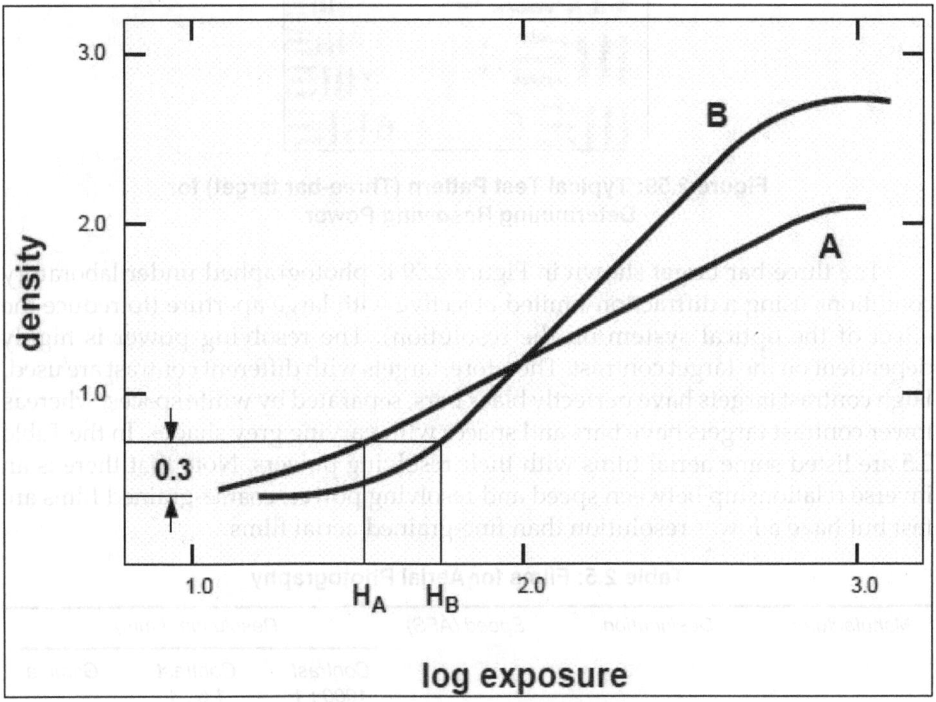

Figure 2.58: Concept of Speed.

4.5 Resolving Power

The image quality is directly related to the size and distribution of the silver halide crystals and the dyes suspended in the emulsion. The crystals are also called *corn*, and the corn size corresponds to the diameter of the crystal. *Granularity* refers to the size and distribution, *concentration* to the amount of light-sensitive material per unit volume. Emulsions are usually classified as fine-, medium-, or coarse-grained.

The resolving power of an emulsion refers to the number of alternating bars and spaces of equal width which can be recorded as visually separate elements in the space of one millimeter. A bar and a space is called a line or line pair. A resolving power of 50 l/mm means that 50 bars, separated by 50 spaces, can be discerned per millimeter. Figure 2.59 shows a typical test pattern used to determine the resolving power.

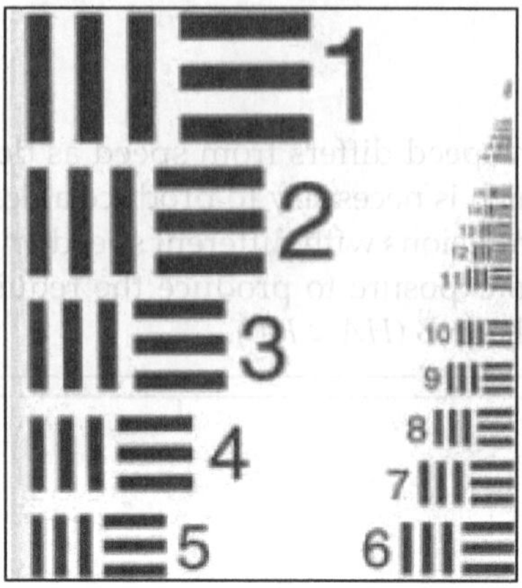

Figure 2.59: Typical Test Pattern (Three-bar target) for Determining Resolving Power.

The three-bar target shown in Figure 2.59 is photographed under laboratory conditions using a diffraction-limited objective with large aperture (to reduce the effect of the optical system on the resolution). The resolving power is highly dependent on the target contrast. Therefore, targets with different contrast are used. High contrast targets have perfectly black bars, separated by white spaces, whereas lower contrast targets have bars and spaces with varying grey shades. In the Table 2.5 are listed some aerial films with their resolving powers. Note that there is an inverse relationship between speed and resolving power: coarse-grained films are fast but have a lower resolution than fine-grained aerial films.

Table 2.5: Films for Aerial Photography

Manufacturer	*Designation*	*Speed (AFS)*	*Resolution (l/mm)*		
			Contrast 1000 : 1	*Contrast 1.6 : 1*	*Gamma*
Agfa	Aviophot Pan		133		1.0-1.4
Kodak	Plus-X Aerographic	160	100	50	1.3
Kodak	High Definition	6.4	630	250	1.3
Kodak	Infrared Aerographic	320	80	40	2.3
Kodak	Aerial Color	6	200	100	

5. Digital Cameras

5.1 Overview

A new trend in aerial cameras for photogrammetry is the development of digital cameras. These are really not new since digital imagery has been used for mapping for years. The popular term "digital camera" is rather informal and may even be misleading because the output is in many cases an analog signal. A more generic term is *solid-state camera*. Other frequently used terms include *CCD camera* and *solid-*

Figure 2.60: Example of 1D Sensor Element and 2D Array Sensor.

state camera. Though these terms obviously refer to the type of sensing elements, they are often used in a more generic sense.

The chief advantage of digital cameras over the classical film-based cameras is the instant availability of images for further processing and analysis. This is essential in real-time applications (*e.g.*, robotics, certain industrial applications, bio-mechanics, etc.).Another advantage is the increased spectral flexibility of digital cameras. The major drawback is the limited resolution or limited field of view.

Digital cameras have been used for special photogrammtric applications since the early seventies. However, vidicon-tube cameras available at that time were not very accurate because the imaging tubes were not stable. This disadvantage was eliminated with the appearance of solid-state cameras in the early eighties. The charge-coupled device provides high stability and is therefore the preferred sensing device in today's digital cameras.

The most distinct characteristic of a digital camera is the image sensing device. Because of its popularity we restrict the discussion to solid-state sensors, in particular to charge coupled devices (CCD). The sensor is glued to a ceramic substrate and covered by a glass. Typical chip sizes are 1/2 and 2/3 inches with as many as 2048×2048 sensing elements. However, sensors with fewer than $1K \times 1K$ elements are more common. Figure 2.60 depicts a line sensor (a) and a 2D sensor chip (b). The dimension of a sensing element is smaller than 10 μm, with an insulation space of a few microns between them. This can easily be verified when considering the physical dimensions of the chip and the number of elements.

5.1.1 Camera Overview

Figure 2.61 depicts a functional block diagram of the major components of a solid-state camera. The optics component includes lens assembly and filters, such as an infrared blocking filter to limit the spectral response to the visible spectrum. Many cameras use a C-mount for the lens. Here, the distance between mount and image plane is 17.526 mm. As an option, the optics subsystem may comprise a

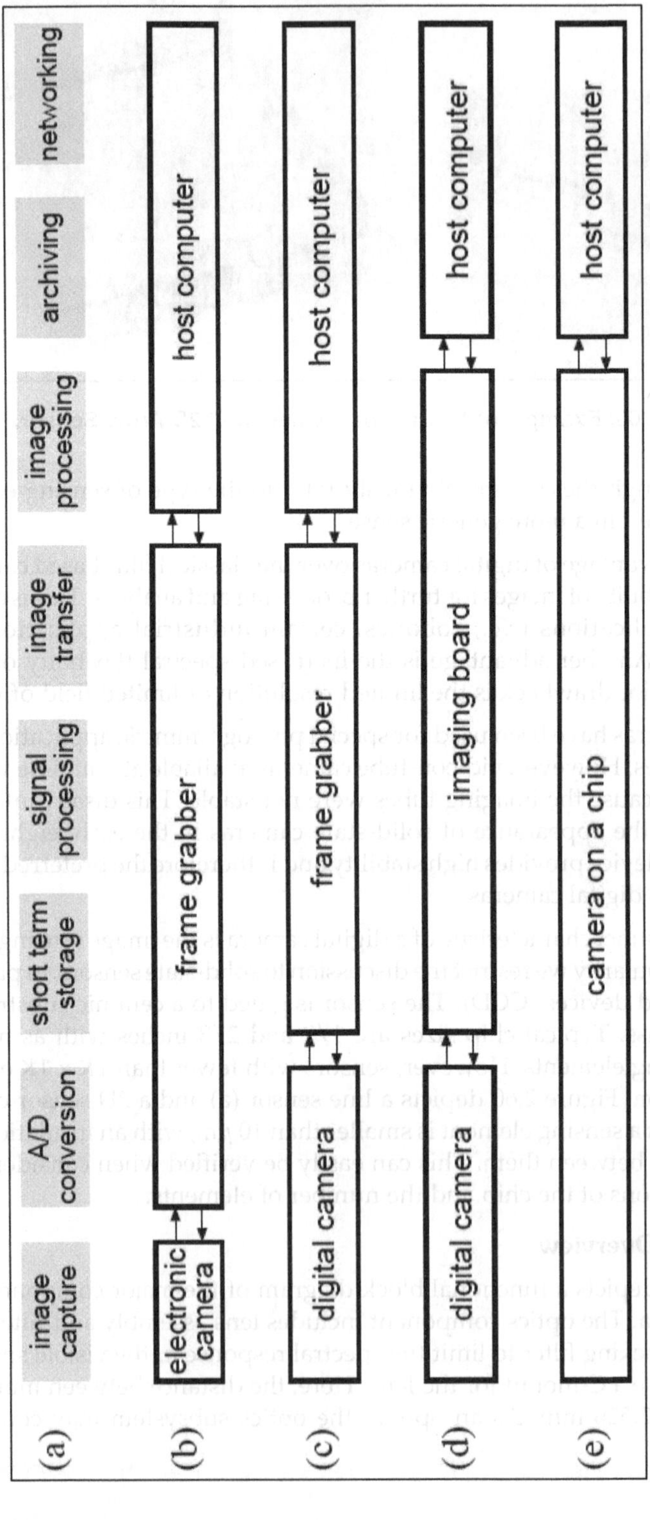

Figure 2.61: Functional Block Diagram of a Solid-State Camera. A real camera may not have all components. The diagram is simplified, *e.g.*, external signals received by the camera are not shown.

shutter. The most distinct characteristic of an electronic camera is the image sensing device.

The solid-state sensor, positioned in the image plane, is glued on a ceramic substrate. The sensing elements (pixels) are either arranged in a linear array or a frame array. Linear arrays are used for aerial cameras while close range applications, including mobile mapping systems, employ frame array cameras.

The accuracy of a solid-state camera depends a great deal on the accuracy and stability of the sensing elements, for example on the uniformity of the sensor element spacing and the flatness of the array. From the manufacturing process we can expect an accuracy of 1/10th of a micron. Considering a sensor element, size 10 μm, the regularity amounts to 1/100. Camera calibration and measurements of the position and the spacing of sensor elements confirm that the regularity is between 1/50th and 1/100th of the spacing.

The voltage generated by the sensor's read out mechanism must be amplified for further processing, which begins with converting the analog signal to a digital signal. This is not only necessary for producing a digital output, but also for signal and image processing. The functionality of these two components may range from rudimentary to very sophisticated in a real camera.

You may consider the first two components (optics and solid-state sensor) as image capture, the amplifiers and ADC as image digitization, and signal and image processing as image restoration. A few examples illustrate the importance of image restoration. The dark current can be measured and subtracted so that only its noise signal component remains; defective pixels can be determined and an interpolated signal can be output; the contrast can be changed (Gamma correction); and image compression may be applied.

5.1.2 Multiple Frame Cameras

The classical film-based cameras used in photogrammetry are often divided into aerial and terrestrial (close-range) cameras. The same principle can be applied for digital cameras. A digital aerial camera with a resolution comparable to a classical frame camera must have on the order of 15,000×15,000 sensing elements. Such image sensors do not (yet) exist. Two solutions exist to overcome this problem: *line cameras* and *multiple cameras*, housed in one camera body. Figure 2.62 shows an example of a multi-camera system (UltraCam from Vexcel). It consists of 8 different cameras that are mounted in a common camera frame. The ground coverage of each of these frame cameras slightly overlaps and the 8 different images are merged together to one uniform frame image by way of image processing.

5.1.3 Line Cameras

An alternative solution to frame cameras are the so called *line cameras* of which the *3-line camera* is the most popular one. The 3-line camera employs three linear areas which are mounted in the image plane in fore, nadir and aft position. With this configuration, triple coverage of the surface is obtained. Examples of 3-line cameras include Leica's ADS40. It is also possible to implement the multiple line

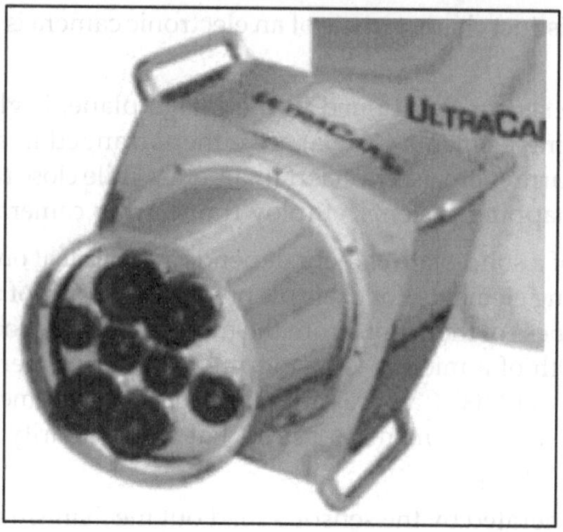

Figure 2.62: Example of a Multi-Camera System (Vexcel UltraCam), Consisting of 8 different Cameras that are Mounted in a Slightly Convergent Mode to Assure Overlap of the Individual Images.

concept by having convergent lenses for every line. A well-known example of a one line-camera is SPOT. The linear array consists of 7,000 sensing elements. Stereo is obtained by overlapping strips obtained from adjacent orbits.

Digital Modular Camera

Z/I-Imaging has introduced the Digital Modular Camera (DMC) system (Figure 2.63). What is unique about this system is the use of one to eight CCD area camera modules that can be mounted together in different configurations. This enables the user to acquire up to four high resolution panchromatic images with the capability of adding up to four multispectral images with reduced resolution.

All CCD cameras are mounted inside the optics frame. Since the DMC camera is the same size as the RMKTOP analog camera, it will fit into the Zeiss gyro-stabilized platform. A schematic diagram of the component parts of the DMC are shown in Figure 2.63. Normally, airborne GPS is integrated with the DMC system. Additionally, the flight management system (FMS) is used to operate the camera system. Finally, an inertial measurement system can be installed in the camera frame.

The electronics module provides the control for the DMC system. It controls the camera module. Not only does it operate the camera shutters and collect the digital imagery, it also communicates with the control unit. The control unit is the heart of the electronics system. It provides the communications with the external components to the camera system, monitors data flow and sends the captured imagery to the data storage RAID.

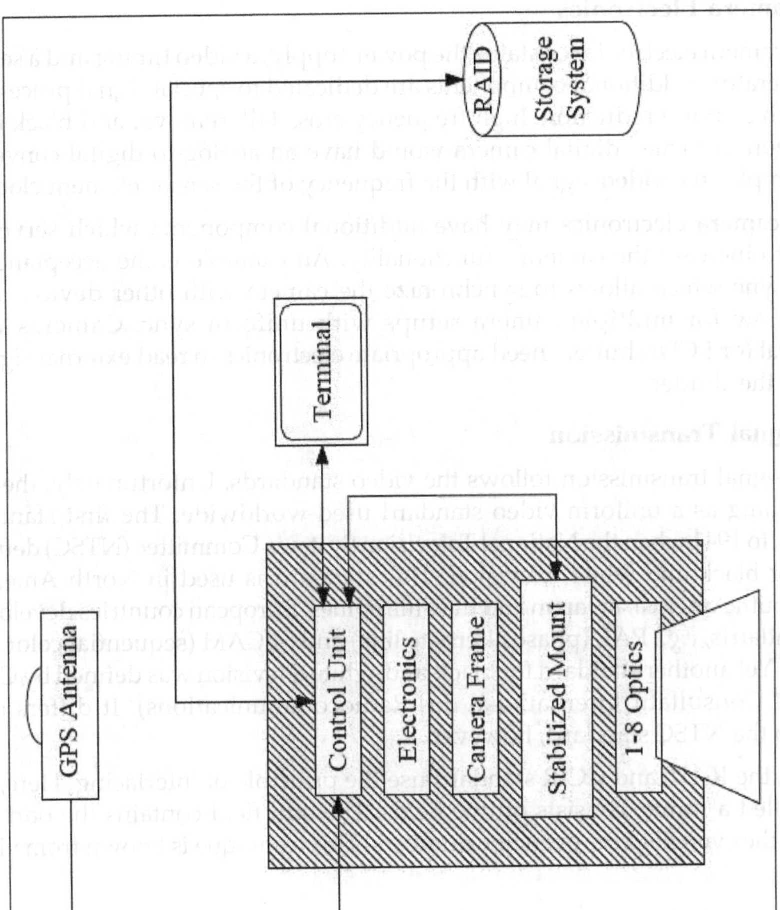

Figure 2.63: DMC System.

5.1.4 Camera Electronics

The camera electronics contains the power supply, a video timing and a sensor clock generator. Additional components are dedicated to special signal processing tasks, such as noise reduction, high frequency cross-talk removal and black level stabilization. A "true" digital camera would have an analog-to-digital converter which samples the video signal with the frequency of the sensor element clock.

The camera electronics may have additional components which serve the purpose to increase the camera's functionality. An example is the acceptance of external sync which allows to synchronize the camera with other devices. This would allow for multiple camera setups with uniform sync. Cameras with mechanical (or LCD) shutters need appropriate electronics to read external signals to trigger the shutter.

5.1.5 Signal Transmission

The signal transmission follows the video standards. Unfortunately, there is no such thing as a uniform video standard used worldwide. The first standard dates back to 1941 when the National Television Systems Committee (NTSC) defined RS-170 for black and-white television. This standard is used in North America, parts of South America, in Japan and the Philippines. European countries developed other standards, *e.g.*, PAL (phase alternate line) and SECAM (sequential color and memory). Yet another standard for black-and-white television was defined by CCIR (Commité Consultatif International des Radiocommunications). It differs only slightly to the NTSC standard, however.

Both, the R-170 and CCIR standard use the principle of interlacing. Here, the image, called a frame, consists of two fields. The odd field contains the odd line numbers, the even field the even line numbers. This technique is known from video monitors.

5.1.6 Frame Grabbers

Frame grabbers receive the video signal, convert it, buffer data and output it to the storage device of the digital image. The analog front end of a frame grabber preprocesses the video signal and passes it to the AD converter. The analog front end must cope with different signals (*e.g.*, different voltage level and impedance).

5.2 CCD Sensors: Working Principle and Properties

The *charge-coupled device (CCD)* was invented in 1970. The first CCD line sensor contained 96 pixels; today, chips with over 50 million pixels are commercially available. Figure 2.64 illustrates the astounding development of CCD sensors over a period of 25 years. The sensor size in pixels is usually loosely termed *resolution*, giving rise to confusion since this term has a different meaning in photogrammetry.

5.2.1 Working Principle

Figure 2.65(a) is a schematic diagram of a semiconductor capacitor-the basic building block of a CCD. The semiconductor material is usually silicon and the insulator is an oxide (MOS capacitor). The metal electrodes are separated from the

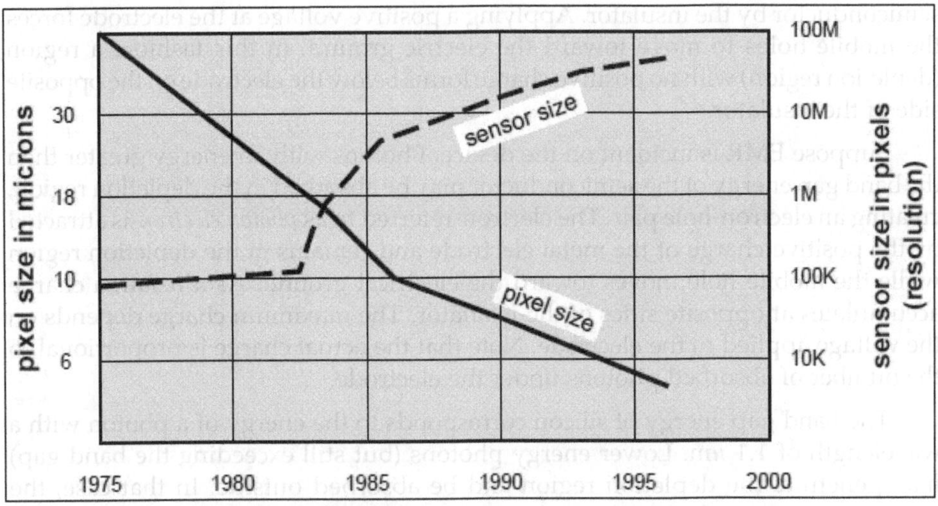

Figure 2.64: Development of CCD Arrays over a Period of 25 Years.

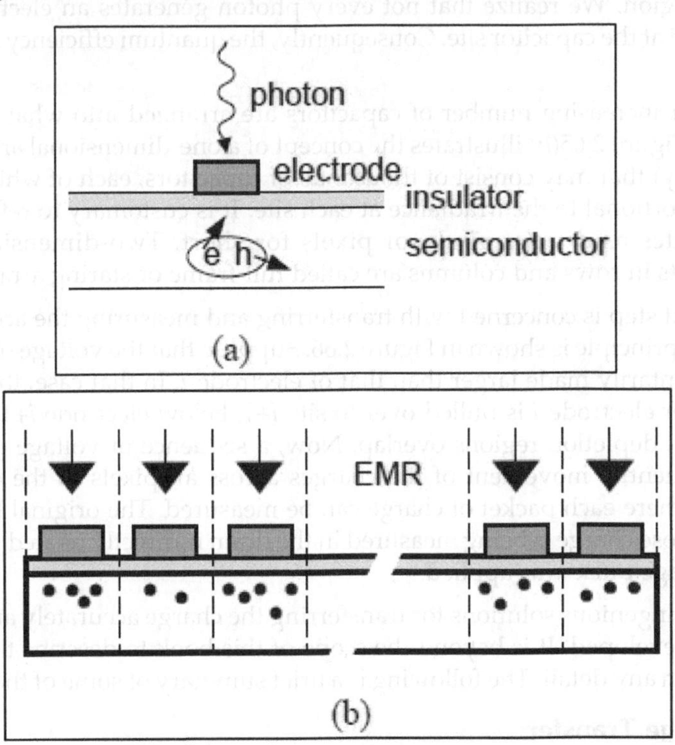

Figure 2.65: Schematic Diagram of CCD Detector.

In (a) a photon with an energy greater than the bandgap of the semiconductor generates an electron-hole pair. The electron e is attracted by the positive voltage of the electrode while the mobile hole moves toward the ground. The collected electrons together with the electrode form a capacitor. In (b) this basic arrangement is repeated many times to form a linear array.

semiconductor by the insulator. Applying a positive voltage at the electrode forces the mobile holes to move toward the electric ground. In this fashion, a region (depletion region) with no positive charge forms below the electrode on the opposite side of the insulator.

Suppose EMR is incident on the device. Photons with an energy greater than the band gap energy of the semiconductor may be absorbed in the depletion region, creating an electron-hole pair. The electron-referred to as *photon electron*-is attracted by the positive charge of the metal electrode and remains in the depletion region while the mobile hole moves toward the electrical ground. As a result, a charge accumulates at opposite sides of the insulator. The maximum charge depends on the voltage applied to the electrode. Note that the actual charge is proportional to the number of absorbed photons under the electrode.

The band gap energy of silicon corresponds to the energy of a photon with a wavelength of 1.1 μm. Lower energy photons (but still exceeding the band gap) may penetrate the depletion region and be absorbed outside. In that case, the generated electron-hole pair may recombine before the electron reaches the depletion region. We realize that not every photon generates an electron that is accumulated at the capacitor site. Consequently, the quantum efficiency is less than unity.

An ever increasing number of capacitors are arranged into what is called a *CCD array*. Figure 2.65(b) illustrates the concept of a one-dimensional array (called a linear array) that may consist of thousands of capacitors, each of which holds a charge proportional to the irradiance at each site. It is customary to refer to these capacitor sites as *detector pixels*, or pixels for short. Two-dimensional pixel arrangements in rows and columns are called full-frame or staring arrays.

The next step is concerned with transferring and measuring the accumulated charge. The principle is shown in Figure 2.66. Suppose that the voltage of electrode i+1 is momentarily made larger than that of electrode i. In that case, the negative charge under electrode i is pulled over to site i+1, below electrode i+1, provided that adjacent depletion regions overlap. Now, a sequence of voltage pulses will cause a sequential movement of the charges across all pixels to the drain (last electrode) where each packet of charge can be measured. The original location of the pixel whose charge is being measured in the drain is directly related to the time when a voltage pulse was applied.

Several ingenious solutions for transferring the charge accurately and quickly have been developed. It is beyond the scope of this book to describe the transfer technology in any detail. The following is a brief summary of some of the methods.

5.2.2 Charge Transfer

Linear Array with Bilinear Readout

As sketched in Figure 2.67, a linear array (CCD shift register) is placed on both sides of the single line of detectors. Since these two CCD arrays are also light sensitive, they must be shielded. After integration, the charge accumulated in the

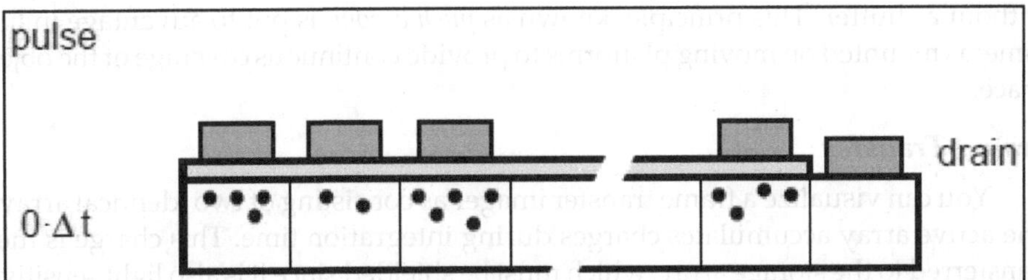

Figure 2.66: Principle of Charge Transfer.

The top row shows a linear array of accumulated charge packets. Applying a voltage greater than V1 of electrode 1 momentarily pulls charge over to the second electrode (middle row). Repeating this operation in a sequential fashion eventually moves all packets to the final electrode (drain) where the charge is measured.

active detectors is transferred to the two shift registers during one clock period. The shift registers are read out in a serial fashion as described above. If the readout time is equal to the integration time, then this sensor may operate continuously

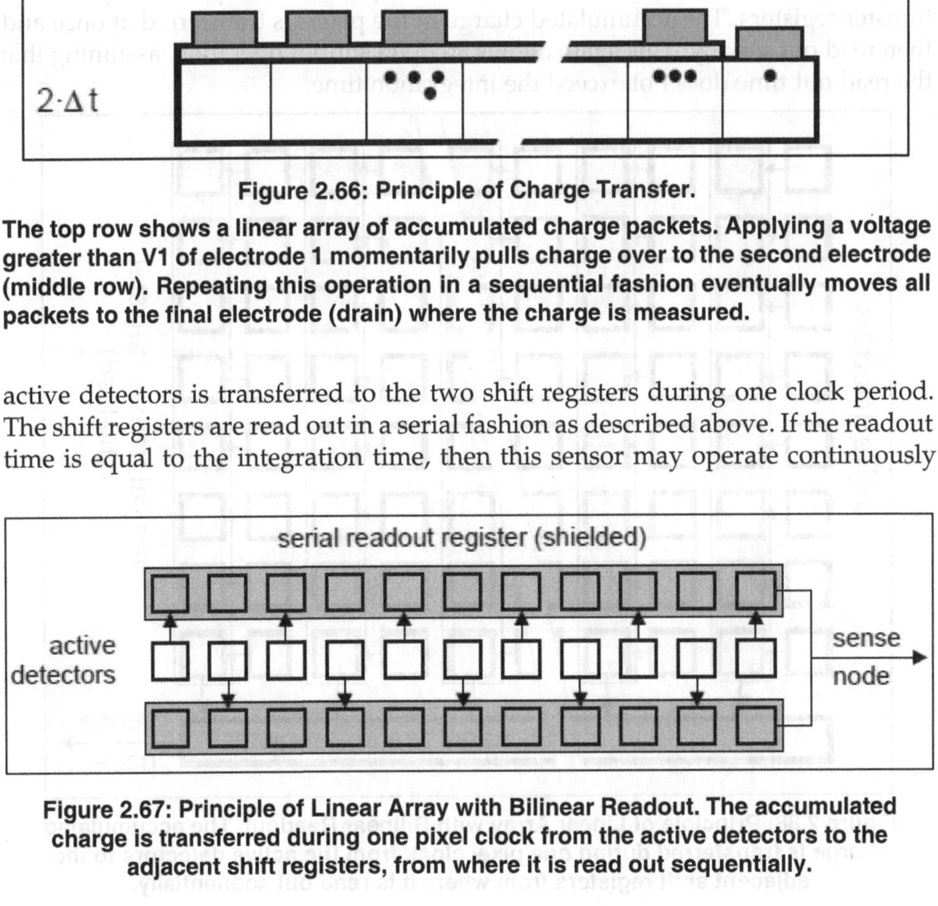

Figure 2.67: Principle of Linear Array with Bilinear Readout. The accumulated charge is transferred during one pixel clock from the active detectors to the adjacent shift registers, from where it is read out sequentially.

without a shutter. This principle, known as *push broom*, is put to advantage in line cameras mounted on moving platforms to provide continuous coverage of the object space.

Frame Transfer

You can visualize a frame transfer imager as consisting of two identical arrays. The active array accumulates charges during integration time. This charge is then transferred to the storage array, which must be shielded since it is also light sensitive. During the transfer, charge is still accumulating in the active array, causing a slightly smeared image.

The storage array is read out serially, line by line. The time necessary to read out the storage array far exceeds the integration. Therefore, this architecture requires a mechanical shutter. The shutter offers the advantage that the smearing effect is suppressed.

Interline Transfer

Figure 2.68 on the following page illustrates the concept of interline transfer arrays. Here, the columns of active detectors (pixels) are separated by vertical transfer registers. The accumulated charge in the pixels is transferred at once and then read out serially. This again allows an open shutter operation, assuming that the read out time does not exceed the integration time.

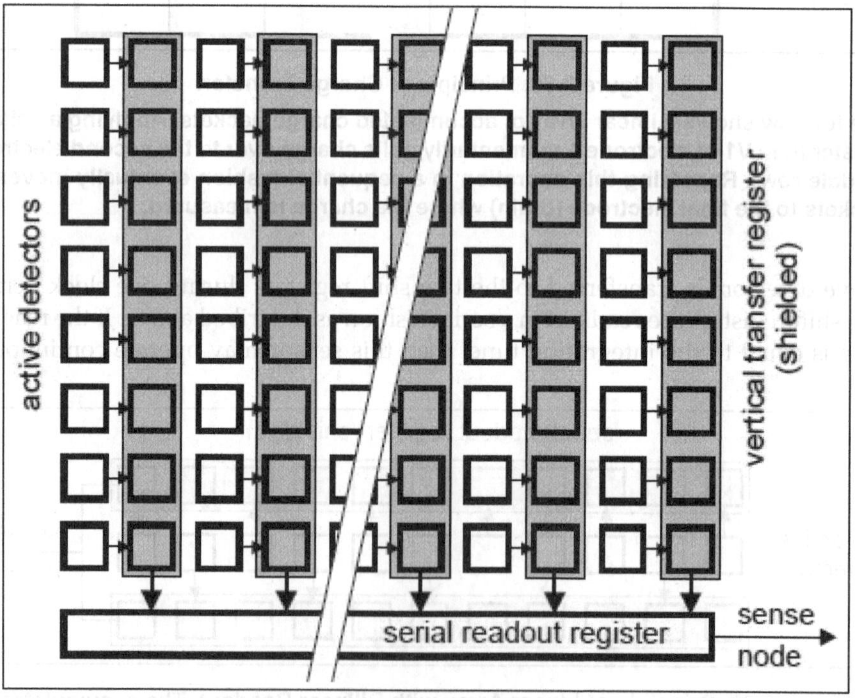

Figure 2.68: Principle of Linear Array with Bilinear Readout. The accumulated charge is transferred during one pixel clock from the active detectors to the adjacent shift registers from where it is read out sequentially.

Since the CCD detectors of the transfer register are also sensitive to irradiance, they must be shielded. This, in turn, reduces the effective irradiance over the chip area. The effective irradiance is often called *fill factor.*

The interline transfer imager as described here has a fill factor of 50 per cent. Consequently, longer integration times are required to capture an image. To increase the fill factor, microlenses may be used. In front of every pixel is a lens that directs the light incident on an area defined by adjacent active pixels to the (smaller) pixel.

5.2.3 Spectral Response

Silicon is the most frequently used semiconductor material. In an ideal silicon detector, every photon exceeding the band gap ($\lambda < 1.1\ \mu$m) causes a photon electron that is collected and eventually measured. The quantum efficiency is unity and the spectral response is represented by a step function. As indicated in Figure 2.69, the quantum efficiency of a real CCD sensor is less than unity for various reasons. For one, not all the incident flux interacts with the detector (*e.g.*, reflected by the electrode in front illuminated sensors). Additionally, some electron-hole pairs recombine. Photons with longer wavelengths penetrate the depletion region and cause electron-hole pairs deep inside the silicon. Here, the probability of recombination is greater and many fewer electrons are attracted by the capacitor. The drop in spectral response toward blue and UV is also related to the electrode material that may become opaque for $\lambda < 0.4\ \mu$m.

Sensors illuminated from the back avoid diffraction and reflection problems caused by the electrode. Therefore, they have a higher quantum efficiency than

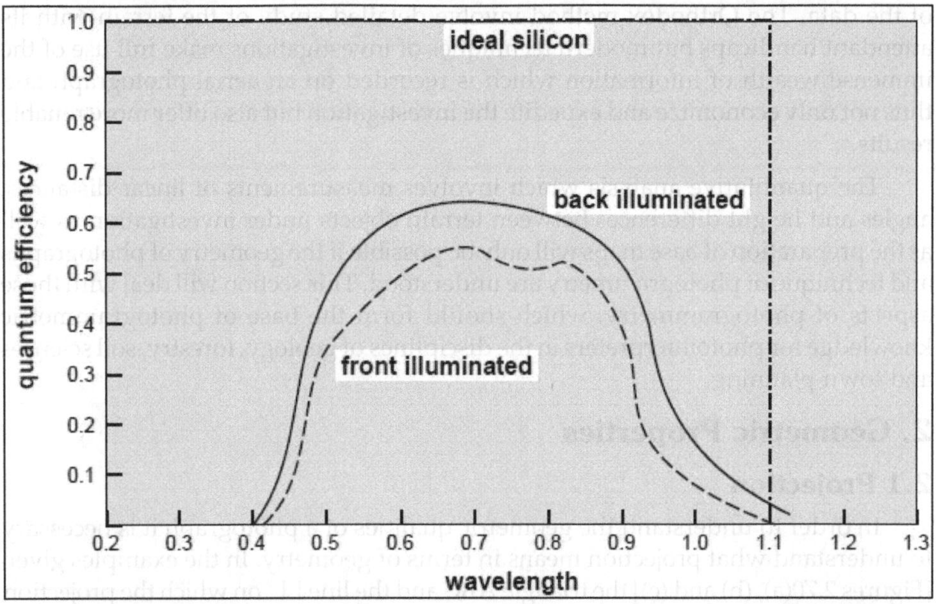

Figure 2.69: Spectral Response of CCD Sensors. In an ideal silicon detector all photons exceeding the band gap energy generate electrons. Front illuminated sensors have a lower quantum efficiency than back illuminated sensors because part of the incident flux may be absorbed or redirected by the electrodes.

front illuminated sensors. However, the detector must be thinner, because high energy photons are absorbed near the surface-opposite the depletion region-and the chances of electron/hole recombination are lower with shorter diffusion length.

In order to make the detector sensitive to other spectral bands (mainly IR), detector material with the corresponding bandgap energy must be selected. This leads to hybrid CCD arrays where the semiconductor and the CCD mechanism are two separate components.

Geometric Characteristics of Aerial Photographs

1. Introduction

Photographs taken from an aircraft commonly termed as *aerial photographs* have come to play an ever-increasing role in the execution of cartographic mapping on various scales and in evaluation of natural resources of a region. Uses of aerial photographs in other fields are also manifold, in fact the scope seems limitless.

The aerial photograph offers possibilities of detailed of the terrain and its culture suited to the need of the investigator, be he a geologist, forester, soil scientist, town planner or any other kind of specialist. The only stipulation is that the specialist must know what he is looking for and how the information, which he is seeking, appears in the aerial photograph. The result of this qualitative analysis will depend upon the specialist's level of specialization and experience.

All studies of natural resources basically involve qualitative examination of the terrain, the correct correlation of the observed data and finally the evaluation of the data. The Orthodox method involve detailed study of the terrain with its attendant handicaps but modern techniques of investigations make full use of the immense wealth of information which is recorded on an aerial photograph and thus not only economize and expedite the investigation but also offer more reliable results.

The quantitative analysis which involves measurements of linear distances, angles and height differences between terrain objects under investigation as well as the preparation of base maps will only be possible if the geometry of photographs and technique of photogrammetry are understood. This section will deal with those aspects of photogrammetry, which should form the base of photogrammetric knowledge for photointerpreters in the disciplines of geology, forestry, soil sciences and town planning.

2. Geometric Properties

2.1 Projection

In order to understand the geometric qualities of a photograph it is necessary to understand what projection means in terms of geometry. In the examples given [Figures 2.70(a), (b) and (c)] the triangle ABC and the line LL′ on which the projection is made are in the same plane.

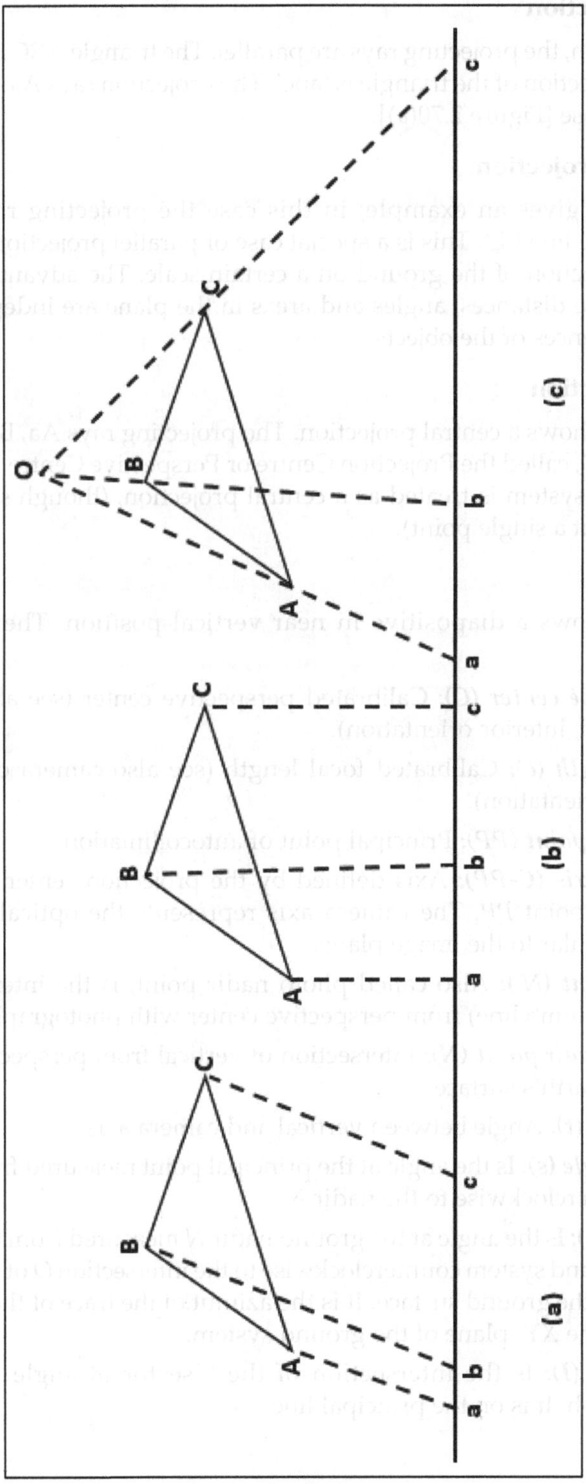

Figure 2.70: Projection.

(a) Parallel Projection

In this projection, the projecting rays are parallel. The triangle ABC is projected on the LL′. The projection of the triangle is 'abc'. The projection rays Aa, Bb, Cc, are all parallel in this case [Figure 2.70(a)].

(b) Orthogonal Projection

Figure 2.70(b) gives an example; in this case the projecting rays are all perpendicular to the line LL′. This is a special case of parallel projection Maps are an orthogonal projection of the ground on a certain scale. The advantage of this projection is that the distances, angles and areas in the plane are independent of the elevation differences of the objects.

(c) Central Projection

Figure 2.70(c) shows a central projection. The projecting rays Aa, Bb, Cc, pass through one point O, called the Projection Centre or Perspective Centre. The image projected by a lens system is treated as a central projection, (though stringy it is not, as the lens is not a single point).

2.2 Definitions

Figure 2.71 shows a diapositive in near vertical position. The following definitions apply:

- ☆ ***Perspective center (C):*** Calibrated perspective center (see also camera calibration, interior orientation).
- ☆ ***Focal length (c):*** Calibrated focal length (see also camera calibration, interior orientation).
- ☆ ***Principal point (PP):*** Principal point of autocollimation.
- ☆ ***Camera axis (C-PP):*** Axis defined by the projection center C and the principal point *PP*. The camera axis represents the optical axis. It is perpendicular to the image plane.
- ☆ ***Nadir point (N′):*** Also called photo nadir point, is the intersection of vertical (plumb line) from perspective center with photograph.
- ☆ ***Ground nadir point (N):*** Intersection of vertical from perspective center with the earth's surface.
- ☆ ***Tilt angle (t):*** Angle between vertical and camera axis.
- ☆ ***Swing angle (s):*** Is the angle at the principal point measured from the *+y-axis* counterclockwise to the nadir *N*.
- ☆ ***Azimut (α):*** Is the angle at the ground nadir *N* measured from the *+Y-axis* in the ground system counterclockwise to the intersection *O* of the camera axis with the ground surface. It is the azimut of the trace of the principal plane in the *XY* -plane of the ground system.
- ☆ ***Isocenter (I):*** Is the intersection of the bisector of angle *t* with the photograph. It is on the principal line.

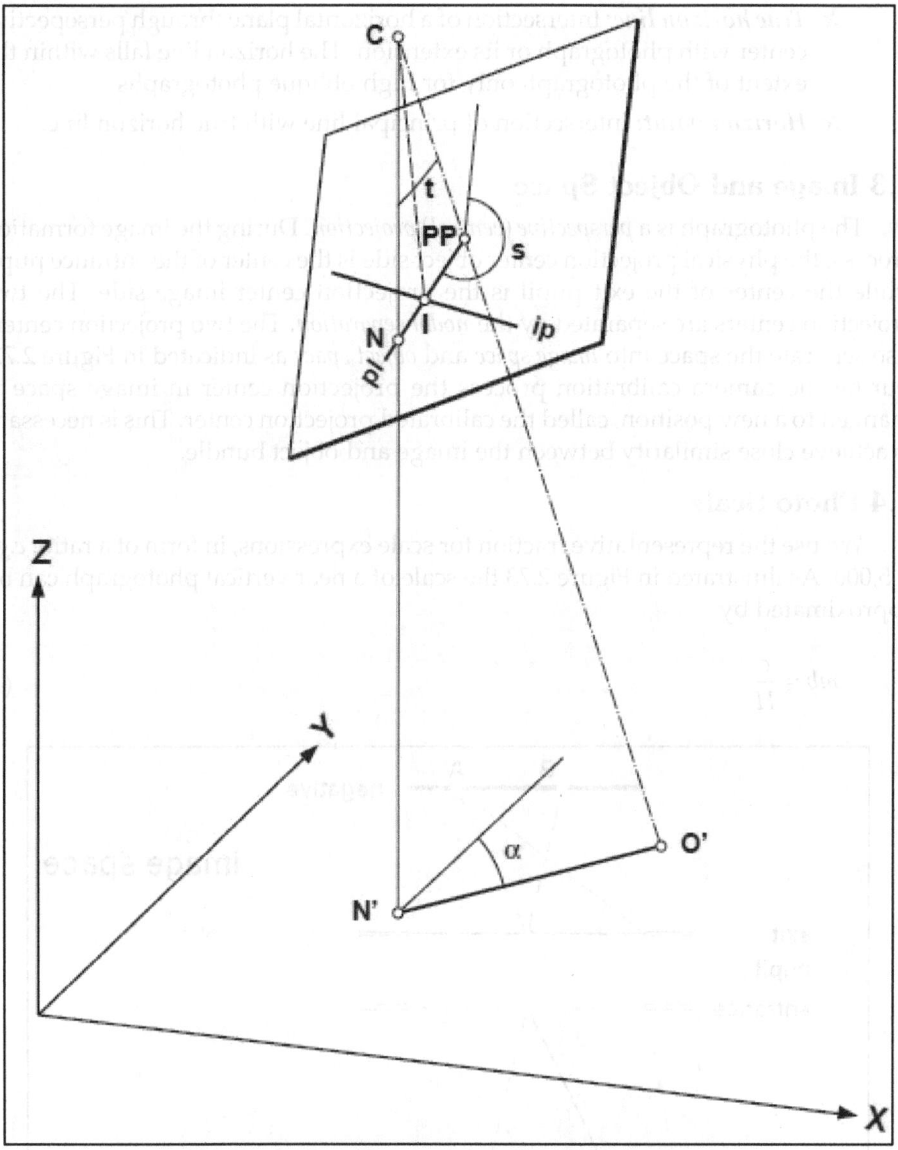

Figure 2.71: Tilted Photograph in Diapositive Position and Ground Control Coordinate System.

☆ ***Principal line (pl):*** Intersection of plane defined by the vertical through perspective center and camera axis with the photograph. Both, the nadir *N* and the principal point *PP* are on the principal line. The principal line is oriented in the direction of steepest inclination of the tilted photograph.

☆ ***Isometric parallel (ip):*** Is in the plane of photograph and is perpendicular to the principal line at the isocenter.

- ☆ ***True horizon line:*** Intersection of a horizontal plane through persepective center with photograph or its extension. The horizon line falls within the extent of the photograph only for high oblique photographs.
- ☆ ***Horizon point:*** Intersection of principal line with true horizon line.

2.3 Image and Object Space

The photograph is a *perspective (central) projection*. During the image formation process, the physical projection center object side is the center of the entrance pupil while the center of the exit pupil is the projection center image side. The two projection centers are separated by the *nodal separation*. The two projection centers also separate the space into *image space* and *object space* as indicated in Figure 2.72. During the camera calibration process the projection center in image space is changed to a new position, called the calibrated projection center. This is necessary to achieve close similarity between the image and object bundle.

2.4 Photo Scale

We use the representative fraction for scale expressions, in form of a ratio, *e.g.*, 1: 5,000. As illustrated in Figure 2.73 the scale of a near vertical photograph can be approximated by

$$mb = \frac{c}{H} \tag{1}$$

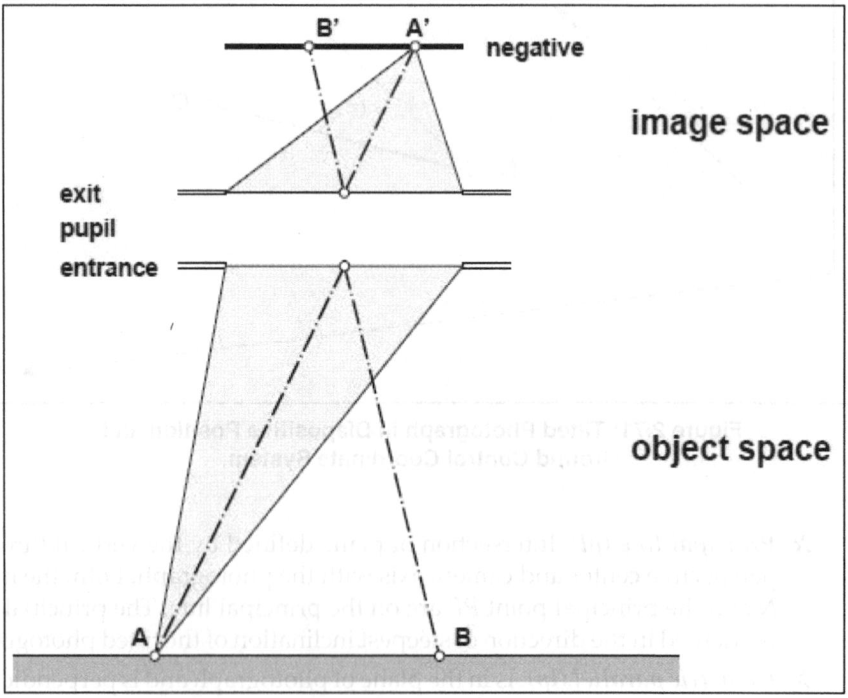

Figure 2.72: The Concept of Image and Object Space.

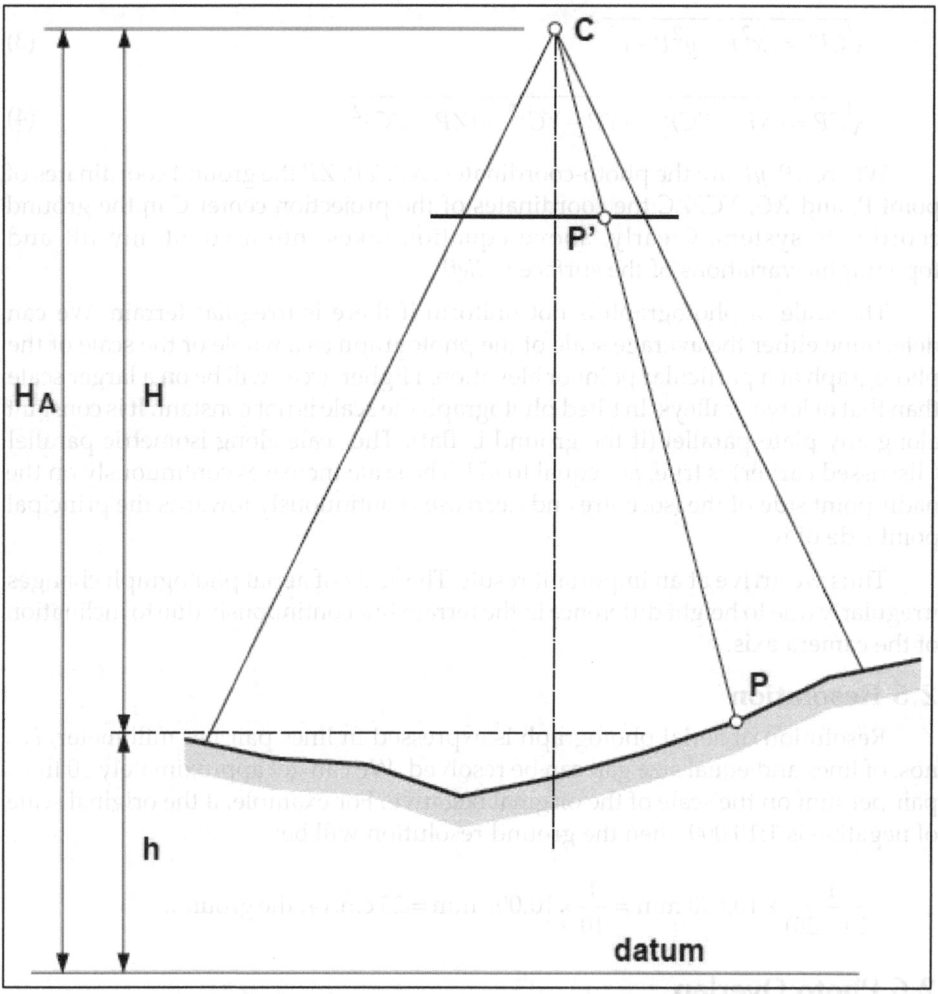

Figure 2.73: Flight Height, Flight Altitude and Scale of Aerial Photograph.

Where *mb* is the *photograph scale number*, *c* the calibrated focal length, and *H* the *flight height* above mean ground elevation. Note that the flight height *H* refers to the average ground elevation. If it is with respect to the datum, then it is called *flight altitude*

$H_{A'}$, with $H_A = H + h$.

The photograph scale varies from point to point. For example, the scale for point *P* can easily be determined as the ratio of image distance *CP'* to object distance *CP* by:

$$m_P = \frac{CP'}{CP} \tag{2}$$

$$\sqrt{CP' = x^2P + y^2P + c^2} \quad (3)$$

$$\sqrt{CP = (XP - XC)^2 + (YP - YC)^2 + (ZP - ZC)^2} \quad (4)$$

Where *xP, yP* are the photo-coordinates, *XP, YP, ZP* the ground coordinates of point P, and *XC, YC, ZC* the coordinates of the projection center C in the ground coordinate system. Clearly, above equation takes into account any tilt and topographic variations of the surface *(relief)*.

The scale of photograph is not uniform if there is irregular terrain. We can determine either the average scale of the photograph as a whole or the scale of the photograph at a particular point or elevation. Higher areas will be on a larger scale than that of lower valleys. In tilted photograph the scale is not constant. It is constant along any plate parallel (if the ground is flat). The scale along isometric parallel (discussed earlier) is true, *i.e.*, equal to *f/H*. The scale increases continuously on the nadir point side of the isocentre and decreases continuously towards the principal point side of it.

Thus we arrive at an important result: The scale of aerial photograph changes irregularly due to height difference in the terrain but continuously due to inclination of the camera axis.

2.5 Resolution

Resolution of aerial photograph is expressed in lines pair per millimeter, *i.e.*, nos. of lines and equal size gap can be resolved. We can get approximately 20 lines pair per mm on the scale of the original negative. For example, if the original scale of negative is 1:10,000, then the ground resolution will be:

$$\frac{1}{(20+20)} \times 10{,}000 \text{ mm} = \frac{1}{10} \times 10{,}000 \text{ mm} = 25 \text{ cm. on the ground.}$$

2.6 Photo Overlap

Aerial photo projects for all mapping and most image analyses require that a series of exposures be made along each of the multiple flight lines. To guarantee stereoscopic coverage throughout the site, the photographs must overlap in two directions: in the line of flight and between adjacent flights.

2.6.1 Endlap

Endlap, also known as forward overlap, is the common image area on consecutive photographs along a flight strip. This overlapping portion of two successive aerial photos, which creates the three-dimensional effect necessary for mapping, is known as a stereomodel or more commonly as a "model." Figure 2.74 shows the endlap area on a single pair of consecutive photos in a flight line.

Practically all projects require more than a single pair of photographs. Usually, the aircraft follows a predetermined flight line as the camera exposes successive overlapping images.

Normally, endlap ranges between 55 and 65 per cent of the length of a photo, with a nominal average of 60 per cent for most mapping projects. Endlap gain, the distance between the centers of consecutive photographs along a flight path, can be calculated by using Equation (5).

$$g_{end} = sp \times w \times [(100 - o_{end})\ 100] \quad (5)$$

where,

g_{end} = Distance between exposure stations (feet)

sp = Photo scale denominator (feet)

o_{end} = Endlap (percent)

w = Width of exposure frame (inches)

When employing a precision aerial mapping camera with a 9 a 9 in. exposure format and a normal endlap of 60 per cent, the formula is simpler. In this situation, two of the variables then become constants:

w = 9 in.

o_{end} = 60 per cent

Then, the expression w*[(100 per cent – o_{end})/100] becomes a constant equal to 3, and Equation 5 may be supplanted by Equation (6).

$$q\ s_{\text{end}}\ p = -3\ 6 \quad (6)$$

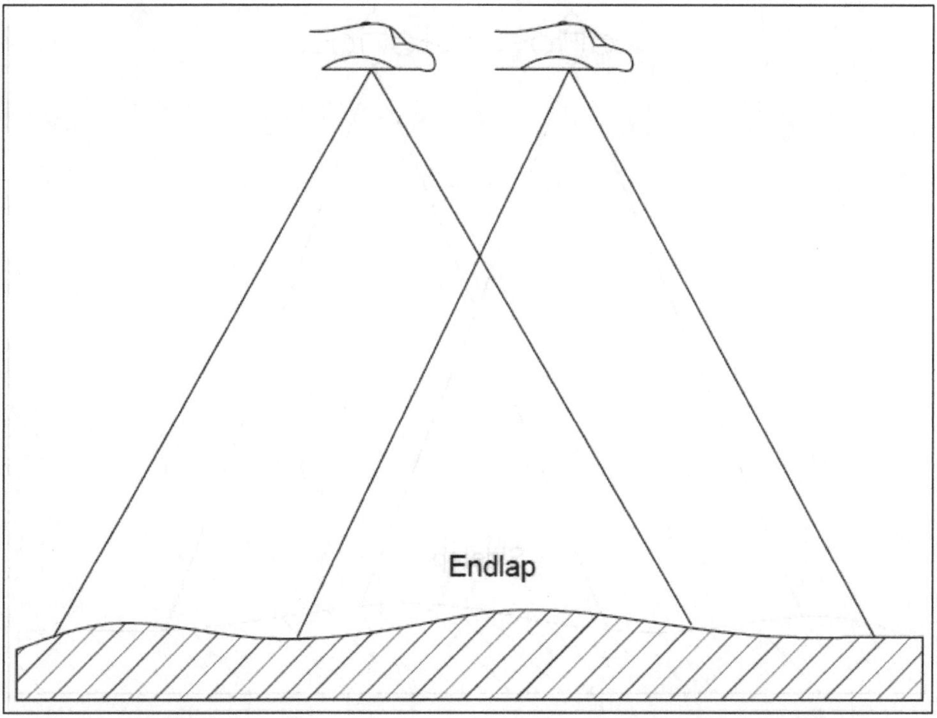

Figure 2.74: Endlap on Two Consecutive Photos in a Flight Line.

When utilizing a camera other than a 9 × 9 in. format and/or an endlap other than 60 per cent, Equation (5) must be employed.

2.6.2 Sidelap

Sidelap, sometimes called side overlap, encompasses the overlapping areas of photographs between adjacent flight lines. It is designed so that there are no gaps in the three-dimensional coverage of a multiline project. Figure 2.75 and 2.76 shows the relative head-on position of the aircraft in adjacent flight lines and the resultant area of exposure coverage.

Usually, sidelap ranges between 20 and 40 per cent of the width of a photo, with a nominal average of 30 per cent. Figure 2.76 portrays the sidelap pattern in a project requiring three flight lines. Sidelap gain, the distance between the centers of adjacent flight lines, can be calculated by using Equation (7).

$$q_{side} = sp - w''[(100 - o_{side})\ 100] \tag{7}$$

where,

g_{side} = Distance between flight line centers (feet)

sp = Photo scale denominator (feet)

o_{side} = Sidelap (percent)

w = Width of exposure frame (inches)

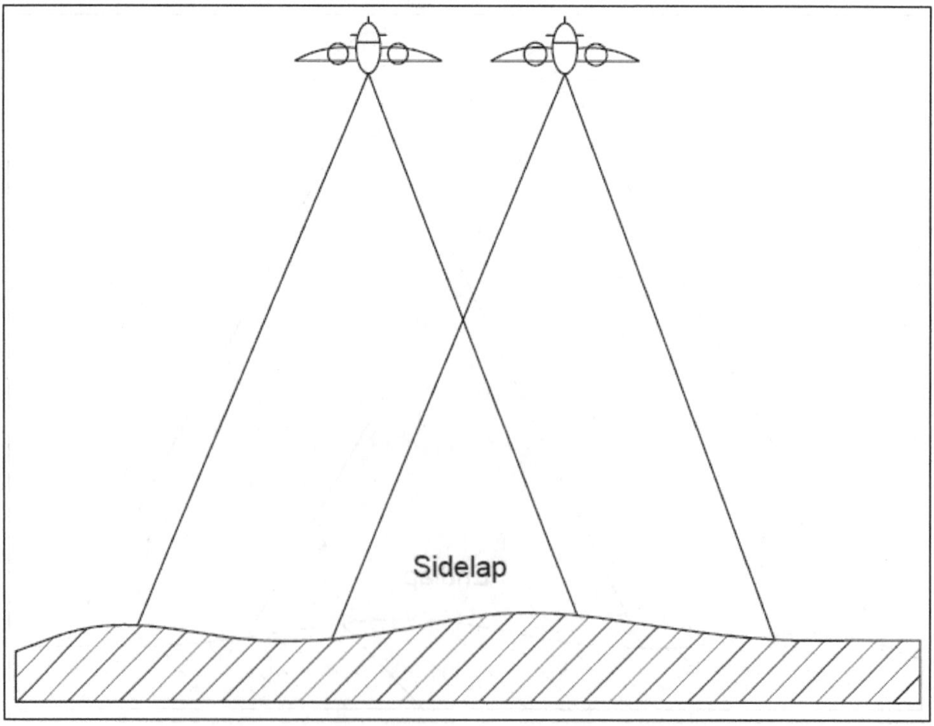

Figure 2.75. Sidelap between Two Adjacent Flight Lines.

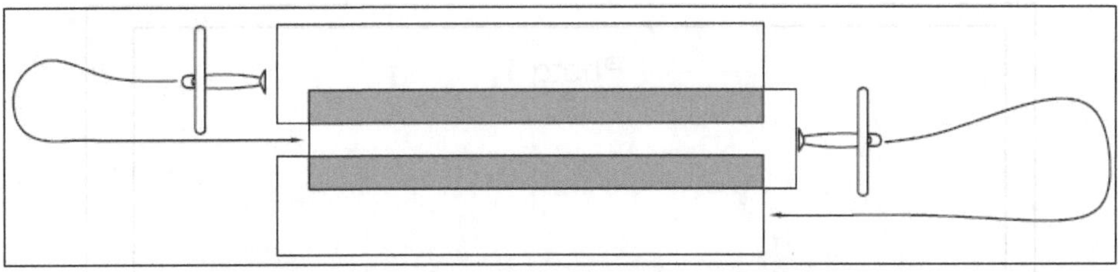

Figure 2.76: Sidelap on Three Adjacent Flight Lines.

When employing a precision aerial mapping camera with a 9×9 in. exposure format and a normal sidelap of 30 per cent, the formula is simpler. In this situation, two of the variables then become constants:

w = 9 in.

o_{side} = 30 per cent

Then, the expression w*[(100 per cent – o_{side})/100] becomes a constant equal to 3, and Equation 7 may be supplanted by Equation (8).

$$q_{side} = Sp''6.3 \quad (8)$$

When utilizing a camera other than a 9 × 9 in. format and/or a sidelap other than 30 per cent, Equation (7) must be employed.

2.7 Stereomodel

From the foregoing discussion of overlap, it is evident that consecutive photos in a flight strip overlap. When focusing each eye on a particular image feature that was viewed by the camera from two different aspects, the mind of the observer is convinced that it is seeing a lone object with three dimensions. Put simply, the threedimensional effect is an optical illusion. This phenomenon of observing a feature from different positions is known as the parallax effect. Although used to describe other facets of photogrammetry, parallax is defined as a change in the position of the observer. This situation allows a viewer, when using appropriate stereoscopic instruments, to observe a pair of two-dimensional photos and see a single three-dimensional image.

Photogrammetrists envision a model as the "neat" area that a single stereopair contributes to the total project. This allows for the endlap and sidelap with surrounding photos. A mapping model is shown as the crosshatched area in Figure 2.77. For a given photograph, several parameters can be found:

- Flight height (above mean ground level)
- Photo center interval
- Flight line spacing
- Acres per model (neat area)

It must be realized that the scale of individual photographs in a project is not a constant. Due to undulations in the aircraft flight and terrain relief, the distance

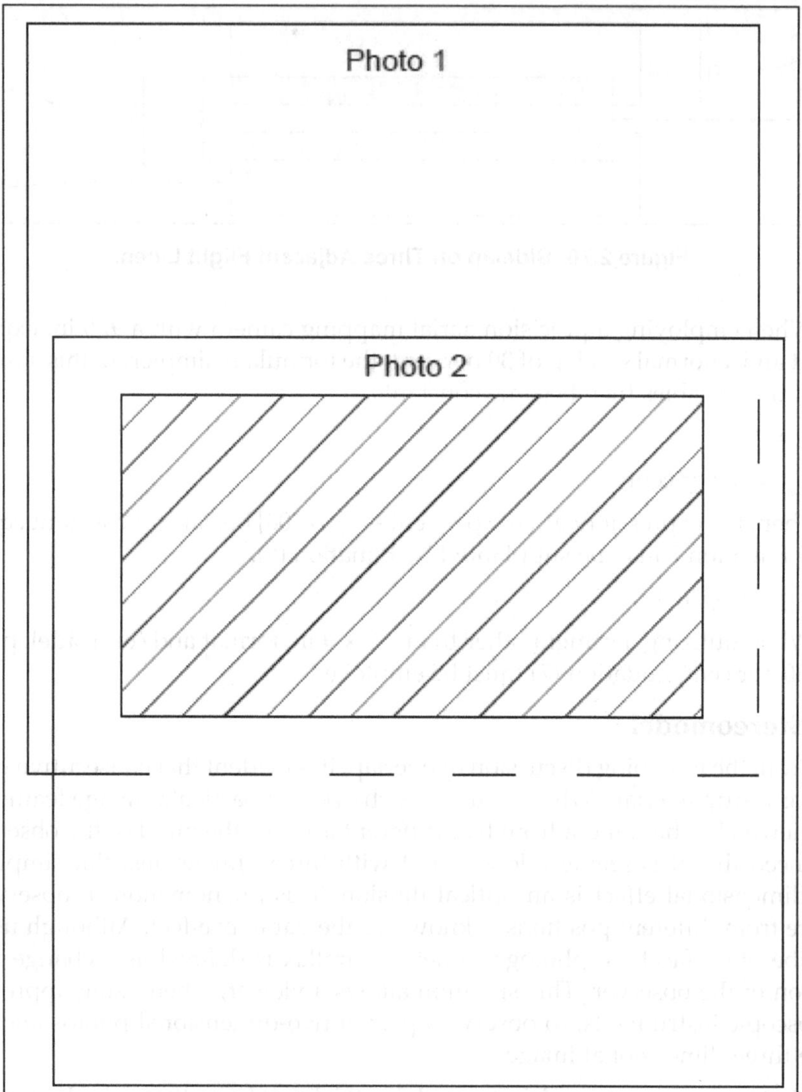

Figure 2.77: Neat Area of a Stereomodel.

between the camera and the ground differs from one exposure to another. Therefore, photo scale must be considered as an average scale for the total project.

2.8 Image Displacement

On a planimetric map all features/details are shown in their correct horizontal position on a certain scale. This is not so in the case of aerial photographs due to image, displacement or distortion. A disturbance of the principle of geometry is called displacement/distortion. There are three major sources of displacement/distortion, which are due to: Optical or photographic deficiencies, *i.e.*, lens distortion and aberration; relief variation of the object photographed and tilt of the camera axis at the moment of exposure.

(a) Lens Distortion

Figure 2.78 shows distortion due to a lens. Object point O is imaged at I′ instead of its correct position I on the image plane. d is the image displacement in this case. In the modern aerial camera lens this type of distortion is negligible.

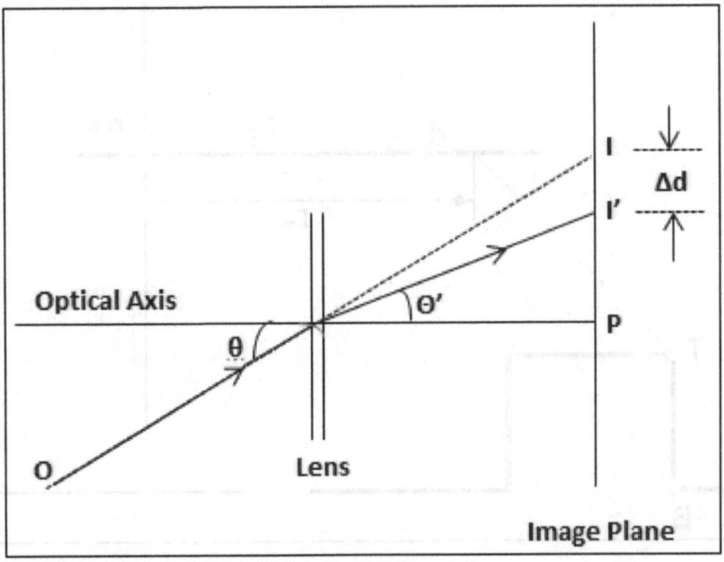

Figure 2.78: Lens Distortion.

2.9 Relief Displacement

The surface of the earth is not smooth and flat. As a consequence, there is a natural phenomenon that disrupts true orthogonality of photo image features. In this respect, an orthogonal image is one in which the displacement has been removed, and all of the image features lie in their true horizontal relationship.

The effect of relief does not only cause a change in the scale but can also be considered as a component of image displacement. Figure 2.79 illustrates this concept. Suppose point *T* is on top of a building and point *B* at the bottom. On a map, both points have identical *X*, *Y* coordinates; however, on the photograph they are imaged at different positions, namely in *T′* and *B′*. The distance *d* between the two photo points is called *relief displacement* because it is caused by the elevation difference Δh between *T* and *B*.

The magnitude of relief displacement for a true vertical photograph can be determined by the following equation:

$$d = \frac{r\Delta h}{H} = \frac{r'\Delta h}{H - \Delta h} \tag{9}$$

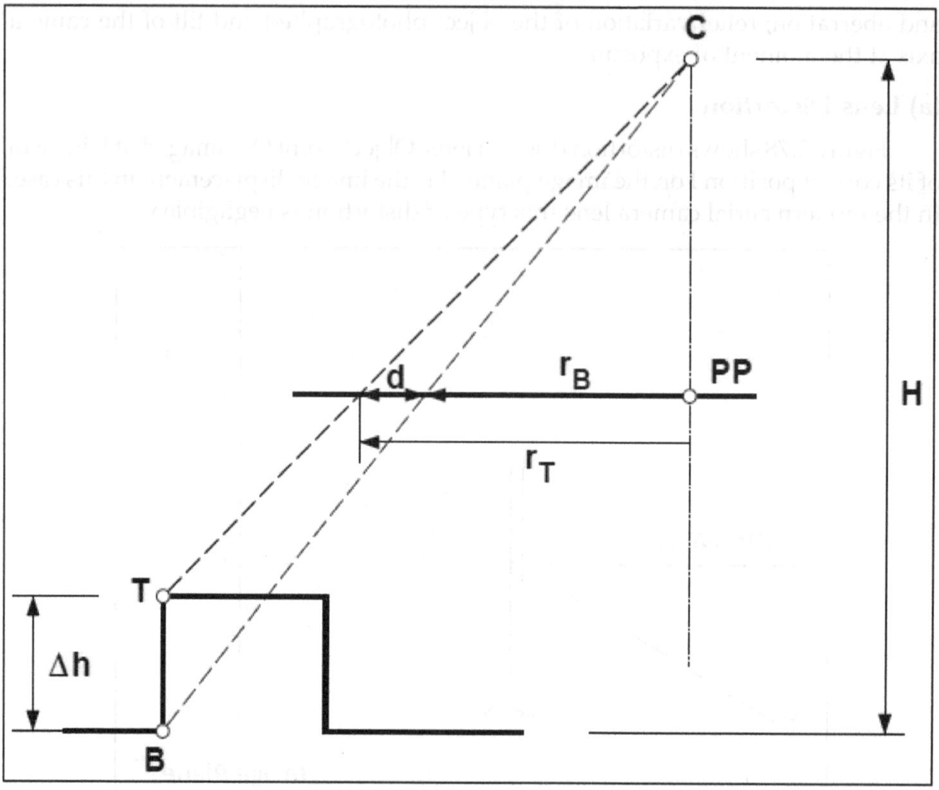

Figure 2.79: Relief Displacement.

Where $r = \sqrt{x^2T + y^2T}$, $r' = \sqrt{x^2B + y^2B}$ and Δh the elevation difference of two points on a vertical. Equation (5) can be used to determine the elevation Δh of a vertical object

$$h = \frac{dH}{r} \tag{10}$$

The direction of relief displacement is radial with respect to the nadir point N', independent of camera tilt.

2.9.1 Causes of Displacement

Camera tilt, earth curvature, and terrain relief all contribute to shifting photo image features away from true geographic location. Camera tilt is greatly reduced or perhaps eliminated by gyroscopically-controlled cameras.

Earth curvature is of little consequence on large-scale photography. The relatively small amount of lateral distance covered by the exposure frame introduces only a minimal amount of curvature, if any. Topographic relief can have a great effect on displacing image features. The amount of image displacement increases on high-degree slopes. Feature displacement also increases radially away from the photo center.

2.9.2 Effects of Displacement

An aerial photograph is a three-dimensional scene transferred onto a two-dimensional plane. Hence, the photographic process literally squashes a three-dimensional feature onto a plane that lacks a vertical dimension, and image features above or below mean ground level are displaced from their true horizontal location. Figure 2.80 illustrates this phenomenon. Assuming that the stack rises straight into the air from the ground, both the top and the base possess the same horizontal (XY) placement.

This diagram belies that fact, because the base and the top are in displaced positions (labelled *"d"* in Figure 2.80) on the negatives. This separation will not be of the same magnitude on successive photos. Figure 2.81 illustrates the radial displacement of an object in an aerial photograph.

Just as images of fast-rising features are displaced, so are the changes in ground elevations, though not as visibly apparent in the photographs. Figure 2.82 illustrates relief displacement on a straight utility clearing that crosses rolling hills. The clearing is identified as the wavy open strip running diagonally through the woods on the

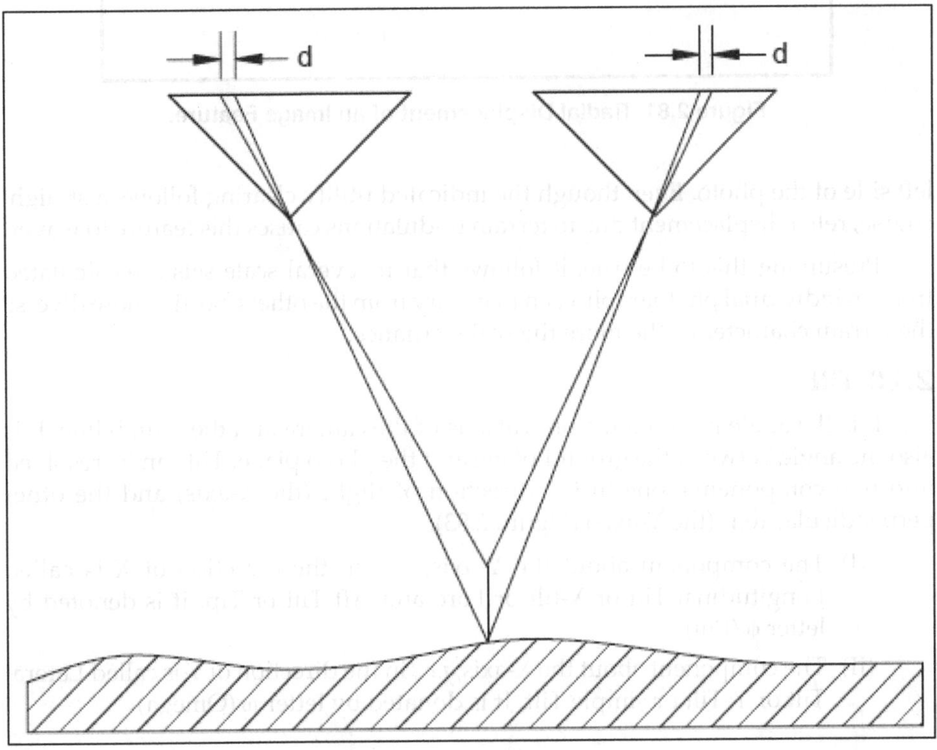

Figure 2.80: Image Displacement.

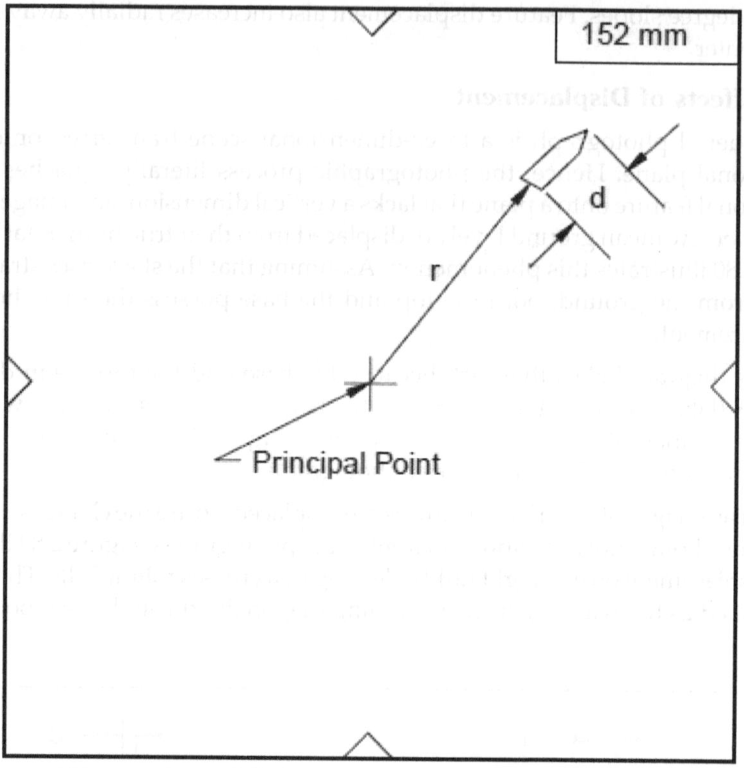

Figure 2.81: Radial Displacement of an Image Feature.

left side of the photo. Even though the indicated utility clearing follows a straight course, relief displacement due to terrain undulations causes this feature to waver.

Presuming this to be true, it follows that if several scale sets are calculated from an individual photograph, each may vary from the others. So, the more diverse the terrain character is, the more the scale variance.

2.10 Tilt

It is the angle between the optical axis of the camera and the plumb line. It is also the angle between the ground plane and the photo plane. Tilt can be resolved into two components, one in the direction of flight (the X-axis) and the other perpendicular to it (the Y-axis) (Figure 2.83).

(I) The component about the Y-axis, *i.e.*, in the direction of X is called Longitudinal Tilt or X-tilt or Fore and Aft Tilt or Tip. It is denoted by letter ϕ (Phi).

(II) The component about the X-axis, *i.e.*, in the direction of Y is called Lateral Tilt or Y-Tilt or simple Tilt. It is donated by letter ω (Omega)

Figure 2.82: Image Displacement on a Utility Clearing.

In Figure 2.83 the vertical ON through the perspective center meets the photo plane at point *'n'* called the Photo nadir point and the ground plane at point N called the Ground nadir point. These points are also called Plumb Point.

The foot of the perpendicular (*p*) from *O* on the photo plane is called **Principle Point.** The length of this perpendicular (*op*) is called **Principle Distance** (Figure 2.84).

The approximate position of the principal point of a photograph is determined by joining the opposite fiducial marks (or collimating marks) (Figure 2.85). Line joining opposite fiducial marks is known as fiducial axis. The point of intersection of the fiducial axes is called fiducial centre (*f*) and is, for practical purposes, coincident with the principal point (*p*) in a well-adjusted camera.

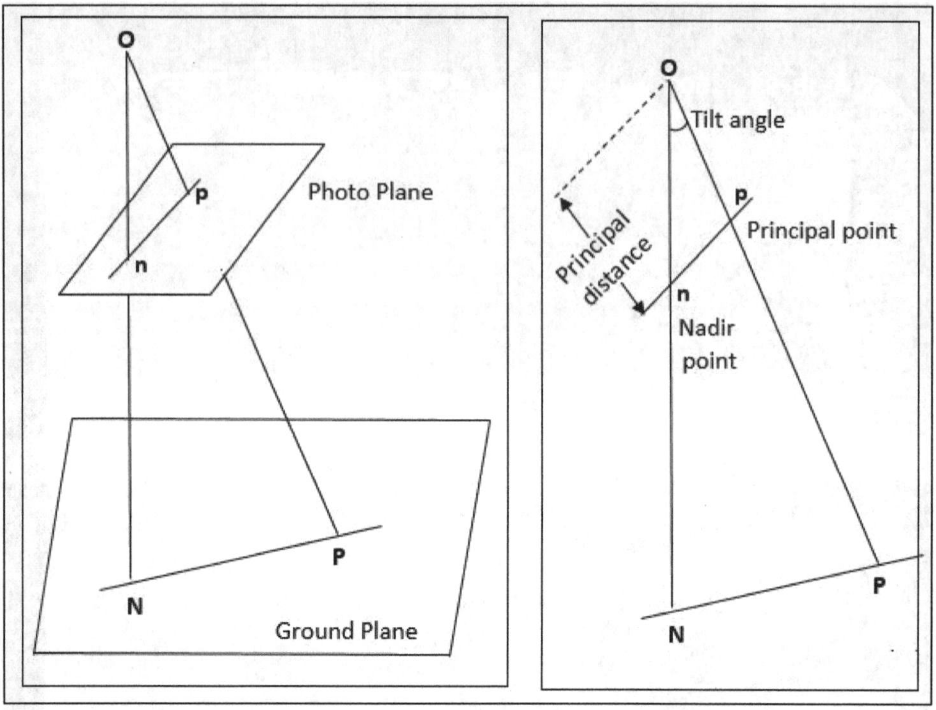

Figure 2.83: Tilt.

2.10.1 Reasons for Photo Tilt

- ☆ Atmospheric conditions (air pockets or currents)
- ☆ Human error of the pilot fails to maintain a steady flight,
- ☆ Imperfections in the camera mounting, etc.

2.11 Swing

Swing is the angle measured in the plane of the photograph between the fiducial axis in the direction of flight and the actual flight line [Figure 2.85(B)]. The angle is denoted by (Kappa).

3. Elementary Mathematical Concepts

An aerial photograph, as already discussed, is a central perspective. In an ideal case of an absolutely vertical photograph of a completely flat terrain the aerial photograph will be geometrically the same as the corresponding map of the area. However, because of tilt of the photograph and relief variation of ground photographed, an aerial photograph differs geometrically from the map of the corresponding area.

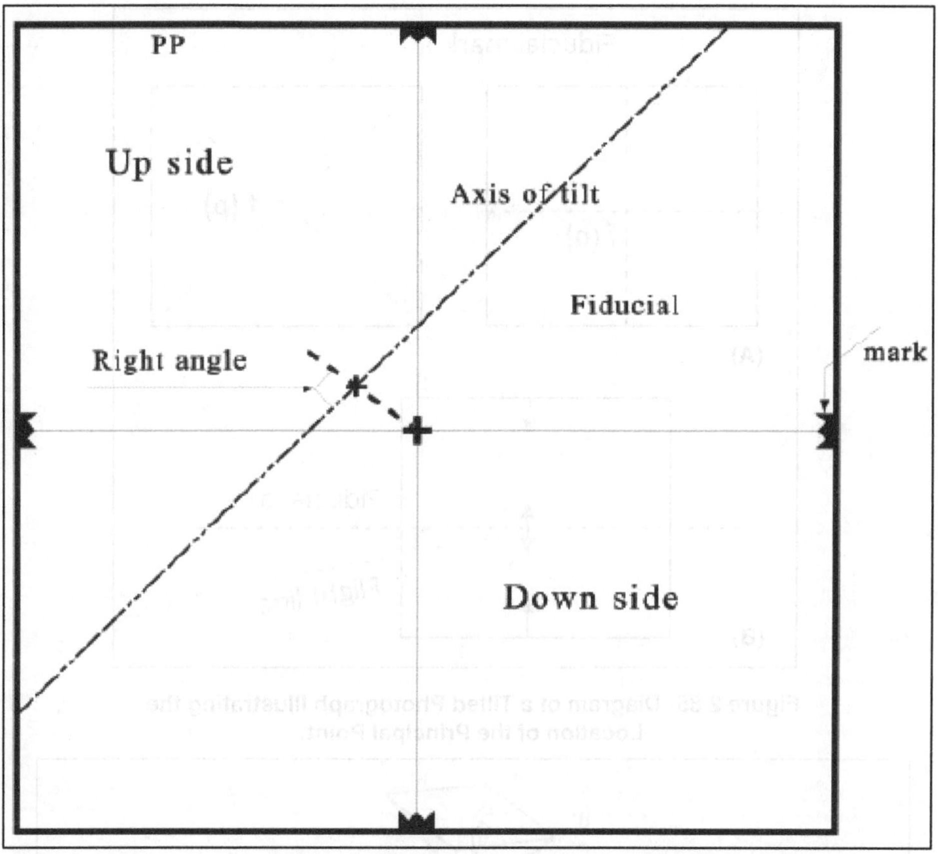

Figure 2.84: Diagram of a Tilted Photograph Illustrating the Location of the Principal Point (PP), the Nadir (n), the Isocenter (i), the Axis of Tilt, and the Direction of Tilt (Up and down sides).

The central perspective (a case of central projection) is characterized by the fact that all straight lines joining corresponding points, *i.e.*, straight lines joining object points to their corresponding images, pass through one point. This point is known as the perspective centre. Figure 2.86 illustrates this relationship.

Straight lines *Aa′*, *Bb′* etc., joining corresponding points, *e.g.*, A on the ground and a′ its image in the image plane (negative plane), are known as perspective rays and pass through the perspective centre.

A plane between the perspective centre and the object is known as a positive plane. The consideration of a positive plane does not involve any significant geometrical change in the relationship.

The purpose of Photogrammetry is to produce an orthogonal projection of the image of the object from its central projection (*i.e.*, from perspective pictures-

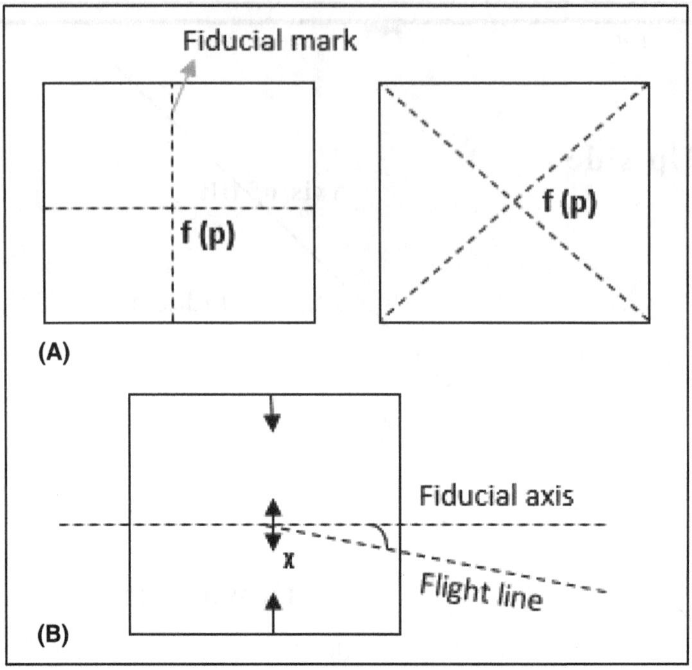

Figure 2.85: Diagram of a Tilted Photograph Illustrating the Location of the Principal Point.

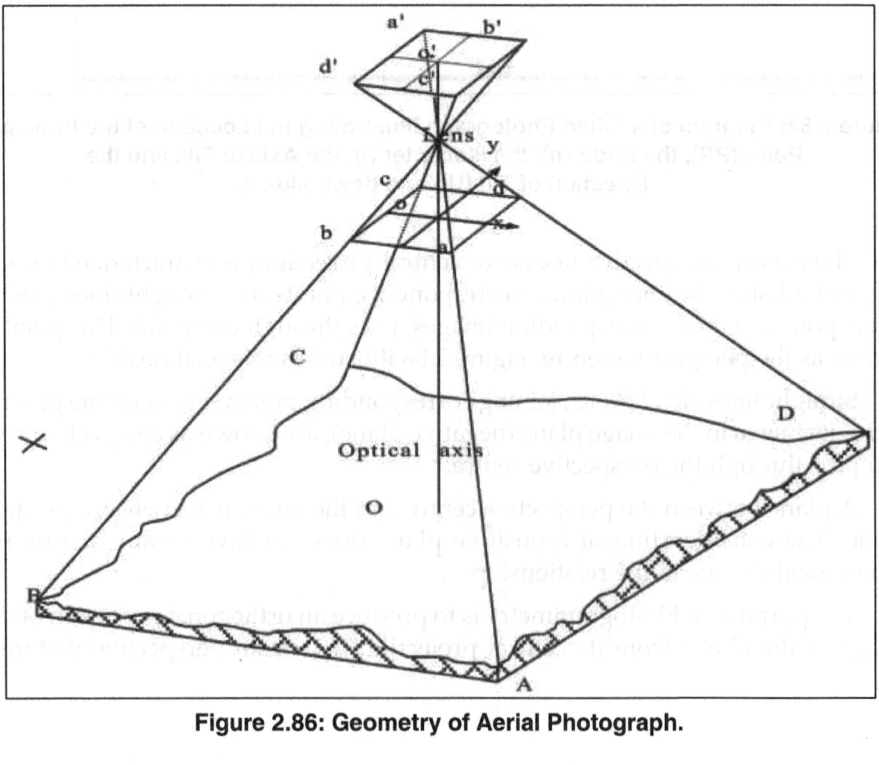

Figure 2.86: Geometry of Aerial Photograph.

photographs) by using the geometrical links between the object and its photo-image at the moment of exposure.

3.1 Properties of Central Projection

To study the properties of an aerial photograph it is necessary to understand the geometry of central (perspective projection). Some properties of this projection will be dealt with in what follows: Plane I can be considered as ground plane and Plane II as positive plane (photograph), see Figure 2.87.

(a) *AB*, the line of intersection of the object and image planes is known as the axis of homology. It is also known as the axis of perspective.

(b) *SN* and *Sp* are the perpendiculars from *S* on to the planes I and II intersecting plane *I* at *N* and *P* and plane II at n and p respectively. The bisector of the angle *PSN* meets the principle line (see subpara (*f*) *EF* in plane I and *EH* in plane II in *I* and *i*. These points are ISOCENTRE is that angle measured at *I* in plane I are the same as corresponding angles measured at *i* in plane II. It is very important to remember that angles are true at the isocentres only when the ground is flat.

(c) A plane parallel to plane I and passing through the perspective centre 'S' cuts the plane II in a line *CD* that is known as the horizon line. Horizon Line and Axis of homology are always parallel to each other. All horizontal lines parallel to the horizon line or axis of homology are termed as plate parallel. The plate parallel passing through the isocentre 'i' is termed the isometric parallel or isoline. It can be proved that this is the only parallel along which the scale = *f/H*, *i.e.*, the same as in the case of a vertical photograph. A line through the perspective centre perpendicular to the principal plane (*f*) is called axis of tilt.

(d) Image of all objects infinitely distant on the right of *AB* will be formed on the horizon line. Points on the horizon line are known as vanishing points. Images of parallel lines in the ground plane converge to a vanishing point. Thus, images of all lines parallel to the principle line in the object plane will converge to a vanishing point *H*, the point of intersection of horizon line and the principal line in the photoplane. It is a fundamental property of perspective projection that a line in one plane projects as a line in the other plane, the two lines meeting at the axis of homology, *e.g.*, line *XY* in ground plane projects as *X'Y'* in photo plane.

(e) A plane, which is perpendicular to both the planes and passes through the perspective centre is known as the principal plane. The lines of intersection of this plane with the two planes are known as Principal lines. *EF* and *EH* are principal lines. Figure 2.87 represent the principal plane. The angle '0' between the principal lines is the angle between the perspective planes. When this angle '0' is equal to Zero the plane II can be considered as a vertical photograph. In normal vertical photography this angle should not exceed 3 or 4 grades.

(f) Any point in plane I such as *X* has a corresponding position X′ in plane II. Such pairs are called homologous points.

(g) If two planes are projectively related as in Figure 2.88 certain important relationships exist between the corresponding details in plane I and II. Lines *O′1, O′2, O′3* and *O′4,* on plane II are the images of lines *O1, O2, O3* and O4 in plane I. Let there be another line UV in plane I which cuts the lines *O1, O2,* etc. in 1', 2', 3' and 4' then it can be shown that

$$\frac{12/13}{24/34} = \frac{1'2'/1'3'}{2'4'/3'4'} = r$$

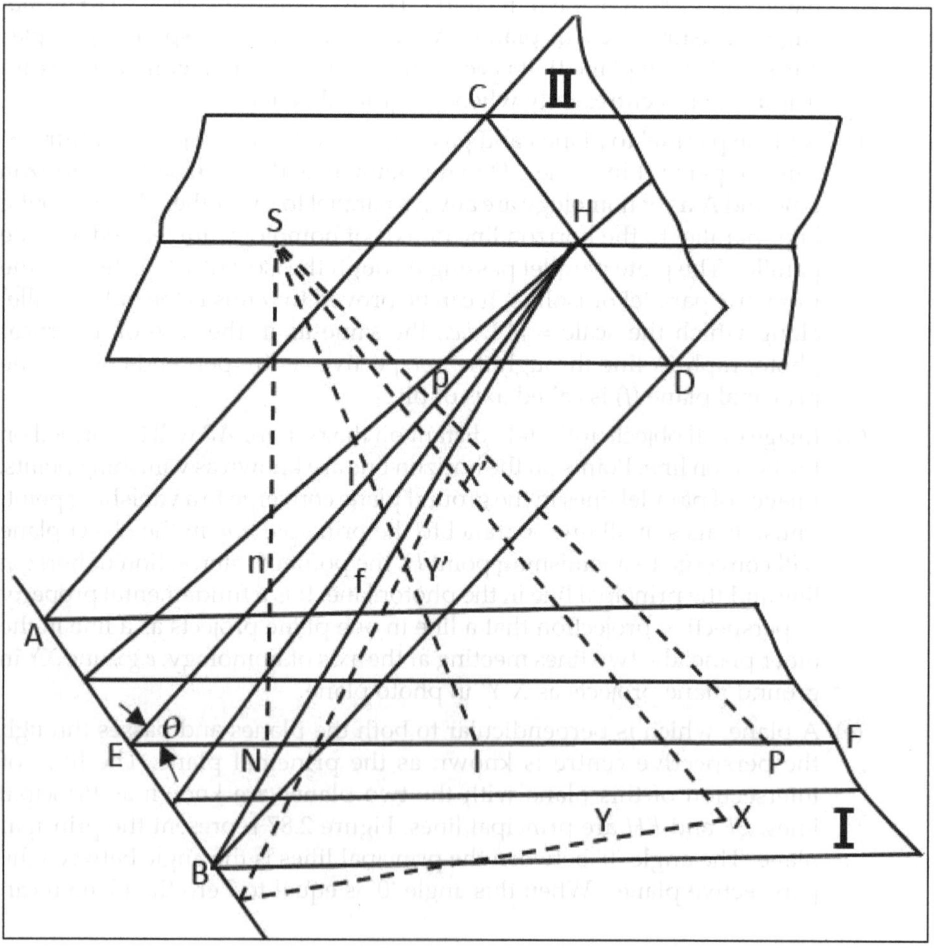

Figure 2.87: Geometry of a Tilted Photograph.

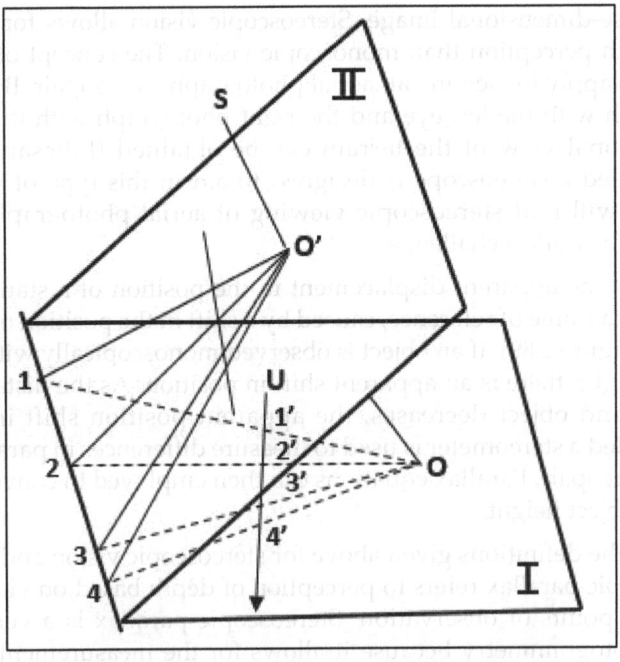

Figure 2.88: Projectively Related Two Planes.

This ratio is known as the Anharmonic ratio (or Cross-ratio) of the four distances. Because of the constancy of this ratio, a unique position can be found for the line *UV* in plane II as well so that 1' falls on line *O'*1, 2' falls on line *O'*2 and so on. This property is used in graphical rectification.

3.2 Distortion vs. Displacement

Often, the term distortion is considered to be synonymous with displacement. Distortion implies aberration. It is caused by discrepancies in the photographic, processing, and reproduction systems. This condition is not correctable in the compilation of a stereomodel.

Displacement is a normal inherent condition. Since mapping instruments work with a three-dimensional spatial image formed by a pair of overlapping two-dimensional photos, predictable displacement can be compensated for in the mapping process. Rather than being a fault in the image structure, displacement is the means by which it is possible to extract spatial information from photographs.

Measurement of Height and Parallax

Methods of visually judging depth may be classified as either stereoscopic or monoscopic. Monoscopic vision refers to viewing surrounding objects with only one eye. Depth is perceived primarily based on the relative sizes of objects and hidden objects; distant objects appear smaller and behind closer objects. However, depth perception is poor with monoscopic vision, and distance estimation can be difficult. In stereoscopic vision, objects are viewed with both eyes, producing a composite three-dimensional image. Stereoscopic vision allows for a far greater degree of depth perception than monoscopic vision. The concept of stereoscopic vision can also apply to viewing an aerial photography stereopair. By viewing the left photograph with the left eye and the right photograph with the right eye, a three-dimensional view of the terrain can be obtained (Lillesand, 1994). An instrument called a stereoscope is designed to aid in this type of visualization. Many people will find stereoscopic viewing of aerial photography without a stereoscope to be quite a challenge.

Parallax is the apparent displacement in the position of a stationary object, with respect to a frame of reference, caused by a shift in the position of observation (Wolf, 1974) (Figure 2.89). If an object is observed monoscopically with the left eye and then the right, there is an apparent shift in position. As the distance between the observer and object decreases, the apparent position shift increases. An instrument called a stereometer is used to measure differences in parallax between objects in a stereopair. Parallax equations can then employed to convert measured parallax into object height.

Based on the definitions given above for stereoscopic vision and parallax, the term stereoscopic parallax refers to perception of depth based on viewing objects from separate points of observation. Stereoscopic parallax is a very important concept in photogrammetry because it allows for the measurement of height of objects appearing a stereopair.

1. Height Measurement

There are a number of methods for measuring the heights of objects using aerial photography. Here briefly discussed three types that are important for photo interpreters. Two methods deals with the measurement of height from a single photo; while the third method deals with the determination of height by measuring parallax differences. The types of parallax being measured are:

1. Absolute Parallax; and,
2. Differential Parallax.

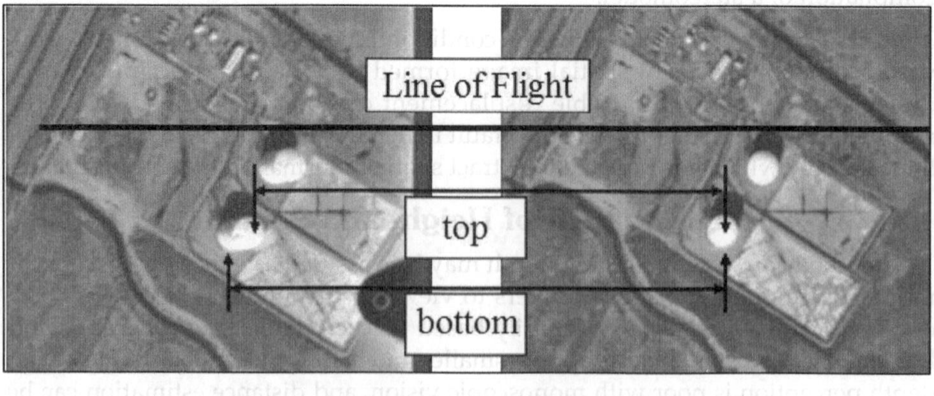

Figure 2.89: Displacement between the Top and Base of the Storage Towers in Photo Stereo-Pair.

1.1 Single Photo Methods of Height Determination

1.1.1 Shadow Height Method

Basically, if the shadow cast by an object can be measured and the sun angle causing the shadow is known or can be derived (from latitude, date and time) then the height of the object can be calculated using simple trigonometry, as follows:

$$h = Ls \times \tan a$$

where,

tan a = The tangent of the sun angle from the ground surface. and
Ls = Length of the shadow.

Here we assume that the shadow on which the ground falls is level and that the object is vertical. The objects top must be sharply defined so that it creates a distinct image. There are many sources of error here.

1.1.2 Displacement Method

In this method of height determination from a single aerial photo we:

1. Accept the principle point as the photo nadir (were assuming a "true" vertical photo);
2. Must precisely know or be able to determine the altitude from which the photo was acquired.
3. Both the top and the bottom of the object to be measured should be clearly visible.
4. The degree of image displacement must be great enough to be accurately measured with available equipment.

If the above conditions are met the formula for the displacement method of height determination from a single aerial photo can be written as:

$$H_0 = \frac{H_a \times D}{R}$$

where,

H_o = Height of the object;
H_a = Altitude above the surface where the photo is taken;
D = Length of the displaced image;
R = Radial distance from the photo nadir to the top of the object.

1.2 Parallax

The feasibility of finding height differences of objects with the help of measurement on photographs of the area concerned is the most important quality of photographs. This is achieved by measuring parallaxes on the photographs.

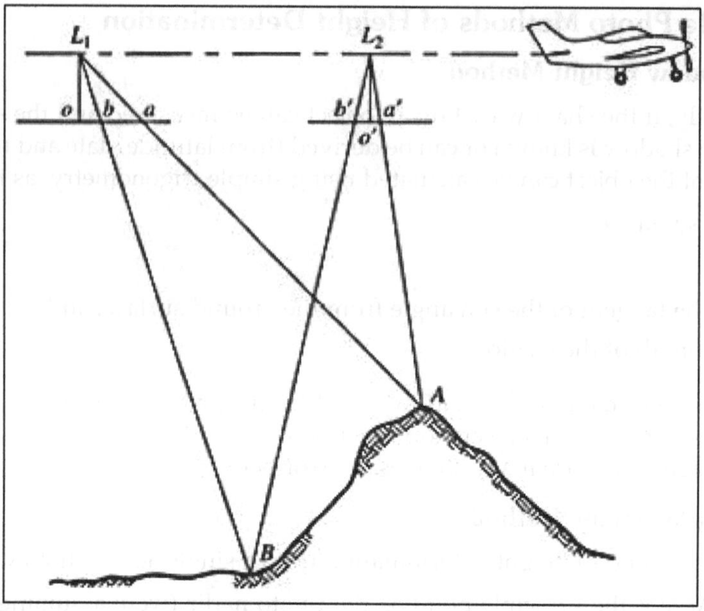

Figure 2.90: Stereoscopic Parallax of Vertical Aerial Photographs.

An aerial camera exposing overlapping photographs at regular intervals time obtains a record of positions of images at the instants of exposure. The change in position of an image from one photograph to next caused by the aircraft's motion is termed as strereo scopic parallax x, or simply parallax. Parallax exists for all images appearing on successive overlapping photographs. In Figure 2.90, for example, images of object points A to B appear on a pair of overlapping vertical aerial photographs which were taken from exposure stations L_1 and L_2. Points A to B are imaged at a and b on the left-hand photograph. Forward motion of the aircraft between exposures, however, caused the images to move laterally across the camera focal plane parallel to the flight line, so that on the right-hand photo they appear a' and b'. Because point A is higher (close to the camera) than point B, the movement of image a across the focal plane was greater than the movement of image b; in the words, the parallax of point A is greater than the parallax of point B. This calls to attention two important aspects of stereo parallax: (1) the parallax of any point is directly related to the elevation of the point and (2) parallax is greater for high points than for low points. Variation of parallax with elevation provides the fundamental basis for determining elevation of points from photographic measurements.

Figure 2.91 shows the two photographs of Figure 2.90 in superposition. Parallax of object points A and B are P_a and P_b respectively.

Stereoscopic parallax for any point such as A who images appear on two photos of a stereopair, expressed in terms of "flight line" photographic coordinetes, is

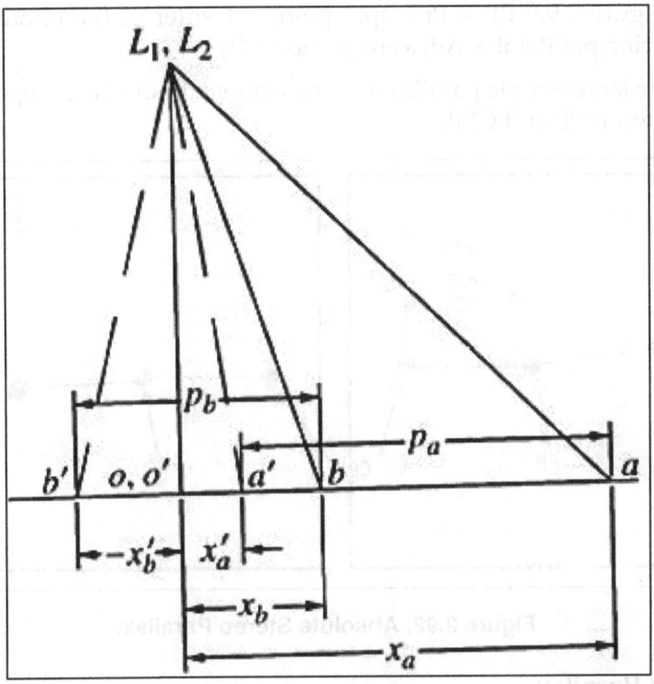

Figure 2.91: The Two Photographs of Figure 1 Shown in Superposition.

$$P_a = X_a - X_a'$$

$$P_b = X_b - X_b' = X_b - (-X_b') = X_b + X_b' \quad (1)$$

(therefore, proper algebraic signs must be cosidered)

where,

P_a = Stereoscopic parallax of object A

P_b = Stereoscopic parallax of object B

X_a = Photocoordinate of image a on the left photograph of stereopair.

X_a'= Photocoordinate of image a′ on the right photograph of stereopair.

X_b = Photocoordinate of image b on the left photograph of stereopair.

X_b'= Photocoordinate of image b′ on the right photograph of stereopair.

The photo coordinates are not measured with respect to the fiducial axis system. Rather, they are measured with respect to the "flight line" axis system.

Absolute Stereo Parallax

Considering a pair of aerial photographs of equal principle distance, the absolute parallax is the difference of a point is the algebraic difference of the distances of the two images from their respective photo nadirs measured in a horizontal plane and parallel to the air base.

From Figure 2.92: PP = Principal point = Center of the photo and CPP = Conjugate Principal Point = Adjacent photo's PP

Absolute stereoscopic parallax a' = the average photo base length = Average distance between PP and CPP.

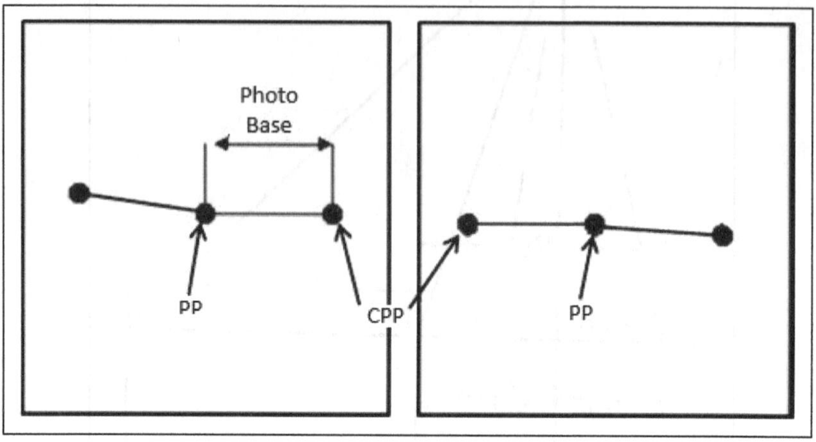

Figure 2.92: Absolute Stereo Parallax.

Differential Parallax

Differential parallax is defined as the difference between the stereoscopic parallax at the top and base of the object. This is usually employed in the determination of the differences in the elevation (Figure 2.93).

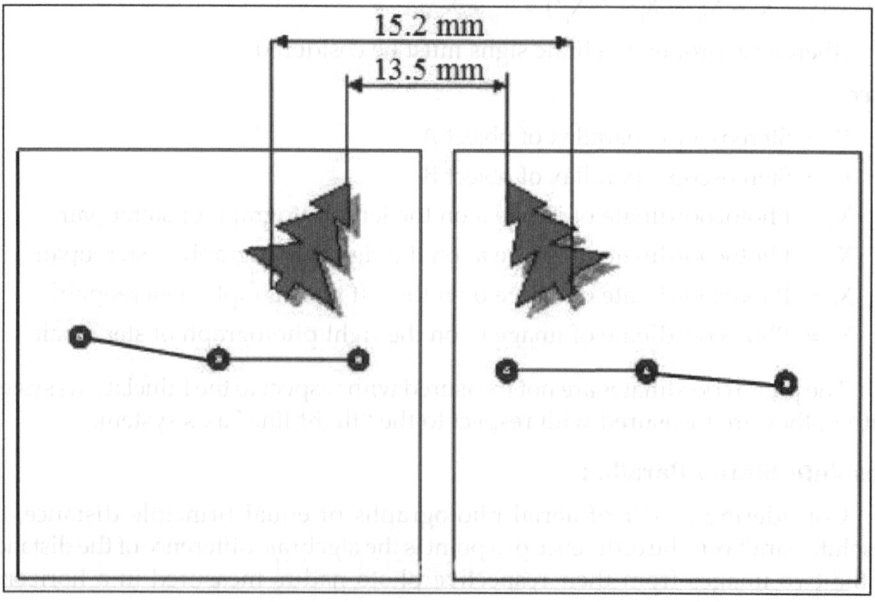

Figure 2.93: Differential Parallax.

(From the figure Differential parallax = 15.2 mm – 13.5 mm = 1.7 mm).

X- and Y-Parallaxes

In Figure 2.94, p_1, p_2' and p_2 p_1' are the photo bases of the left hand and right hand photographs respectively. a_1 and a_2 are the corresponding images of an object point A p_1 a_1 and p_2 a_2 can be resolved into two mutually perpendicular directions - one along the direction of flight (X-direction) and the other perpendicular to it (Y-direction). Then, if X_1, Y_1 and X_2, Y_2 are the resolved parts of p_1 a_1 and p_2 a_2 respectively in the two directions.

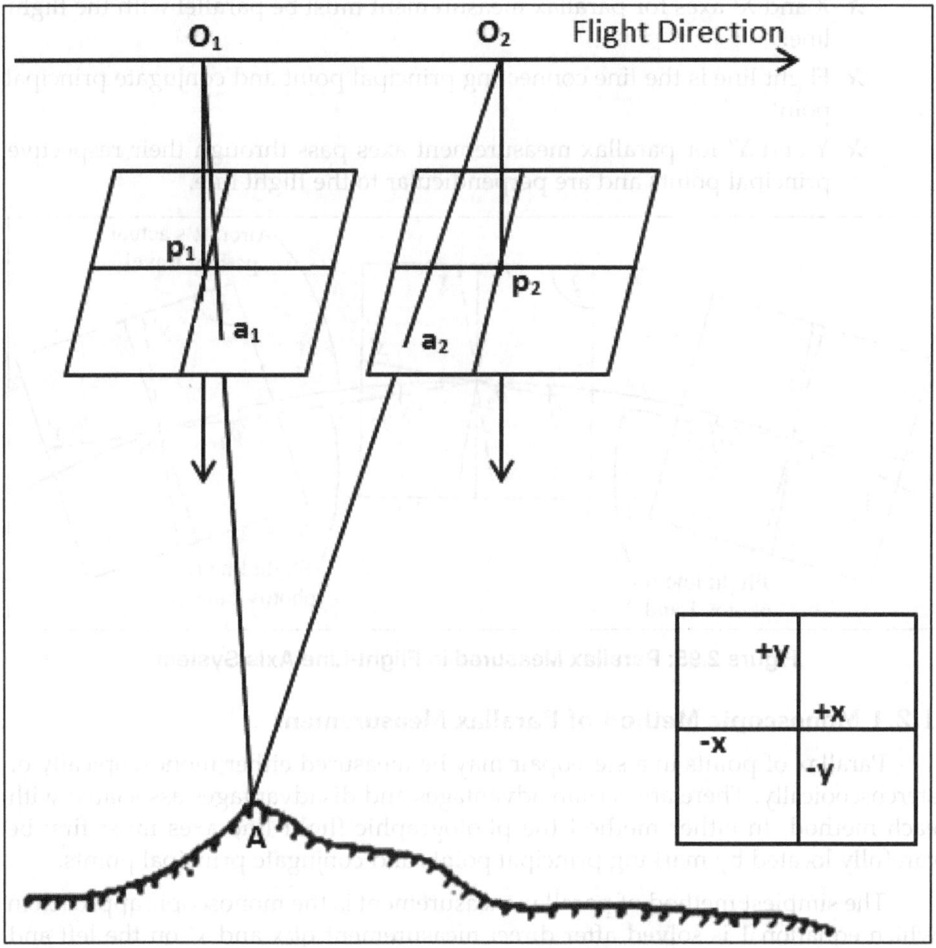

Figure 2.94: Parallax of Principal Points.

$X_1 - (-X_2) = (X_1 + X_2)$ is the X-parallax or absolute stereoscopic parallax or horizontal parallax and is defined as the algebraic difference in the direction of the air base of the distances of the two images of an object from their respective principal points. (Note: Minus sign is given to X_2 as the distance from the principal point is measure in the negative X-direction, *i.e.*, opposite the flight direction). Similarly $(Y_1 - Y_2)$ is called Y-Parallax or Vertical Parallax. If the paired photographs are assumed to be vertical and taken from equal altitude above the datum, the Y-Parallax is absent.

Photographic "Flight Line" Axes for Parallax Measurement

An example is shown in Figure 2.95, where photographs 1 through 3 were exposed as shown.

- ☆ Parallax occurs parallel to the direction of flight.
- ☆ X and X′ axes for parallax measurement must be parallel with the flight line.
- ☆ Flight line is the line connecting principal point and conjugate principal point.
- ☆ Y and Y′ for parallax measurement axes pass through their respective principal points and are perpendicular to the flight line.

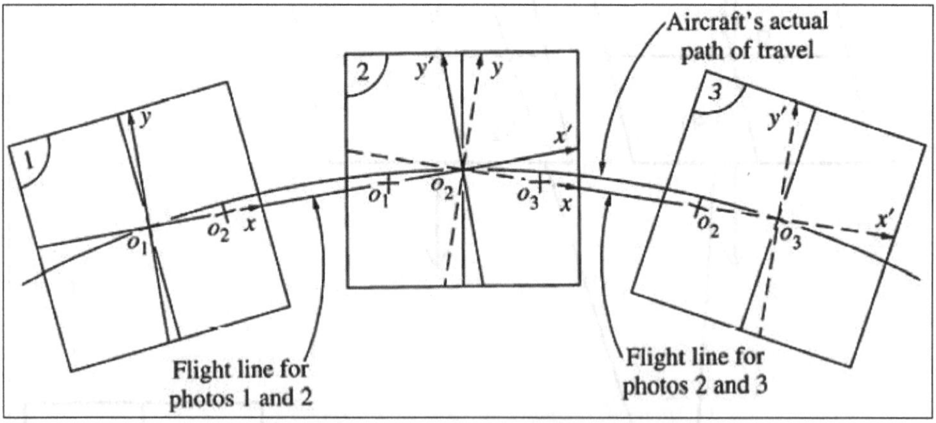

Figure 2.95: Parallax Measured in Flight-Line Axis System.

1.2.1 Monoscopic Method of Parallax Measurement

Parallax of points in a stereopair may be measured either monoscopically or stereoscooically. There are certain advantages and disadvantages associated with each method. In either method the photographic flight line axes must first be carefully located by marking principal points and conjugate principal points.

The simplest method of parallax measurement is the monoscopic approach in which equation 1 is solved after direct measurement of x and x′ on the left and right photos respectively. A disadvantage of this method is that two measurements are required for each point.

Another method approach to parallax measurement is to fasten the photographs down on a table or base material as shown in Figure 2.96. In this method the photographic flight lines $O_1 O_2$ and $O_1' O_2'$ are marked as usual. A long straight line, AA′, is drawn on the base material and two photos are carefully mounted as shown so that the photographic flight lines are coincident with this

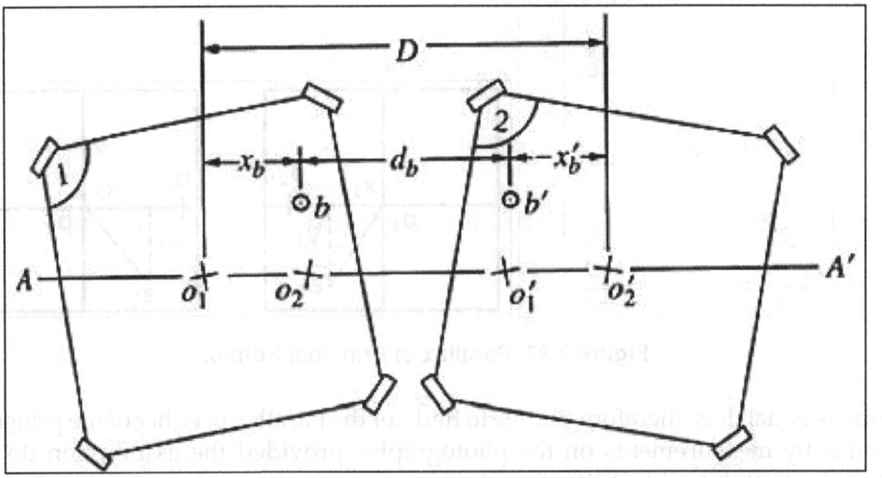

Figure 2.96: Parallax Measurement Using Simple Scale.

line. Having fastened the photos down, the distance D between principal points is a constant which can be measured.

$$P_b = D - d_b \text{ (or) } P_b = X_b - (X_b') \quad (2)$$

where,

D = Distance between principal points (constant)

D = Distance between its image on the left and right photos.

Parallax of Principal Points

Geometric center of the photograph, and the intersection of the X and Y-axes is called as Principal Point. The intersection of the North-South and East-West fiducial marks.

If we transfer the principal point P_1 of the left hand photograph of a stereo pair on the right hand photograph at p_1' (Figure 2.97), then by definition $p_2\,p_1'$ is the parallax of principal point of the left photograph and is the distance between the exposure stations (air-base) on the scale of the right hand photograph.

Similarly, $p_1\,p_2'$ is the parallax of the principal point of the right hand photo and represents the air-base on the scale of the left hand photograph.

Parallax of left ground principal point is photo base b′ measured on right photo, and the parallax of right ground principal point is photo base b measured on left photo

- ✰ For moderate relief, b ≈ b′
- ✰ Photo base is average of two values

If the terrain is flat and flight is level and flight altitude does not change the scale of the two photos will be exactly same and hence the two photo-bases will be

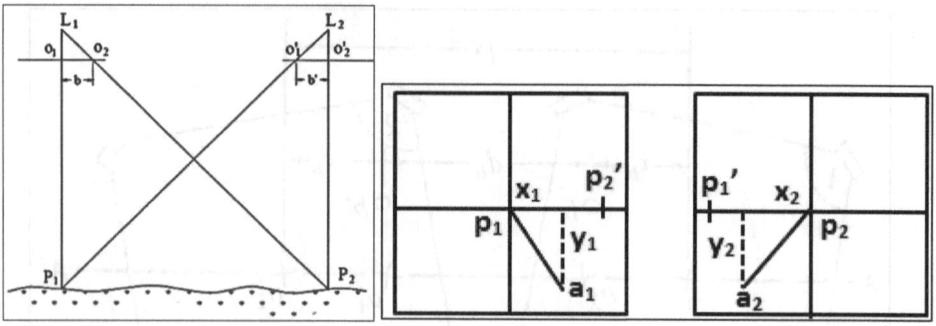

Figure 2.97: Parallax of Principal Points.

exactly equal. It is, therefore, simple to find out the parallax of either of the principal points by measurements on the photographs, provided the assumption do not deviate much from ideal situation, *viz.*

(i) Focal length in both cases same, which is always the case
(ii) Flying height is the same
(iii) Optical axis vertical

In practice we tolerate tilts of about 3 degrees; of course, flying heights are within reasonable limits.

1.2.2 Stereoscopic Method of Parallax Measurement

Through the principal of the floating mark, parallaxes of points may be measured stereoscopically. This method employs a stereoscope in conjunction with an instrument called a parallax bar and also frequently called a stereometer. A parallax bar consists of a metal bar to which are fastened two half marks. The right half mark may be moved with respect to the left mark by turning a micrometer screw. Readings from the micrometer are taken with the floating mark set exactly on points whose parallax is desired. From the micrometer readings, parallax or differences in parallax are obtained. A parallax bar is shown lying beneath a lens/ mirror stereoscopic in Figure 2.98.

Figure 2.99 is schematic diagram of the operating principal of the parallax bar. Having oriented the photos and positioned the fixed half mark as just described, the parallax bar constant, C, for the setup is determined.

$$P_a = X_a - X_a' = D - (K - r_a) = (D - K) + r_a$$

$$P_a = C + r_a \qquad (3)$$

where,

C = Parallax constant

D = Distance between principal points (constant)

K = Distance from fixed mark to index mark (constant)

r_a = Micrometer reading

Figure 2.98: Parallax Bar Attached to Lens/Pocket Stereoscope.

$P = c + r_a$ (forward-reading type, readings increase with increasing parallax)

$P = c - r_a$ (backward-reading type, readings decreased with decreasing parallaxes)

$$C = p - r \tag{4}$$

The parallax bar constant should be determined on the basis of micrometer readings and parallax measurements for two points. Then the mean of the two values of C may be adopted. Any two points may be selected for this purpose;

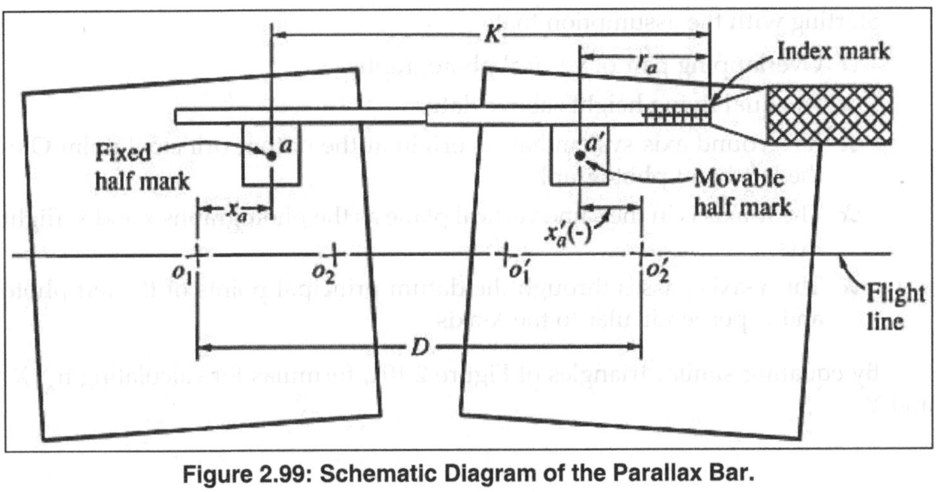

Figure 2.99: Schematic Diagram of the Parallax Bar.

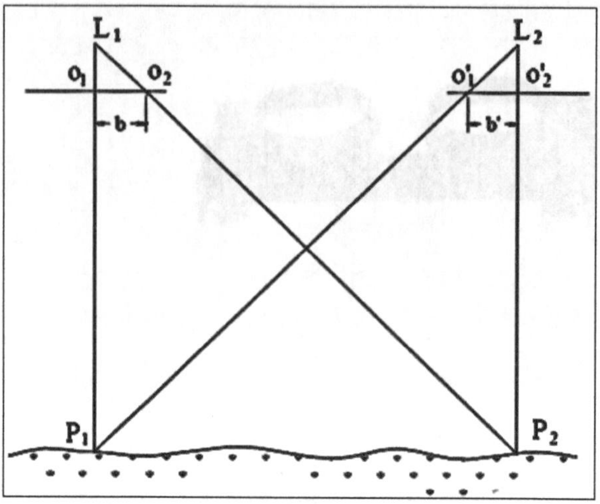

Figure 2.100: Parallax of Principal Points.

however, the two principal points are convenient and often used. Figure 2.100 is a section through a pair of overlapping vertical photos.

By equation:

$(P_b = X_b - X_b' = X_b - (-X_b') = X_b + X_b')$

$P_{o1} = X_{o1} - (-x_{o1}') = o - (-b') = b$

$P_{o2} = X_{o2} - x_{o2}' = b - o = b'$

$C_1 = P_{o1} - r_{o1} = b' - r_{o1}$

$C_2 = P_{o2} - r_{o2} = b - r_{o2}$

$C = (C_1 + C_2)/2$

1.2.3 Parallax Formula

Starting with the assumption that:

- ☆ Overlapping pair of vertical photographs
- ☆ At equal flying heights above datum
- ☆ XY ground axis system has its origin at the datum principal point O of the left-hand photograph
- ☆ The X-axis is in the same vertical plane as the photographs x and x′ flight axis
- ☆ The Y-axis passes through the datum principal points of the left photo and is perpendicular to the X-axis

By equating similar triangles of Figure 2.101, formulas for calculating h_A, X_A and Y_A

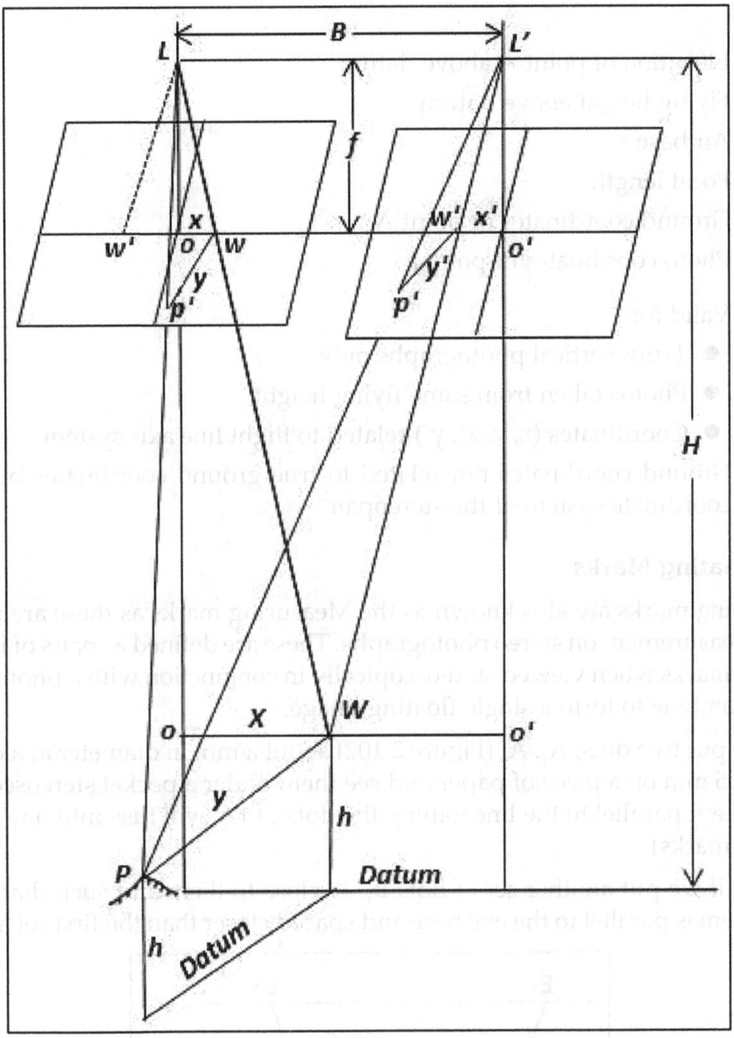

Figure 2.101: Geometry of an Overlaying Pair of Vertical Photographs.

$$h_A = H - \frac{Bf}{p_a} \quad (5)$$

$$X_A = B - \left[\frac{x_a}{p_a}\right] \quad (6)$$

$$Y_A = B - \left[\frac{y_a}{p_a}\right] \quad (7)$$

where,

h_A = Elevation of point A above datum
H = Flying height above datum
B = Air base
f = Focal length
X_A, Y_A = Ground coordinates of point A
X_a, Y_a = Photo coordinates of point a

☆ Valid for
- Truly vertical photographs only
- Photos taken from same flying height
- Coordinates (x, y, x′, y′) related to flight line axis system

☆ Ground coordinates not related to true ground coordinates but to the coordinate system of the stereopair

1.2.4 Floating Marks

Floating marks are also known as the Measuring marks as these are used for precise measurement on stereo photographs. These are defined as pairs of identical reference marks when viewed stereoscopically in conjunction with a photographic overlap combine to form a single floating image.

If we put two dots, A_1, A_2 (Figure 2.102) about a mm in diameter at a distance of about 65 mm on a piece of paper and see them under a pocket stereoscope (the eye base kept parallel to the line joining the dots), they will fuse into one dot (See reference marks).

Now if we put another set of dots B_1, B_2 close to them and such that the line joining them is parallel to the eye base and spaced closer than the first set and seen

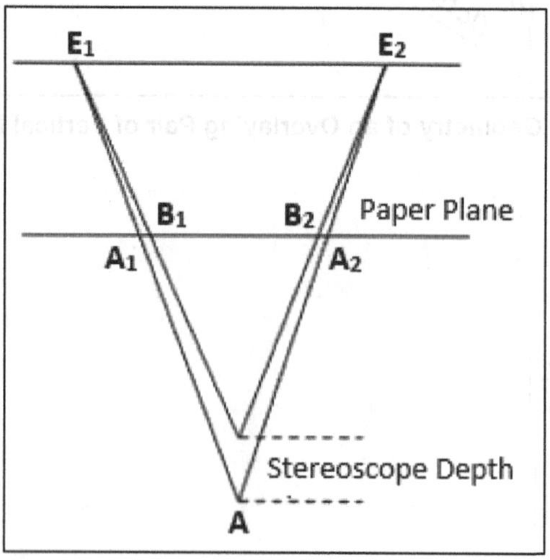

Figure 2.102: Principle of Floating Marks.

under stereoscope, we find that this set also fuses into one mark, but floating *i.e.*, higher above the first one. The vertical distance 'AB' is known as stereoscopic depth. That means if the mark has a different parallax from that of its surroundings, it will appear higher or lower. But if no parallax exists between the dots and the object images, the fused dot appears in contact with the fused image. At this moment quite accurate measurement may be made of the distance between the

1.2.5 Parallax Difference

Assuming that there is no tilt and flight is level, two photographs are taken with image of an object point A a_1 and a_2 on them respectively. If the two photographs are put on top of the other with their principal point's p_1 and p_2 and flight direction in coincidence, then by definition a_2 a_1 is the absolute stereoscopic parallax [Figure 2.103(a)].

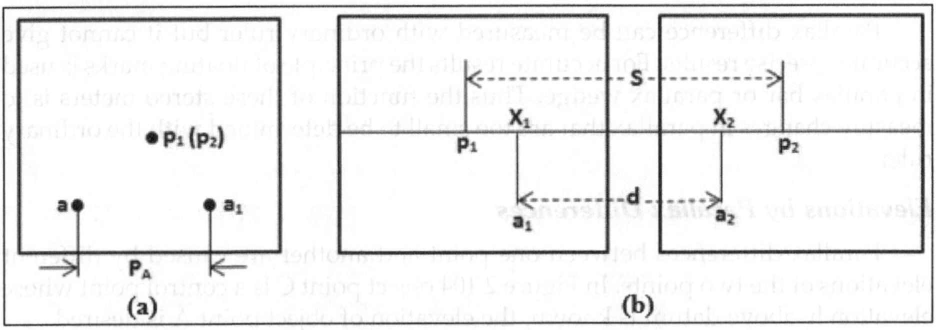

Figure 2.103: Parallax Difference.

If now we put the pair of photographs under a stereoscope for fusion, they will have to be separated at a convenient distance p_1p_2 say a distance represented by 'S' [Figure 2.103(b)].

The parallax of A, $P_A = p_1 p_2 - a_1 a_2$

$= S - a_1 a_2$

Similarly parallax of another point Q,

$PQ = S - q_1 q_2$

Considering 'A' as the reference point, the parallax difference between 'A' and 'Q' is

$\Delta p = PQ - PA$

$= a_1 a_2 - q_1 q_2$

In practice, direct measurement of parallax is seldom done, instead we measure the parallax difference (Δ p) with the help of parallax bar or parallax wedge.

Generally, graduations on parallax bar are marked in such a way that if the separation between corresponding images decreases [*i.e.*, 'd' decreasing – Figure 109(b)], the reading on parallax bar increases – the point with larger parallax gives a higher reading, and correspond to a point of higher elevation. In such a case the parallax difference Δp = (Parallax bar reading for Q - Parallax bar reading for A)

$= q_1 q_2 - a_1 a_2$

Measurement of Parallax Difference

Parallax differences may be determined in any of the following ways:

1. By monoscopic measurement of parallax followed by subtraction.
2. By taking differences in parallax bar readings. The validity of this approach is seen if parallax determined from parallax bar readings are subtracted, as

   $\Delta p = p - p_c = (C + r_c) - (C + r_c) = r - r_c$
3. By parallax wedge.

Parallax difference can be measured with ordinary ruler but it cannot give accurate/precise results. For accurate results the principle of floating marks is used in parallax bar or parallax wedge. Thus the function of these stereo meters is to measure changes in parallax that are too small to be determined with the ordinary rulers.

Elevations by Parallax Differences

Parallax differences between one point and another are caused by different elevations of the two points. In Figure 2.104 object point C is a control point whose elevation h_c above datum is known, the elevation of object point A is desired.

$$h_a = h_c + \frac{\Delta p(H - H_c)}{p_a} \quad (8)$$

Approximate Equation for Elevation from Parallax Differences

The following approximate equation for elevation differences is obtained from equation 8 by (1) substitute the photo base b of the stereopair for p_a; (2) substituting average flying height above ground, H′, for $(H - h_c)$; and (3) letting $\Delta h = h_A - H_c$;

$$\Delta h = \frac{\Delta p\, H'}{b} \quad (9)$$

where,

Δh= Difference in elevation between two points whose parallax difference is Δp

B = Photo base ≈ 90 mm (for 9- inch- square format with 60 per cent end lap)

H′ = Average flying height above ground

Parallax Correction Graph

Errors in measured parallax due to photographic tilt, imperfect alignment of the photos for parallax measurements, shrinkage or expansion of the photo peppers,

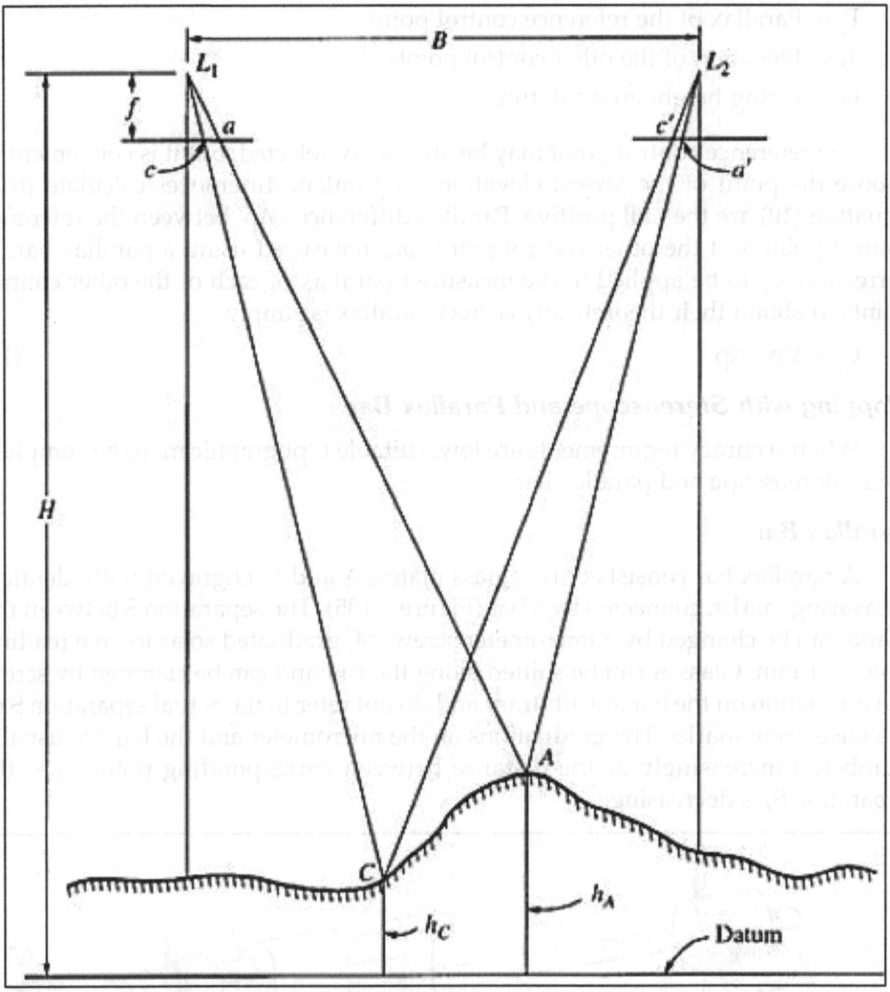

Figure 2.104: Elevations of Parallax Differences.

and camera lens distortions can be effectively compensate for by constructing a parallax correction graph.

Based upon selected reference control point, parallax differences Δp that theoretically must exist for the other control points is calculated using following rearranged fern Equation (8).

$$\Delta p = \left[\frac{h - h_c}{H - h_c}\right] p_c \tag{10}$$

where,

h_c = Elevation of the reference control point

P_c = Parallax of the reference control point

h = Elevation of the other control points

H = Flying height above datum

The reference control point may be arbitrarily selected, but it is convenient to choose the point of the lowest elevation, as parallax differences calculate from Equation (10) are then all positive. Parallax differences Δp′ between the reference control point and the other control points are measured using a parallax bar. A correction, c_p, to be applied to the measured parallax of each of the other control points to obtain their theoretically correct parallax is simply

$$C_p = \Delta'p - \Delta p \qquad (11)$$

Mapping with Stereoscope and Parallax Bar

When accuracy requirements are low, suitable topographic maps be compiled using stereoscope and parallax bar.

Parallax Bar

A parallax bar consists of two glass plates, A and B, engraved with identical measuring marks, connected by a bar (Figure 2.105). The separation S between the marks can be changed by a micrometer screw. M, graduated so as to give reading upto 0.01 mm. Glass A can be shifted along the rod and can be clamped by screw C. Graduation on the bar are arbitrary and do not refer to the actual separation S of the measuring marks. The graduations on the micrometer and the bar are usually numbered increasingly as the distance between corresponding points, *i.e.*, the separation S, is decreasing.

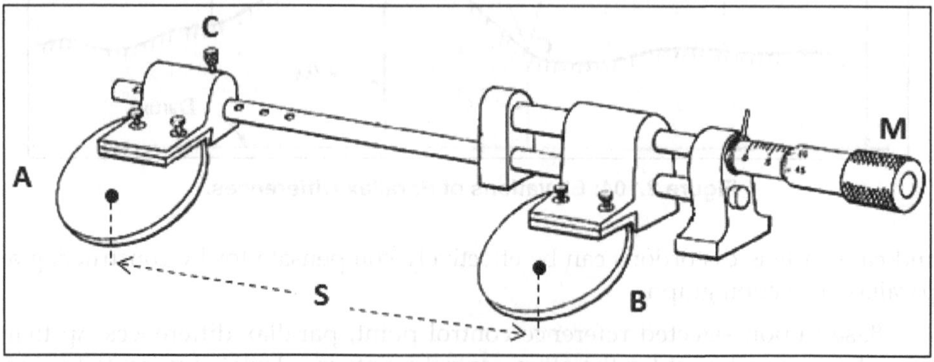

Figure 2.105: Parallax Bar.

Elevation of contour points are preferably determined using parallax difference equation. These points can all be plotted in a planimetrically correct manner according to their coordinates. Elevation contours can then be interpolated between contour points. With the proper parallax bar micrometer readings, contours may be traced directly by keeping the floating mark in contact with the ground as it is moved about the stereomodel. Parallaxes bar setting required to trace each contour are readily computed.

Parallax Wedge

This is a sheet of transparent material with no converging rows of dots. The wedge is slid backwards and forwards in 'Y' direction until two dots fuse as one dot on the ground, the reading then being noted. The dots are numbered in accordance with corresponding parallax values (Figure 2.106).

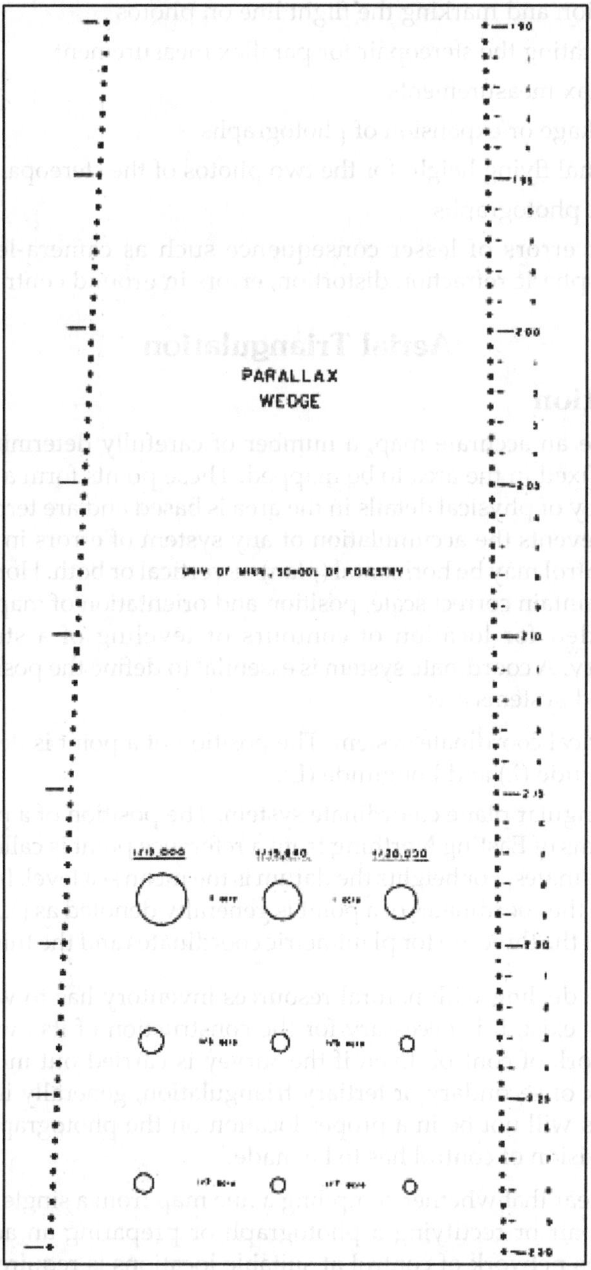

Figure 2.106: Parallax Wedge.

Error Evaluation

Answers obtained using the various equations will inevitably contain errors. It is important to be aware of the presence of these errors and to be able to assess their magnitude. Some of the sources of error in computed answers using parallax equation are as follows:

- ☆ Location and marking the flight line on photos
- ☆ Orientating the stereopair for parallax measurement
- ☆ Parallax measurements
- ☆ Shrinkage or expansion of photographs
- ☆ Unequal flying height for the two photos of the stereopair
- ☆ Tilted photographs
- ☆ Other errors of lesser consequence such as camera-lens distortion, atmospheric refraction distortion, errors in ground control, etc.

Aerial Triangulation

1. Introduction

To produce an accurate map, a number of carefully determined points are required to be fixed in the area to be mapped. These points form a framework on which the survey of physical details in the area is based and are termed as control. This control prevents the accumulation of any system of errors in the measuring operations. Control may be horizontal (plan) or vertical or both. Horizontal control is needed to maintain correct scale, position and orientation of map while vertical control is needed for location of contours or leveling of a stereo model in photogrammetry. A coordinate system is essential to define the position of a point. Commonly used systems are:

(a) Spherical coordinate system: The position of a point is denoted in terms of latitude (λ) and Longitude (L).

(b) Rectangular plane co-ordinate system: The position of a point is defined in terms of Easting Northing from a reference point is called the origin of co-ordinates. For heights the datum is the mean sea level. In mathematical terms the coordinates of a point is generally denoted as (X, Y, Z). The first two of the three are for planimetric coordinates and the third is for height.

If a service dealing with natural resources inventory has to work in an area where no maps exist, it is necessary for the construction of its own base map to have a framework of control. Even if the survey is carried out in an area with a denser network of secondary or tertiary triangulation, generally it will be found that pass-points will not be in a proper location on the photograph or model. In both cases provision of control has to be made.

It is now clear that whether compiling a line map from a single photograph or from a stereo pair or rectifying a photograph or preparing an aerial mosaic, a sufficiently dens network of control at suitable locations is required to orient the photograph or the stereo model. Provision of such a dense control by ground methods would be very expensive and time consuming. Fortunately, methods have been developed for the determination of this control by photogrammetric methods called erotriangulation bridging or control extension methods

2. Definition

Aerial triangulation is the simultaneous space resection and space intersection of image rays recorded by an aerial mapping camera. The spatial direction of each image ray is determined by projecting the ray from the front nodal point of the camera lens through the image on the positive photograph. Conjugate image rays projected from two or more overlapping photographs intersect at the common ground points to define the three-dimensional space coordinates of each point. The entire assembly of image rays is fit to known ground control points in an adjustment process. Thus, when the adjustment is complete, ground coordinates of unknown ground points are determined by the intersection of adjusted image rays. Simultaneously with the ground point intersections, the exterior orientation of each photograph is determined by image ray resection through the camera.

3. Purpose and Applications

Aerial triangulation is a complex operation. Its purpose is to simultaneously determine the exterior orientation of all exposure stations in a project (block) and to compute points in object space from measurements in overlapping photographs. Figure 2.107 depicts an overview of the different methods and approaches in aerial triangulation. The shaded boxes refer to manual operations. For example, the leftmost sequence of operations are all carried out manually, including the block adjustment. The manual way of performing aerial triangulation was practiced before the age of computers.

It is adequately termed *analog aerial triangulation*–obviously a thing of the past. The processes depicted in the center are known as *semi-analytic aerial triangulation*, characterized by measuring model coordinates on a stereoplotting instrument, followed by a block adjustment with independent models. In digital photogrammetry we are concerned with *analytical aerial triangulation*. Here, the only manual process is the measurement of photo coordinates. Hence, the smallest unit that enters the block adjustment is a photograph, measured on an analytical plotter, or comparator, for example. In semi-analytic aerial triangulation, the basic unit is a measured stereomodel and in analog aerial triangulation it is a measured strip, accomplished on stereoplotting instruments that support the conjunction of successive photographs.

The main application of aerial triangulation is to provide control information for the absolute orientation of stereomodels as required in subsequent photogrammetric operations, such as the compilation of topographic maps or the generation of DEMs. Traditionally, the absolute orientation is accomplished with control points obtained by aerial triangulation. Using the exterior orientation parameters is a quicker way, however. After the interior orientation is established

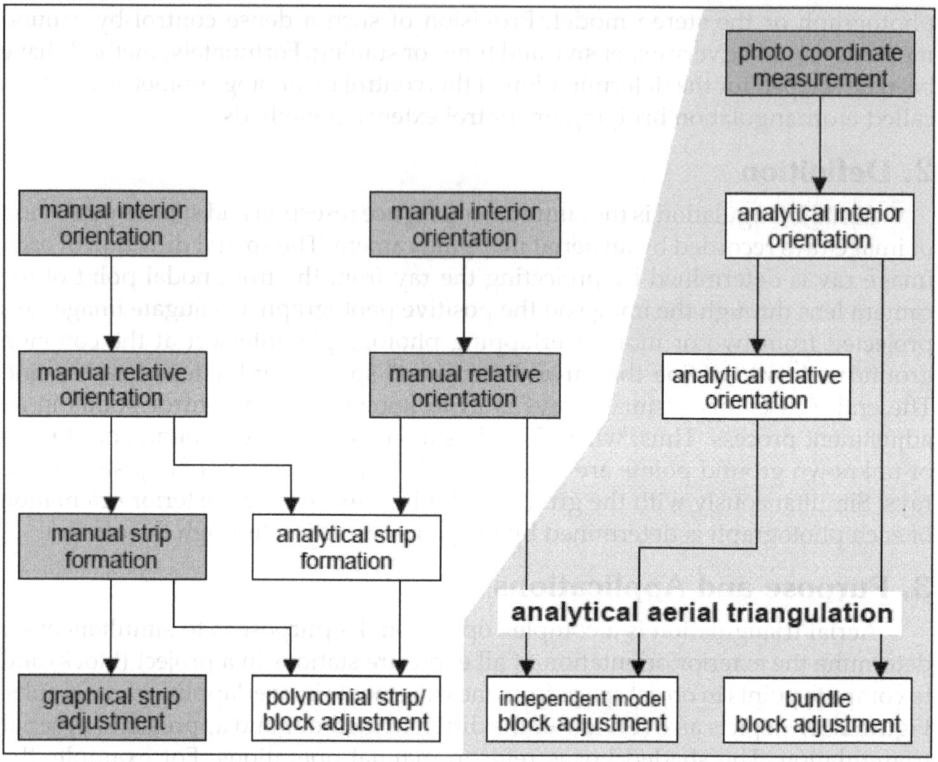

Figure 2.107: Overview of the different Aerial Triangulation Methods and Procedures. The shaded boxes refer to manual processes.

on analytical plotters, the exterior orientation is imported and the model can be automatically oriented.

Aerial triangulation is also used extensively for many other purposes. As a viable method to densify precise ground control networks it finds applications in areas such as property surveying, construction surveys, and subdivision layout. The principle of simultaneously orienting imagery and establishing points in object space is also successfully employed in closerange photogrammetry, for example as a quality control tool in manufacturing.

4. Historical Development

Figure 2.108 illustrates the development of analytical aerial triangulation during the past thirty years expressed as progress over time. In this context, progress includes economical factors (*e.g.*, increase in throughput time, cost per model) and technical aspects (*e.g.*, accuracy, reliability).

Although the foundation of analytical photogrammetry dates back to the turn of the last century, it was not until the 1950s when H. Schmid developed the principles of analytical aerial triangulation, including error analysis (*e.g.*, Schmid, 1957). Numerous contributions have been made by D. Brown, among them an

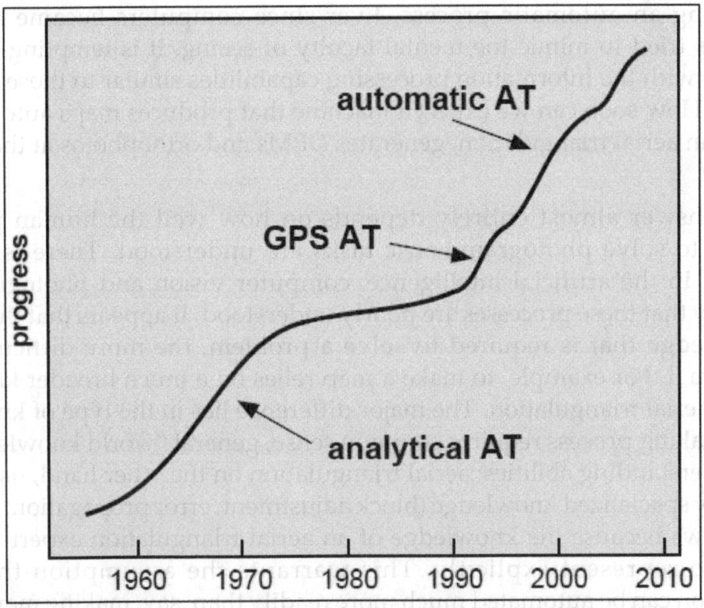

Figure 2.108: Development of Aerial Triangulation during the Past 30 Years.

elegant treatment of least squares adjustment and efficient solutions for large normal equation systems (Brown (1958). In Europe, major research efforts were directed toward the development of block adjustment by independent models. Original work was undertaken by van den Hout and led to the anblock method (van den Hout (1966). A rigorous 3D solution and error propagation studies were developed at the University of Stuttgart (*e.g.*, Ackermann *et al.*, 1973).

The rapid progress in the late sixties and early seventies was mainly in the areas of block adjustment methods and error propagation analysis. The mathematical models were extended by including the interior orientation in the block adjustment or by adding additional parameters to compensate for not-modeled systematic errors. With the availability of analytical plotters some procedural changes took place. For example, one of the most significant result was the great simplification of the point transfer problem. Not surprisingly, GPS had also a great impact on aerial triangulation.

Not only did it spark new research interest but it changed significantly the way aerial triangulation projects are carried out. Once a sensitive problem, the determination of good control points is now a snap. Using differential GPS in the aircraft for control points in the block adjustment, sometimes called *airborne GPS control*, or simply *GPS aerial triangulation* greatly helps reducing the number of required ground control points.

Digital photogrammetry and computer vision have brought, and are continuing to bring new changes that will shape aerial triangulation as dramatically as analytical photogrammetry did during the past 30 years. Aerial triangulation is on the verge of becoming an automatic process. Ever since computers became available, researchers tried to mimic the mental faculty of seeing. It is tempting to endow computers with the information processing capabilities similar to those of human operators. How soon can we expect a machine that produces maps automatically, performs an aerial triangulation, generates DEMs and orthophotos at the touch of a button?

The answer almost entirely depends on how well the human processes necessary to solve photogrammetric tasks are understood. There is common agreement in the artificial intelligence, computer vision and photogrammetry community that these processes are poorly understood. It appears that the broader the knowledge that is required to solve a problem, the more difficult it is to encapsulate it. For example, to make a map relies on a much broader knowledge base than aerial triangulation. The major difference lies in the type of knowledge: the map making process requires common sense, general "world knowledge" and image understanding abilities; aerial triangulation on the other hand, uses a great deal of very specialized knowledge (block adjustment, error propagation, etc.). This is good news because the knowledge of an aerial triangulation expert is far less difficult to represent explicitly. This warrants the assumption that aerial triangulation can be automated much more readily than, say, making maps. Oddly enough, it is much more difficult to explain aerial triangulation to a layman than how to make a map.

4.1 History of Aerial Triangulation

Aerial triangulation has gone through different stages in the past; namely, radial, strip, and block triangulation.

4.1.1 Radial Triangulation

Radial triangulation is a graphical approach and based on the fact that angles measured in a photograph at the iso-centre, located in the middle of line connecting the principal point and photo-nadir, are true horizontal angles and can be used for planimetric triangulation. For a vertical photography, the principal point and the photo-nadir coincide, therefore, the fiducial centre is a suitable approximation of isocentre for this method. In this approach, the principal points of the neighboring photographs are transferred and the horizontal rays can then be drawn for each photo. These planimetric bundles can be put together along a strip or in a block using two ground control points. The other points can be defined by multiple intersections. Radial triangulation was used in 1950s in which stereo and slotted templates layouts provided photo-control for mapping purposes. Although this simple technique provided the required accuracy of the day, it was limited by the large physical space required for the layout. Airplane hangars were usually used for this purpose.

4.1.2 Strip Triangulation

Strip triangulation was developed in the 1920s, where a Multiplex instrument was used to recreate the aerial photography mission. It is based on dependent pair relative orientation and scale transfer to ensure uniform scale along the strip. The sequential dependent pair relative orientation plus scale transfer starting from a controlled model is known as cantilever extension which is equivalent to an open traverse in surveying. If ground control points are used at the end or in between, the method is called "bridging" which is similar to a controlled traverse in surveying.

Mechanical or graphical interpolation technique were then used to fit the measured strip coordinates to the ground control. Numerical strip adjustments started in the 1960s when electronic digital computers became available. A number of polynomial interpolation adjustment formulations were developed for this purpose (Schut, 1968). The transition from analog aerial triangulation to analytical procedure was realized with the advent of computers (*e.g.*, analytical relative orientation, absolute orientation, etc.). The input for fully analytical aerial triangulation is photo coordinates measured in mono or stereo mode using comparators.

4.1.3 Block Triangulation

Block triangulation (bundles or independent models) provides the best internal strength compared to strip triangulation (Ackermann, 1975). The available tie points in consecutive strips assists in the roll angle recovery which is one of the weaknesses inherent in strip triangulation. In terms of the computational aspect, aerial triangulation methods are categorized as: analog, semi-analytical, analytical, and digital triangulation.

4.1.3.1 Analog Aerial Triangulation

This method uses a "first order stereo" plotter to carry out relative and approximate absolute orientation of the first model and cantilever extension. The strip or block adjustment is then performed using the resulting strip coordinates.

4.1.3.2 Semi-Analytical Aerial Triangulation

Relative orientation of each individual model is performed using a precision plotter (*e.g.*, Wild A10). The resulting model coordinates are introduced in a rigorous simultaneous independent model block adjustment. Independent models can also be linked together analytically to form strips which are then used for strip adjustment or block adjustment with strips.

4.1.3.3 Analytical Aerial Triangulation

The input for analytical aerial triangulation are image coordinates measured by a comparator (in stereo mode or mono mode plus point transfer device). A bundle block adjustment is then performed by using all image coordinates measured in all photographs. An analytical plotter in comparator mode can also be used to measure the image coordinates.

4.1.3.4 Digital Aerial Triangulation

This method uses a photogrammetric workstation which can display digital images. Selection and transfer of tie points and measurement tasks that are performed manually in analytical triangulation are automated using image matching techniques. The procedure is fully automatic, but allows interactive guidance and interference.

4.1.4 Control Requirements for Photogrammetric Blocks

Any block consisting of two or more overlapping photographs requires that it be absolutely oriented to the ground coordinate system. The 3D spatial similarity transformation with 7 parameters (3 rotations, 3 translations, 1 scale) is usually employed for absolute orientation and requires at least 2 horizontal and 3 vertical control points. Due to some influences caused by transfer errors (*e.g.*, image coordinate measurements of conjugate points), and extrapolation beyond the mapping area, the theoretical minimum control requirement is unrealistic.

Theoretical and practical studies (Ackermann, 1974 and Brown, 1979) showed that only planimetric points along the perimeter of the block and relatively dense chains of vertical points across the block are necessary to relate the image coordinate system to the object coordinate system. These measures also ensure the geometric stability of the block as well as control the error propagation.

4.1.4.1 Auxiliary Data as Control Information

Perimeter control for planimetry has reduced the number of control points and required terrestrial work for a photogrammetric block. However, the dense chains of vertical control demand additional surveys. A number of studies (Ackermann, 1984; Blais and Chapman, 1984; Faig, 1979) have been carried out to reduce the number of control points, especially vertical points, using measured exterior orientation parameters at the time of photography. These studies showed that great savings in the number of vertical control points could be achieved.

4.1.4.2 History of Auxiliary Data in Aerial Triangulation

The use of auxiliary data in aerial triangulation goes back to more than half a century. As mentioned by Zorzychi (1972), statoscope, horizon camera, and solar periscope were used to directly measure the exterior orientation parameters of the camera at the time of photography. A statoscope provides the Z coordinate of the exposure stations using differential altimetry. An horizon camera provides the rotation angles of the mapping camera with respect to the horizon. A solar periscope can also determine the rotation angles but they are referred to the sun's location. Airborne ranging was also developed to determine the horizontal positions of the camera stations in order to support the airborne determination of large horizontal networks (Corten, 1960). Except for the statoscope, these instruments were not accepted for practical applications due to accuracy and economic reasons. The Airborne Profile Recorder (APR) was developed in 1960 and used extensively in Canada. This instrument provides the Z coordinate for a number of identifiable features that can be correlated to the profile. This can be achieved by using a statoscope to measure the movements of the aircraft with respect to an isobaric surface and a continuous record of the profile between terrain and aircraft as shown in Figure 2.109.

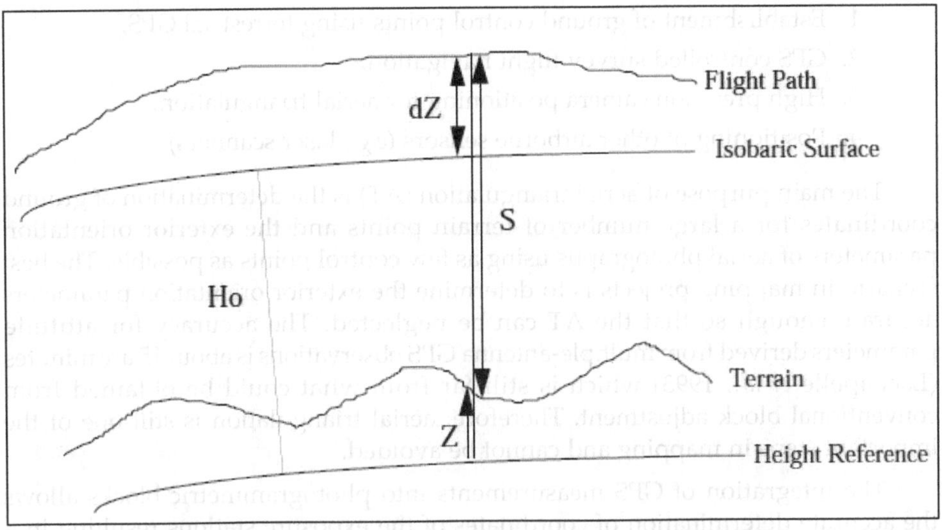

Figure 2.109: Airborne Profile Recorder Concept.

The height information can be derived as:

$$Z = H_0 + dZ - S$$

Blais (1976) used lake leveling information which makes use of the condition that points along the shoreline of a lake have the same elevation. Gyroscopes were employed to determine the exterior orientation parameters but gained little practical acceptance because of accuracy limitations (Corten and Heimes, 1976). With the advances in inertial and satellite positioning technology, the subject of direct measurements of exterior orientation parameters has gained high attentions in recent years. The Global Positioning System (GPS) can provide the position of the exposure stations while Inertial Navigation System (INS) can determine the attitude of the exposure stations. Ackermann (1984) found that positioning data (X, Y, Z) are more effective than attitude data as the latter imply a summation of errors that affect the final object coordinates. The integrated GPS-INS approach to fully recover the exterior orientation parameters is the newest technique (Schwarz *et al.*, 1993) but one of the problems is the high cost of INS.

4.1.5 GPS Assisted Aerial Triangulation

The Navstar Global Positioning System (GPS) has become generally available and has been considered fully operational on a worldwide basis since 1993. It can be used for direct positioning practically anywhere on earth and at any time. GPS has already had a revolutionary impact in various disciplines which are involved with navigation and geodetic positioning. A real time capability is required if GPS is used for navigation purposes. However, it was soon realized that GPS offers a very high accuracy for positioning in combination with post-processing methods.

Since the launch of GPS satellites in early 1980s, photogrammetrists realized the usefulness of GPS for their particular interests (*e.g.*, aerotriangulation). There are four main areas in photogrammetry where GPS can be used (Ackermann, 1994):

1. Establishment of ground control points using terrestrial GPS,
2. GPS controlled survey flight navigation,
3. High precision camera positioning for aerial triangulation,
4. Positioning of other airborne sensors (*e.g.*, laser scanners).

The main purpose of aerial triangulation (AT) is the determination of ground coordinates for a large number of terrain points and the exterior orientation parameters of aerial photographs using as few control points as possible. The best scenario in mapping projects is to determine the exterior orientation parameters accurate enough so that the AT can be neglected. The accuracy for attitude parameters derived from multiple-antenna GPS observations is about 15 arc minutes (Lachapelle *et al.*, 1993) which is still far from what could be obtained from conventional block adjustment. Therefore, aerial triangulation is still one of the important steps in mapping and cannot be avoided.

The integration of GPS measurements into photogrammetric blocks allows the accurate determination of coordinates of the exposure stations resulting in a reduction of the number of ground control points to a minimum. Therefore, the goal is to improve the efficiency of aerotriangulation by avoiding ground control points almost completely. The combined adjustment of photogrammetric data and GPS observations can be carried out by introducing GPS observation equations to the conventional block adjustment.

5. A Broader View: From Aerial Triangulation to Sensor Orientation

The previous section described in general terms the development of analytical aerial triangulation. Progress has been continuous, whether in terms of instrumentation, procedures, or mathematical models and computational techniques. Analytical aerial triangulation has been perfected over the years in several aspects. It has reached a high level of performance, for example, the throughput time per model is about 15 minutes, service companies offer aerial triangulation for $20 per model, and an accuracy of 5 microns can be reached under good conditions.

We now turn to recent developments that have brought significant changes to aerial triangulation. These changes require a broader view of aerial triangulation. As a consequence, Part 4 is called *Sensor Orientation*. During the last decade we have seen major changes in data collection. Let us begin with classical film-based aerial cameras. A noteworthy addition are new gyro-stabilized mounts. Conventional mounts attach the camera to the aircraft body using dampener devices in an attempt to reduce the effect of aircraft vibrations on image motion. The use of new high-resolution film and a desire to widen the weather window by flying in less-than-ideal weather conditions requires longer exposure times. In turn, longer exposure times increase the image blur caused by pitch and roll movements of the aircraft that can reach as much as 10 *o/s* in turbulent conditions. Gyroscopes sense aircraft rotations in the longitudinal (roll), transverse (pitch), and vertical (yaw) axes, send the readings via control unit to servo motors, which keep the camera leveled better than 1. Combined with forward motion compensation (FMC), a superior image quality can be expected. This impacts aerial triangulation in several aspects. Better image quality increases interpretability-crucial in the corners of images where tie points should be identified on several overlapping images. An additional benefit are improved initial estimates for exterior orientation parameters in the block adjustment.

Example

Assume a wide angle camera (f = 153 mm) with an exposure time of 1/250 second, encountering turbulence that causes the airplane to rotate about its principle axes with 6 o/s. This will move an 'ideal' image point in the center by 64 μm and in the corners by 130 μm. Consider now a stabilized system with a remaining angular motion of 0.3 o/s. Repeating this back-of-the-envelope computations reveals that the image blur in the center reduces to 3 μm and in the corners to 6.5 μm.

The introduction of precise navigation and positioning systems to airborne data collection brought another significant change to aerial triangulation. These systems are usually coupled with a flight management system that includes hardware and software and offers broad functionality, ranging from planning and executing photo flights to post-flight processing. The design data of the planned flight lines and waypoints are entered to the control computer on the aircraft. This control unit is linked to GPS, the camera and its mount, and provides the pilot and photographer navigation and graphical guidance. The control unit automatically triggers the camera, records the midpoint of exposure and the data from the gyro-stabilized mount. In this fashion, pinpoint photography is obtained. In addition, position and attitude of the exposure stations become available, which serve as excellent initial estimates in the block adjustment. Adding a GPS receiver and processing software with geodetic accuracy and an inertial measuring unit capable to deliver high-precision navigation information increases the accuracy of the position and attitude of exposure stations to a degree comparable to results obtained by aerial triangulation, prompting the question whether aerial triangulation is still necessary.

The advent of digital photogrammetric cameras brings new challenges to aerial triangulation for several reasons. In order to meet the resolution and ground coverage of classical aerial cameras, digital equivalents employ the technology of using multiple digital cameras housed in a single box, or utilize the push-broom principle. Each of the individual cameras in the multiple camera solution covers a part of the ground. Image processing software assembles the individual images together to a frame image. Implementations of this technology vary. Regardless of the implementation the result is the same: a perspective digital image. Except for proper camera calibration and perhaps considering different models for compensating systematic errors in self-calibrating bundle block adjustment, such frame images do not change aerial triangulation procedures.

Digital photogrammetric cameras employing pushbroom technology, also referred to as *line cameras*, are of course a different story altogether. Obviously, a different mathematical model is required for the interior orientation and the block adjustment. But there is also a considerable impact on other aerial triangulation processes, requiring adequate changes.

Let us move on to spaceborne cameras. High-resolution satellite imagery is becoming increasingly available for mapping purposes-at least for those who can afford it. In three consecutive articles, Petrie provides a world-wide survey on this topic (Petrie, 2004). There are several notable differences between space and aerial imagery regarding their georeferencing processes. Usually hundreds, if not thousands of aerial images make up a triangulation project whereas only a few satellite images require simultaneous processing. A concern with some of the spaceborne image acquisition systems is related to the sensor model. Sensor specifications are often times not made available and prevent exact photogrammetric solutions. Instead, parameters of rational function models must be used. There is yet another trend in airborne and spaceborne data acquisition that requires changes in aerial triangulation methods. With the miniaturization and diversification of sensors it is now tempting to mount several different sensors on the same platform to obtain complimentary data, for example high-resolution imagery, multi/hyperspectral imagery, and range data. This brings along the problem of multisensor orientation.

From satellites to close-range photogrammetry. With the availability of digital cameras in the 1990s the field of architectural photogrammetry with its main goal to document cultural heritage has received considerable attention by researchers. This lead to innovative concepts, such as the combination of orientation and model-based reconstruction, camera calibration, and the use of linear features. In fact, incorporating straight lines as computational entities in orientation and reconstruction processes, known as *line photogrammetry*, is the hallmark of modern architectural photogrammetry.

Finally we should turn our attention to more theoretical developments that broaden our view of aerial triangulation. Most remarkable contributions come from the computer vision community. For example in Triggs et al. (2000), the authors promote methods for bundle block adjustment that are more general than the traditional least-squares approaches. Among other topics, the authors elaborate on the choice of cost functions and robustness, on linearly convergent approximations, updating and recursive methods, and quality control.

6. Aerial Triangulation Principles

The technique of AT is used to reduce the number of control points required, achieve higher accuracy across the entire project area, ensure a good edge-match between photos, and to create additional control points through control point extension. Essentially, the known points (ground control points) are used to establish a geometric relationship between a set of points with unknown ground coordinates that are measured on at least two photo/images known as "tie points". These tie points have an unknown X, Y, and Z coordinate before triangulation, but are calculated during AT to add additional control mathematically. After a successful AT, the following results are available:

- ☆ The exterior orientation parameters of each image.
- ☆ The X, Y and Z tie point coordinates.
- ☆ The residuals of the measured points (control points and tie points).
- ☆ The image coordinates residuals.

All of these results are used to determine the overall accuracy of the block, the statistical robustness of the solution, and as input parameters to do the following:

- ☆ Create stereopairs for 3-D feature collection.
- ☆ To measure distances, angles and elevation directly from the source imagery.
- ☆ To create DTM's.
- ☆ To produce orthophoto.

Nowadays, nearly all organizations involved in photogrammetric applications make use of AT techniques. In this field, each organization has developed its own procedures and techniques according to:

- ☆ The available equipment,
- ☆ The landscape, the possibilities to prepare control points and tie points,
- ☆ The scale of the images,
- ☆ The experience of the operators,
- ☆ The intended purpose, the required accuracy, the required output.

Today, AT can be performed either with the use of conventional photography on analytical systems or with the use of digital imagery on DPWs where automation (partly) of the processes involved can be achieved. The development of DP techniques such as image matching which nowadays, can be considered as operational, has given the possibilities for partly or fully automating the digital acquisition of triangulation, especially, image coordinates of tie-points. Already these automated techniques have been realized and implemented in a number of DPS (Bouloucos, 2001).

6.1 Geometric Principles

The aerotriangulation geometry along a strip of photography is illustrated in Figure 2.110. Photogrammetric control extension requires that a group of photographs be oriented with respect to one another in a continuous strip or block configuration. The exterior orientation of any photograph that does not contain ground control is determined entirely by the orientation of the adjacent photographs. In a given aerotriangulation configuration, each photograph contributes to the exterior orientation of the adjacent photographs through the pass points located in the triple overlap areas.

The term triple overlap refers to the ground area shared by two adjacent stereomodels along the strip. The triple overlap area is imaged on three consecutive photographs. Thus, when end lap is specified to be 60 percent, the alternate photographs will overlap each other by 20 percent at average terrain elevation.

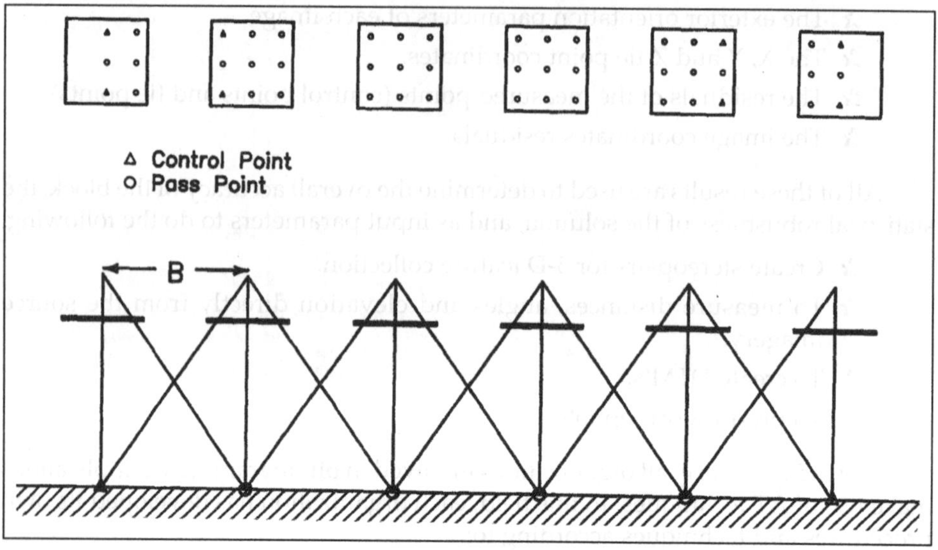

Figure 2.110: Aerotriangulation Geometry.

6.2 Relationship to Ground Control

Aerotriangulation is essentially an interpolation tool, capable of extending control to areas between ground survey control using several contiguous uncontrolled stereomodels. An aerotriangulation solution should never be extended or cantilevered beyond the ground control. Ground control should be located at the ends of single strips and along the perimeter of block configurations. Within a strip or block, ground control is added at intervals of several stereomodels to limit error propagation in the adjusted pass point coordinates. Extending control by aerotriangulation methods is often referred to as bridging since the spatial image ray triangulation spans the gap between ground control.

6.3 Pass Points

A pass point is an image point that is shared by three consecutive photographs (two consecutive stereomodels) along a strip. As the pass point is positioned by one stereomodel, it can be used as control to orient the adjacent stereomodel. Thus, the point "passes" control along the strip. Only three-ray points in the triple overlap area serve this pass point function.

A two-ray point in the stereomodel is an intersection point only and does not contribute to the aerotriangulation function unless it is also a ground control point. A pass point that is shared between two adjacent strips is called a tie point. Pass point requirements are related to type of point used, location, and point transfer and marking requirements.

(a) ***Type of Points:*** Pass points may be artificially marked points, targeted points, or natural images. However, since pass points must lie at or very near the center line of the triple overlap area, artificially marked points designed from the photography taken should be used. Premarked targets are too expensive and too difficult to align with the triple overlap areas. Natural images are not suitable for precise pointing.

(b) ***Marking artificial points***: Artificially marked pass points must be well-defined symmetrical patterns drilled, punched, or otherwise marked in the emulsion using a suitable marking instrument such as a Wild PUG or equivalent. Only the aerotriangulation/compilation positives should be marked. Normally, the original negatives should not be marked.

(c) ***Location:*** A minimum of three pass points must be marked along the center line of each triple overlap area. One pass point must lie near the photo principal point, and one point must lie near each of the corners of the neat model. Pass point locations must be selected by examining the photographic prints with a stereoscope. Pass points must be located on unobscured level ground in accordance with the characteristics for vertical photo control. All pass point locations must be symbolized and labelled on the control photographs.

(d) ***Point transfer for monoscopic measurements:*** Artificial pass points are typically marked in stereoscopic correspondence on all photographs showing the site of the point, using a stereoscopic transfer and point marking device such as a Wild PUG or equivalent. For the minimum three pass points in each triple overlap area, this operation will result in a minimum nine pass points on each photo. This method is required when photo coordinates are to be measured on a monocomparator. This method may be used when pass points are to be measured stereoscopically, either as photo coordinates on a stereocomparator or analytical stereoplotter or as model coordinates on any stereoplotter. Stereoscopic point transfer and marking should be done by a highly experienced operator using utmost care in choosing the site and in the marking of each pass point.

(e) ***Point transfer for stereoscopic measurements***: When pass points are to be measured stereoscopically, either as photo coordinates on a stereocomparator or analytical stereoplotter or as model coordinates on any stereoplotter, artificial pass points need be marked on the center photograph only using a suitable point marking device. Viewing the marked point stereoscopically with adjacent photographs will accomplish the point transfer of the pass point location to the overlapping photo as part of the measurement process. When parallel flight lines of photography are used, tie points should be transferred from one flight line to each adjacent flight line using a stereoscopic transfer and point marking device such as a Wild PUG or equivalent. Artificial points are typically not superimposed on images of targets.

6.4 Ground Control Points

Ground control requirements are related to targeting, control location, and survey accuracy requirements.

(a) ***Targeting***: Aerotriangulation for Class 1 and Class 2 mapping should require ground photo control points to be targeted prior to the flight. Aerotriangulation for Class 3 mapping may use postflight natural photo control points.

(b) ***Control location:*** Final control location and bridging distances used should be the decision of the contractor, but the following guidelines should be applied:

(i) ***Single Strips***

☆ Along a single flight line of photography (Figure 2.111), horizontal control points should be in pairs at the strip ends within the terminal stereomodel, one on each side of the flight line approximately opposite each other. For each single flight line, additional horizontal control points should be located at intervals along the strip that do not exceed the maximum allowable bridging distance. Horizontal basic control points should be no more than one-third of the width of coverage of a photograph from the flight line.

☆ Along a single flight line of photography, vertical control points should be in pairs, one on each side of the flight line approximately opposite each other and at a distance from the flight line of between one-fourth and one-third of the width of coverage of a photograph. For each single flight line, the pairs of vertical control points should be located at the strip ends within the terminal stereomodel and at intervals along the strip that do not exceed the allowable bridging distance for the class of mapping.

(ii) ***Blocks***

A block of photography, consisting of two or more flight lines of photography (Figure 2.111), should have control points spaced approximately equally round the periphery following the same spacing and location guidelines as for strips. There should be at least one horizontal control point and two vertical control points near the center of any block.

☆ Additional horizontal control should be located in the center of the block such that horizontal control falls in alternating strips at an interval not to exceed two times the allowable horizontal bridging distance. There should be at least one horizontal control point and two vertical control points near the center of any block.

☆ Additional vertical control should be located in the center of the block such that vertical control falls in each strip at an interval not to exceed two times the allowable vertical bridging distance.

(c) ***Bridging distance:*** Table 2.5(a) lists typical allowable bridging distances that may be used as a guide in estimating control requirements for a project. These are minimum guidelines, and many contractors will design

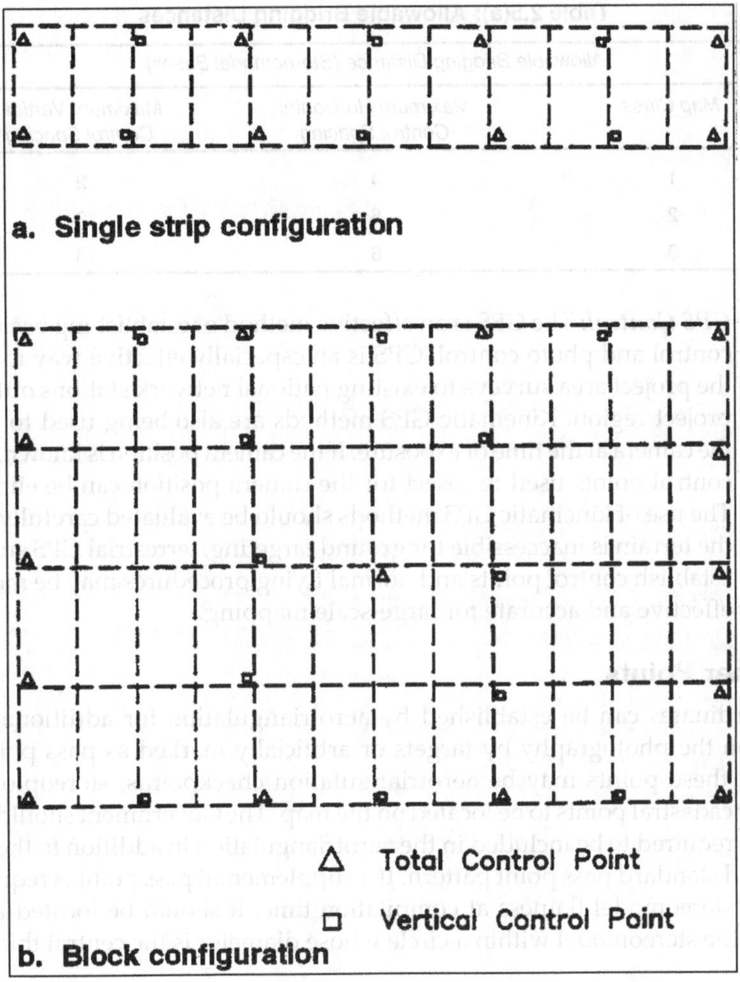

Figure 2.111: Typical Strip and Block Control Configurations.

a more dense control pattern. For example, a horizontal control point every five stereomodels regardless of block size.

(d) ***Ground control accuracy:*** Ground control accuracy for aerotriangulation should be more stringent than for a project fully controlled by field survey points. The aerotriangulation solution will contribute to the propagated error in the pass point ground control values. Since the pass point coordinates should meet the accuracy required for photo control, the photoidentifiable ground control points used to control the aerotriangulation should be more typical of the accuracy of the basic control survey.

Table 2.5(a): Allowable Bridging Distances

Allowable Bridging Distance (Stereomodel Bases)		
Map Class	*Maximum Horizontal Control Spacing*	*Maximum Vertical Control Spacing*
1	4	2
2	5	3
3	6	3

(e) ***GPS Control:*** The GPS is an effective method of establishing basic project control and photo control. GPS is an especially effective way to connect the project area surveys to existing national network stations outside the project region. Kinematic GPS methods are also being used to position the camera at the time of exposure. If the camera position is known, ground control points used to resect for the camera position can be eliminated. The use of kinematic GPS methods should be evaluated carefully. Unless the terrain is inaccessible for ground targeting, terrestrial GPS surveys to establish control points and normal flying procedures may be more cost-effective and accurate for large-scale mapping.

6.5 Other Points

Coordinates can be established by aerotriangulation for additional points located on the photography by targets or artificially marked as pass points. For example, these points may be aerotriangulation checkpoints, stereoplotter test points, or cadastral points to be located on the map. The Government should specify all points required to be included in the aerotriangulation in addition to the control points and standard pass point pattern. If a supplemental pass point is required for checking stereomodel flatness at compilation time, it should be located near the center of the stereomodel within a circle whose diameter is the central third of the air base.

7. Typical Workflow of Aerial Triangulation

For comparing traditional analytical aerial triangulation with digital methods in subsequent chapters, it is useful to recall the major steps.

- ☆ **Project planning:** comprises the planning of aerial photography and the ground control, the selection of instruments and adequate procedures to achieve the desired result, and the estimation of costs and delivery schedules.
- ☆ **Flight mission:** involves the execution of the flight plan, the film development, and the generation of a photo flight report.
- ☆ **Preparation:** on every photograph a suitable number of well distributed points are selected. The location must be chosen such that the same point appears on as many neighboring photographs as possible. Selected points are appropriately annotated.

- ☆ **Point transfer:** the points selected in the previous phase must be marked on every photograph, either directly by drilling a hole into the emulsion, or indirectly by the stereovision ability of a human operator.
- ☆ **Point mensuration:** all selected points, including control points, are precisely measured on analytical plotters or softcopy workstations, either manually by a skilled operator or automatically by image matching methods.
- ☆ **Block adjustment:** is performed by one of the many highly efficient aerial triangulation programs. The success of aerial triangulation depends on highly skilled personnel. With the increased use of GPS technology for providing control information, the point transfer phase is now arguably the single most critical factor.

8. Block Adjustment

There are two block adjustment methods used in analytical aerial triangulation. In the *bundle block adjustment* the exterior orientation parameters of all images involved in a block are simultaneously determined together with the tie points. The basic unit of the *independent model method* is a stereomodel as measured on an instrument, or computed during relative orientation.

8.1 Concept of Blocks

Most mapping projects require many models; large projects may involve thousands and medium sized projects hundreds of photographs. Figure 2.112 illustrates the concept of a block, consisting of photographs, models and strips. Projects with single strips are quite common in highway and railroad projects.

8.1.1 Overlap Considerations

Obviously, for block adjustment to work common points between photographs or models are required. Such points are known as *tie points, pass points*, or *bridging points* (usually in conjunction with independent model block adjustment). Common points between adjacent strips are called *strip tie points*. In order to find tie points, photographs, models, and strips must overlap. Figure 2.113(a) shows the overlap configuration of an ideal block. The forward overlap (overlap in strip direction) is 60 per cent, the side overlap 20 to 30 per cent.

Common points are found in the overlapping areas. As illustrated, the degree of overlap varies within the block. If you move from left to right, you find the typical pattern of 1, 2, 3, 2, 3,. where the numbers refer to the number of photographs covering the same area. This sequence of numbers doubles where adjacent strips overlap. With this standard overlap configuration, the maximum overlap is 6. Consecutive models overlap by 20 per cent.

Figure 2.113(b) depicts the overlap configuration of a photograph within the block. Again, the numbers refer to the number of photographs covering the same area in object space.

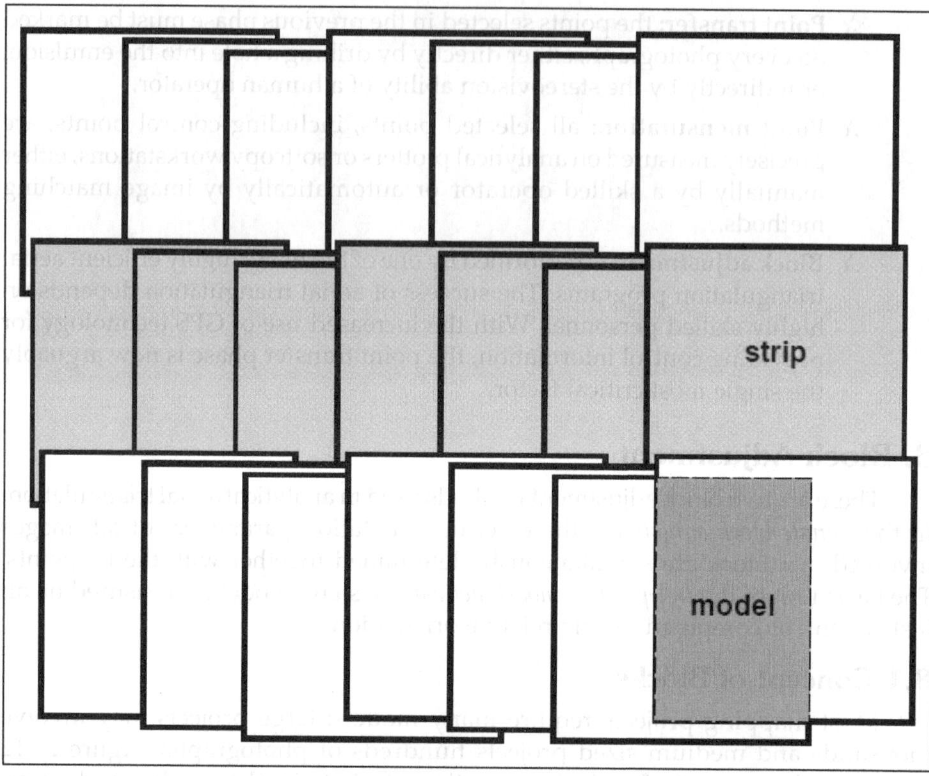

Figure 2.112: Concept of Models, Strips and Blocks.

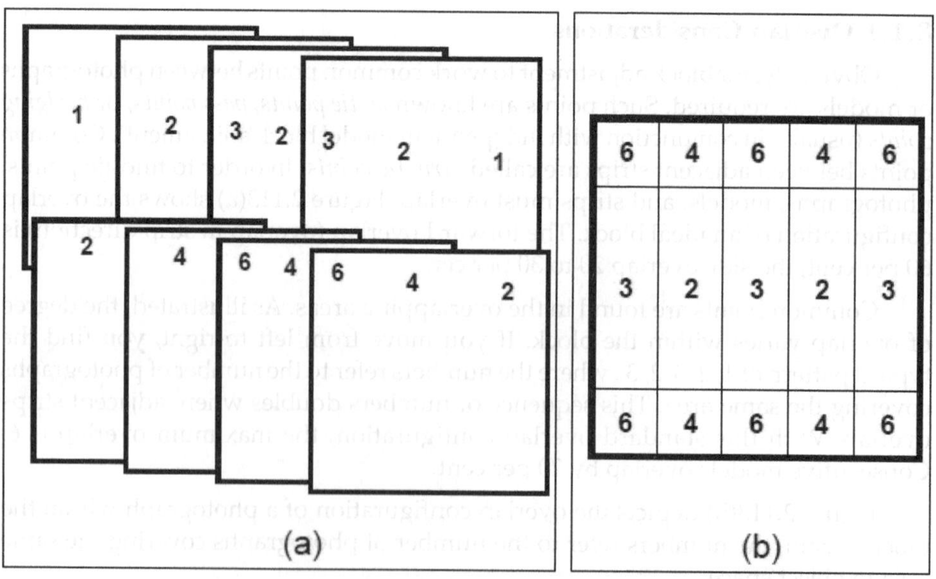

Figure 2.113: Overlap Configuration of an Adeal Block (a).
The photograph in (b) shows the zones with their associated overlaps.

8.1.2 Tie Points and Block Points

Tie points, or common points in photographs, models or strips are measure on all photographs or modes they occur. Tie points are introduced as unknowns in the block adjustment; their position in object space will be known after the adjustment. In order to differentiate between measured tie points in photographs or models and their estimated position in object space I refer to the latter as *block points*, short for tie point in object space. For estimating the size of normal equations or for discussing redundancy budgets it is helpful to know the ratio of tie points to block points.

Consider the typical 9-point pattern used in traditional aerial triangulation, illustrated in Figure 2.114. The pattern is optimal because it satisfies economical constraints (minimum number of points), simplifies point transfer and chooses points in the high overlap areas.

To facilitate the point transfer, three points are selected on every photograph as shown in Figure 2.114. Assuming stereoscopic measuring instruments are used, the point transfer for photographs within a strip is accomplished by the measuring instrument. Hence, a separate process for transferring points from one photograph to the next one is not necessary. However, strip tie points must be transferred separately, prior to the measuring process.

Let us now consider the ratio of tie points to block points for the 9- point pattern. Examining Figure 2.113(b) closely reveals that six points are in 6-overlap areas and three points are in 3-overlap areas. It follows that the ratio tie points to block points

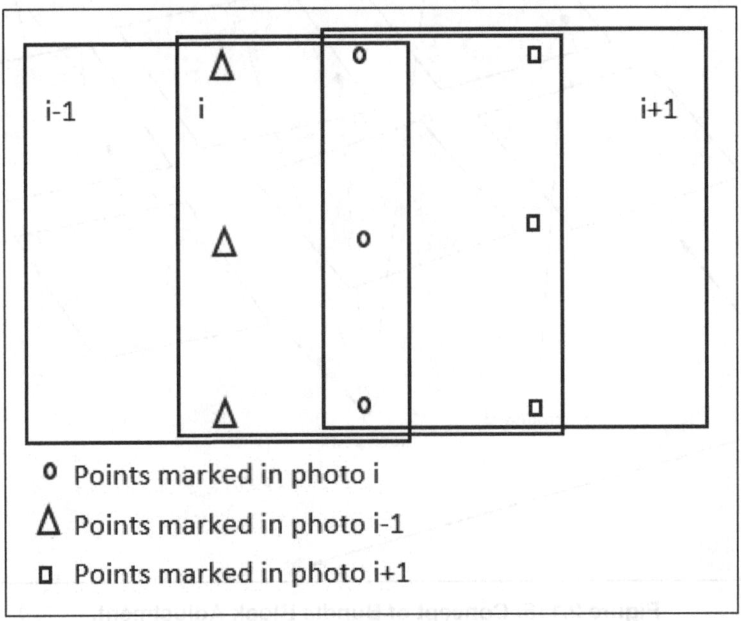

Figure 2.114: The Traditional 9-point Pattern is Ideal because it Simplifies Point Transfer and Achieves a High Redundancy with a Minimum Number of Points.

is 6 × 6 + 3 × 3/9 = 5, that is, in ideal blocks, a tie point of a 9-point pattern occurs on average on five photographs.

In automatic aerial triangulation many more points are used in the block adjustment. Suppose we have *n* tie points regularly distributed in the image. How can we estimate the ratio tie point to block points? We can determine the overlap areas shown in Figure 2.113(b) and use them as weights. Bearing in mind the traditional overlap configurations of 60 per cent forward overlap and 20 per cent sidelap, we arrive at the numbers listed in Table 2.5. For example, 24 per cent of a photograph is covered by six other photographs, 16 per cent is covered by 4 photographs and so on. In conclusion, the ratio of tie points to block points for many regularly distributed points is ≈ 3.6.

8.2 Bundle Block Adjustment

A bundle of rays that originates from an object point and passes through the projective centre to the image points (Figure 2.115) forms the basic computational unit of aerial triangulation. Bundle block adjustment means the simultaneous least squares adjustment of all bundles from all exposure stations, which implicitly includes the simultaneous recovery of the exterior orientation elements of all photographs and the positions of the object points (Faig, 1985).

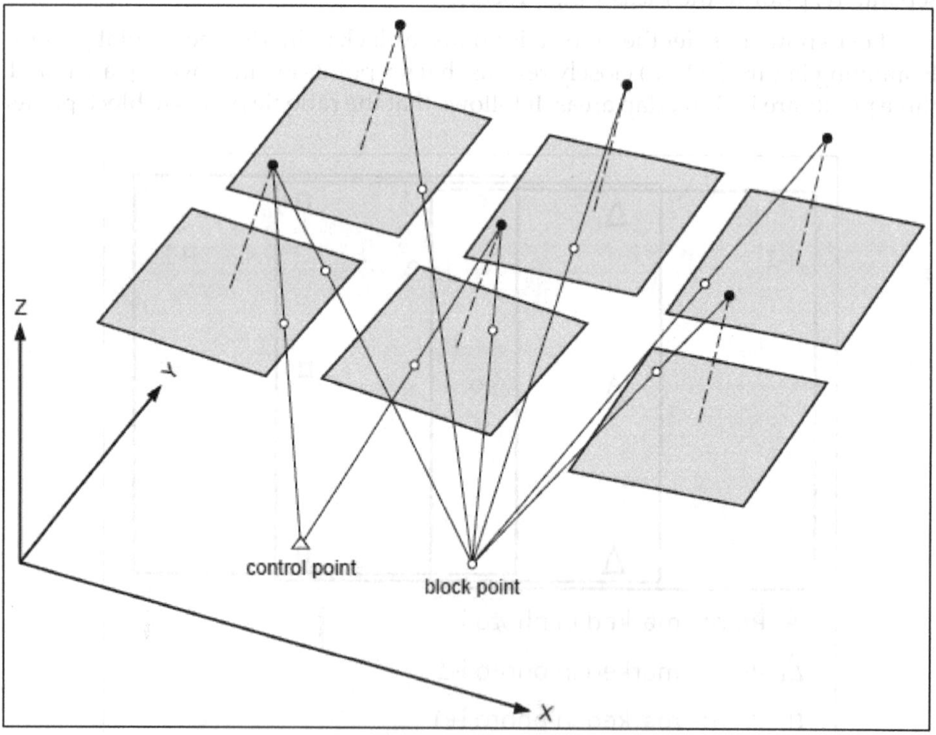

Figure 2.115: Concept of Bundle Block Adjustment.

The exterior orientation parameters of all images involved are simultaneously determined using tie points and control points.

Figure 2.115 illustrates the concept of the bundle block adjustment. The basic unit is the pair of coordinates of a point measured on an image. The six exterior orientation parameters of every image and the 3D positions of all tie points are determined simultaneously by an adjustment procedure.

The fundamental equation of aerial triangulation is the collinearity equation which states that an object point, its homologous image point, and the perspective centre are collinear (Figure 2.116). Usually, a least-squares adjustment model is employed. The functional relationship between measured quantities in image space and parameters in object space is given by the collinearity equations:

$$x_p = -c\frac{(X_p - X_c)r_{11} + (Y_p - Y_c)r_{12} + (Z_p - Z_c)r_{13}}{(X_p - X_c)r_{31} + (Y_p - Y_c)r_{32} + (Z_p - Z_c)r_{33}}$$

$$y_p = -c\frac{(X_p - X_c)r_{21} + (Y_p - Y_c)r_{22} + (Z_p - Z_c)r_{23}}{(X_p - X_c)r_{31} + (Y_p - Y_c)r_{32} + (Z_p - Z_c)r_{33}} \tag{1}$$

Where x_p, y_p are the refined image coordinates1, c is the calibrated focal length, X_C, Y_C, Z_C the coordinates of the perspective center, and r_{11},....,r_{33} are the elements of the 3D orthogonal rotation matrix **R**, containing the three rotation angles ω, φ and κ. The tie points in object space are denoted by X_p, Y_p, Z_p and are referred to as block points.

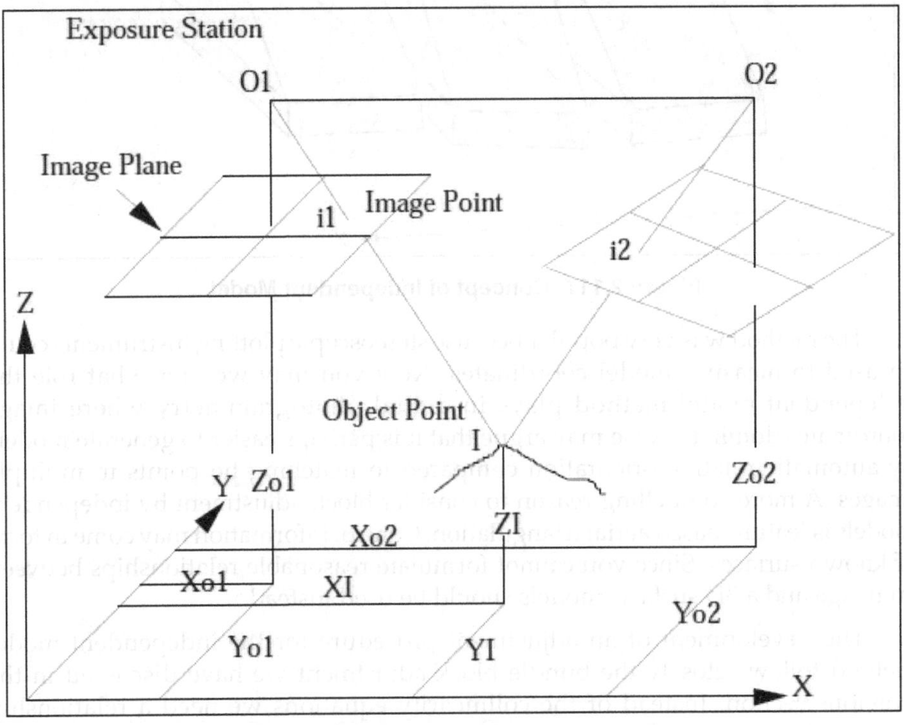

Figure 2.116: Concept of Collinearity Equation.

After linearization of Eq. 1 with respect to the parameters, a Gauss-Markov model for estimating the unknowns is employed as shown below

$$\mathbf{y} = \mathbf{A}\xi + \mathbf{e}, \mathbf{e} <- (0, \sigma^2_0 \mathbf{P}^{-1}) \quad (2)$$

Here, **y** is the observation vector, **A** the design matrix, ξ the parameters to be estimated, and **e** is the error vector. The parameters are estimated such that the collinearity equations are satisfied by minimizing the square sum of the errors.

8.3 Independent Model Method

A stereomodel is the smallest unit that enters the block adjustment by independent models. Figure 2.117 illustrates the concept. Think of it as a simultaneous adjustment of models based on the model tie points, including the perspective centers-in other words, a series of absolute orientations that are all performed at once.

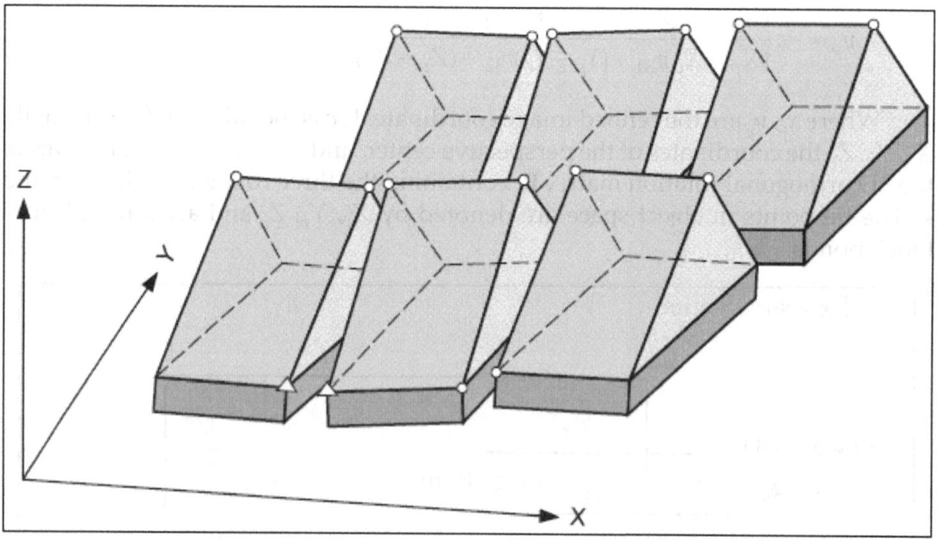

Figure 2.117: Concept of Independent Model.

The method was very popular because stereoscopic plotting instruments could be used to measure model coordinates. Now you may wonder what role the independent model method plays in digital photogrammetry where image coordinates dominate. One may argue that it is perhaps easier to generate models by automatic relative orientation compared to matching tie points in multiple images. A more compelling reason to consider block adjustment by independent models is feature-based aerial triangulation. Control information may come in form of known surfaces. Since you cannot formulate reasonable relationships between an image and a 3D surface, models should be used instead.

The development of an adjustment procedure for the independent model method follows closely the bundle block adjustment we have discussed in the previous section. Instead of the collinearity equations we need a relationship between the model and the object space coordinate systems. This relationship can be conveniently expressed by 3D similarity transformation.

Referring to Figure 2.117, we have

$$\mathbf{p} = s\mathbf{R}\mathbf{m} + \mathbf{t} \tag{3}$$

with $\mathbf{p} = (X_p, Y_p, Z_p)$ the object space coordinates of model tie points, s the scale factor, $\mathbf{R}$ a 3D orthogonal rotation matrix, $\mathbf{m} = (x_m, y_m, z_m)$ the measured (or computed) model coordinates, and $\mathbf{t} = (X_t, Y_t, Z_t)$ the translation vector.

Considering that approximations for the unknown parameters in this non-linear problem are required we may just as well approximate the rotation matrix by its differential form. Hence

$$R \approx \begin{bmatrix} 1 & -\Delta k & \Delta \\ \Delta k & 1 & \Delta\omega \\ -\Delta & -\Delta\omega & 1 \end{bmatrix} \tag{4}$$

With $s = 1 + \Delta s$, the linearization of Equation (3) with respect to the points orientation parameters and the block is given by

$$A_a = \begin{bmatrix} 0 & z_m & -y_m & x_m & 1 & 0 & 0 \\ -z_m & 0 & x_m & y_m & 0 & 1 & 0 \\ y_m & -x_m & 0 & z_m & 0 & 0 & 1 \end{bmatrix} \tag{5}$$

$$\xi_a = [\Delta\omega\ \Delta\ \Delta\kappa\ \Delta s\ \Delta X_t\ \Delta Y_t\ \Delta Z_t]^T \tag{6}$$

$$\mathbf{A}_m = \begin{bmatrix} -1 & 0 & 0 \\ 0 & -1 & 0 \\ 0 & 0 & -1 \end{bmatrix} \tag{7}$$

$$\xi m = [\Delta X_p\ \Delta Y_p\ \Delta Z_p]^T \tag{8}$$

Based on which we estimate the unknown parameters by a least-squares adjustment using the Gauss-Markov model.

9. Instrumentation

Precise photo coordinate measurements are required for fully analytical aerotriangulation. An analytical stereoplotter or a monocomparator must be used. When semianalytical aerotriangulation serves as a preadjustment for fully analytical aerotriangulation, instrumentation and measurement procedures are controlled by the requirements for fully analytical aerotriangulation. When semianalytical aerotriangulation is to be the final adjustment for pass point coordinates, coordinates of independent stereomodels must be measured on a Second-Order analog instrument or better.

10. Aerotriangulation Methods

Aerotriangulation methods can be characterized in several ways:

(*a*) Photogrammetric projection method (analogic or analytical).

(*b*) Strip or block formation and adjustment method (sequential or simultaneous).

(*c*) Basic unit of adjustment (strip, stereomodel, or image rays).

Modern instruments and adjustment methods dictate that an analytical adjustment of stereomodels or image rays be emphasized.

(A) Semianalytical Aerotriangulation

Semianalytical methods use the stereomodel as the basic unit of the aerotriangulation process. Stereomodels may be formed by either analogical or analytical methods. After interior and relative orientation of each stereopair of photographs, stereomodel coordinates are measured in an arbitrary coordinate system defined in the model space. For aerotriangulation purposes, the spatial coordinates of the two exposure stations and each pass point in the stereomodel are required.

(1) Sequential Strip Formation

A strip unit is formed sequentially from the stereomodels by joining them at the exposure station and pass points shared by the new stereomodel and the previous stereomodel already in the strip coordinate system. Each stereomodel added to the strip provides the necessary points to bring the succeeding stereomodel into the strip unit.

The sequential nature of the strip formation procedure allows systematic errors to accumulate as the strip unit is assembled from the individual stereomodels. Residual systematic errors in the interior orientation, the relative orientation, and the measurement system are propagated down the strip causing the horizontal and vertical strip coordinate datums to bow and twist. Although it is difficult to measure and completely remove individual systematic errors in each stereomodel, the total effect of all the errors acting on a strip unit is readily apparent after initially leveling and scaling the assembled strip unit. Discrepancies at the horizontal and vertical control coordinates are calculated by taking the coordinate differences between the known position of each control point and its position in the assembled strip unit.

(2) Strip Adjustment

The sequential strip formation causes the X, Y, and Z coordinate discrepancies at the control points to form smooth error curves. The error curves can be approximated by second- or third-order polynomial functions. Corrections to the strip pass point coordinates are derived by evaluating the polynomial functions at the strip location of each pass point. The corrections are applied to the original strip coordinates to complete the strip adjustment process.

- ☆ The polynomial functions model the error propagation in a single strip unit. Block configurations may be adjusted by using separate polynomials for each strip. The unique strip units in a block are joined together at tie points in the overlap area between strips. The polynomial functions adjusting adjacent strips are constrained to yield the same adjusted coordinates at each tie point in the block.
- ☆ The polynomial adjustment of aerotriangulation is an approximate method that is less accurate than the simultaneous adjustment of stereomodels or photographs. The purpose of polynomial strip adjustment in modern aerotriangulation systems is to perform a preliminary adjustment of all the photogrammetric and ground survey measurements. The sequential strip formation and polynomial adjustment is an excellent way to detect blunders in the data and provide initial approximations for use in simultaneous adjustments.

(B) Simultaneous Adjustment of Stereomodels

The adjustment of independent stereomodels as individual units is the logical extension of the strip unit adjustment discussed in the previous paragraph. Sequentially formed strip units are not used, but rather all stereomodels in an entire strip or block are simultaneously transformed into the ground coordinate system. Stereomodels without ground control points are carried into the ground coordinate system by the pass points and exposure stations shared with adjacent stereomodels. The simultaneous adjustment of stereomodels can rival the simultaneous adjustment of photographs if the individual stereomodels are corrected for all systematic deformations. However, the development of analytical stereoplotters that measure precise photocoordinates warrants that the more rigorous fully analytical aerotriangulation be required for all large-scale and high-accuracy mapping.

(C) Fully Analytical Aerotriangulation

Fully analytical aerotriangulation is a simultaneous adjustment solution of collinearity equations representing all the image rays in a strip or block of photography. The adjustment is often referred to as the "bundle" method. Although the basic unit of adjustment is actually the individual image ray, the solution determines the exterior orientation parameters of the bundle of image rays recorded by each photograph, and the adjusted ground coordinates of each ground intersection point in the ground coordinate system.

11. Analytical Aerial Triangulation

Traditionally, highly skilled operators are employed to perform AT on analytical plotter. The approach can be classified into three-phased:

These three steps are being implemented sequentially and possibly in separate and different work environments.

11.1 Project Preparation and Planning

(a) Selection of Ground Control Points

A block of photography, consisting of two or more flight lines of photography, should have planimetric control points spaced approximately equally around the block perimeter. The height control points form chains of controls in a direction perpendicular to the strips. The accuracy of ground control points for AT should be more stringent than for a project fully controlled by field survey points.

GPS is an effective method of establishing basic project control and photo control to connect the project area surveys to an existing national network stations outside the project region. Kinematic GPS methods are also being used to position the camera at the time of exposure. If the camera position is known, ground control points used to resect for the camera position can be eliminated. The use of kinematic GPS methods should be evaluated carefully; unless the terrain is inaccessible for ground targeting, terrestrial GPS surveys to establish control points and normal flying procedures may be more cost-effective and accurate for large-scale mapping.

(b) Tie Points

Tie points are used to create geometric harmony among the images in a project so they're positioned correctly relative to one another. Automatic tie point collection uses digital image matching techniques to automatically identify and measure tie points across multiple images and strips of imagery. During the preparation stage of conventional AT an operator selects and marks conjugate pass and tie points. Tie points requirements are related to type of point used, location and point transfer and marking requirement.

- ☆ ***Type of points***: tie points may be artificially marked points or targeted points. However, since tie points must lie at or very near the centreline of the triple overlap area, artificially marked points designed from the photography taken should be used. Premarked targets are too expensive and too difficult to align with the triple overlap areas.
- ☆ ***Marking artificial points***: artificially marked tie points must be well-defined symmetrical patterns drilled, punched or otherwise marked in the emulsion using a suitable marking instrument such as a Wild PUG or equivalent. Normally the original negatives should not be marked. This manner provides an unambiguous and permanent identification for the tie points that were selected.
- ☆ ***Location:*** a minimum of three tie points must be marked along the centreline of each triple overlap area. One tie point must lie near the photo principal point, and one point must lie near each of the corners of the neat model. Tie points locations must be selected by examining the photographic prints with a stereoscope. Tie points must not be located on obscured ground level and must be symbolized and labelled on the control photographs.
- ☆ ***Point transfer for stereoscopic measurement:*** by viewing the marked points stereoscopically with adjacent photographs the points are

transferred to the overlapping photos. This is quite crucial, particularly the transfer of strip tie points. The success of an AT project largely depends on the quality of the point transfer. Measuring process can only be proceeded after the points are transferred or clearly identified.

11.2 Point Measurements

During this phase of AT, the coordinates of both the ground control points and the tie points are measured and recorded in the photo coordinate system. This is very tiring especially for large projects since for the traditional AT, operators can only view two images at the same time. Special software such as PAT-MR can then be used for gross error elimination and block adjustment by independence models and PAT-B for bundle block adjustment. The tie points are then used to setup stereo models on analytical stereo plotters through RO and AO processes. With the advent of DPW, EO parameters can be used to set-up stereo models directly on these photogrammetric workstations without the RO and AO steps. It is worth noting that while photogrammetric mapping is moving to the digital domain, not all related firms and agencies are using DAT. Instead, many of them still apply analytical AT for image orientation and move the rest of the processing to DPW equipment.

12. Digital Aerial Triangulation

DAT is increasingly being carried out in photogrammetric production environments as greater efficiency is being aimed for through increased automation. Currently, the major DAT software's on the market are the Helava Automated Triangulation System (HATS) from LH Systems, Match-AT from Inpho and Integraph, and Phodis-AT from Zeiss.

Digital point selection, point transfer, point measurement by image matching procedures and blockadjustment has opened new possibilities for almost complete automation of AT. In an automated production, the DAT is divided into several processing steps, which include data preparation (photos and control points), automatic data import and image magnification, automatic IO, automatic AT measurements, bundle block adjustment, and quality control.

12.1 Data Preparation

The data preparation includes configuration of the photo block (providing images, loading digital images from tape, if not available on disk) and providing control point data (co-ordinates, overview plot, available sketches). GCPs in DP can be collected from existing vector files (*i.e.,* road intersection), orthorectified images, DTMs and scanned topographic and cartographic maps. Camera calibration parameters (principal distance, radial distortion) are also imported into the project file.

12.2 Automatic Data Import and Image Magnification

Before starting the measurements, the digital images are imported into the DPW and an image pyramid is generated automatically for measurements, display and zooming. The GPS photo centre coordinates of each image if available are imported to provide approximate values for the overlaps between images in the blocks. Additionally, the preparations of the triangulation files, which contain all parameters for the automatic measurements, were performed through a number of interfaces.

12.3 Automatic Interior Orientation

Automatic IO of digitised film images is fundamentally straightforward; the only thing needed is to automatically find the pixel coordinate (row and column) values of fiducial marks. This can be done by correlating restored digital images of fiducial marks against corresponding images on the digitized photo. These coordinates can then be used, together with calibrated fiducial coordinates, to compute the needed pixel-to-photo transformation. Automatic fiducial measurement is possible after manually measured the first two fiducial marks.

12.4 Automatic AT Measurements

The processes of AT measurements, as currently used in HATS, are divided into four steps that include Automatic Point Measurement (APM), Interactive Point Measurement (IPM), Simultaneous Solve (Re-measurements) and Blunder Detection.

The underlying principles of all three are similar in that the operator should have to measure the ground control points in at least one image, but all other processes are automatic. HATS request the operator to define a pattern of tie points and image matching proceeds accordingly. Other software possibilities like those applied on (MATCH-AT and PHODIS-AT) prefer to select very large numbers of tie points by means of an interest operator. In all case, the different levels of the image pyramids are used to facilitate the process. Triangulation automates the tedious procedure of selecting and measuring the image coordinates of tie points, and control points. However, interactive point measurements are used to measure the ground control points and the image coordinates of control points and tie points (in case of failure). It automatically flags unacceptable tie points and displays the required images for re-measurement without operator's intervention. It also has the ability to automatically remove unacceptable tie points during the solve process. Here is where the automatic AT offers significant improvements. The point transfer can be combined with simultaneous measurements of conjugate points on all images involved. A distinct advantage is the possibility to simultaneously determine conjugate points that appear on more than two photographs by a process known as multiple image matching.

12.5 Bundle Block Adjustment

The system flags unacceptable tie points and displays the required images for measurement without operator intervention. All the operator has to do is to re-measure these unacceptable points by moving the floating mark to their proper locations, if requested. Once GCPs and tie points are collected, block bundle adjustment can begin. Block bundle adjustment is essential to determining the information required to create orthophotos, DTMs, digital stereo models (DSMs) and 3-D features. The results of block bundle adjustment include the internal and external sensor model information, the 3-D coordinates of tie points and additional parameters that characterize the sensor model. An output report can provide detailed statistical reports outlining the accuracy and precision of the derived data.

13. Digital AT Performance

It is not unusual for traditional analytical AT to consume one hour per stereo model (Xiaopeng and Baker, 2000). Time required for DAT is largely dependent on the image block configuration and on the terrain conditions. When the image block is uniform (with consistent overlap and sidelap) or GPS/INS data is available, the image block configuration is relatively easy to set up. Otherwise, considerable effort is needed for the preliminary orientation of the block. Terrain type and image scale also affect the accuracy that can be achieved and the execution time required. More user intervention is needed for difficult terrain types and large-scale imagery. Ten minutes (operator time) or better per image has been reported by some authors like DeVenecia et al. (1996) and Kersten (1999) for DAT projects.

Kaczynski and Ziobro repoted at OEEPE Workshop in 1999 that labour time needed for DAT is about 3 times less than that needed to perform the same work on the analytical plotter. Under extremely difficult conditions (large height difference, different flying seasons, heavily vegetated terrain), DAT will take more time than AAT or simply will not work at all (Xiaopeng and Baker, 2000). In these cases, AAT may be the only methodology and will still be competitive to DAT in difficult terrain. However, with more improvement and more experience DAT productivity will be enhanced. It is predicted that a triangulation rate of better than 5 minutes per image can be achieved in the future (Kersten, 1999).

These efficiencies can also be achieved if precise direct measurements of the EO elements with GPS/INS are available. This leads to a better performance of the automatic point transfer and mensuration due to the better approximation of the block configuration.

DAT has attracted much attention from both academic (Ackermann and Tsingas, 1994; Agouris and Schenk, 1996) and production areas (Kersten, 1999) since early nineties. A series of tests were designed and carried out to compare different AT scenarios and to obtain experience with the digital systems. Results from these tests and practical experience indicate that the accuracy of DAT is at least at the same level or better when compared to conventional AAT.

13.1 Accuracy and Process

The contract agency is responsible for designing the aerotriangulation scheme that will meet the requirements of the photogrammetric product. Table 2.6 summarizes the guidelines for evaluating aerotriangulation methods. However, since meeting required pass point accuracies is dependent on the photogrammetric system, the contract agency should be allowed some latitude in meeting criteria for these intermediate results.

(a) Photo Coordinate Measurements

Photo coordinate measurement is the most critical factor contributing to the accuracy of aerotriangulation results. The contract agency should be especially careful to control the quality of point transfer, point marking, and point measurement. The measurement stage(s) of the stereoplotter or the monocomparator should have a least count of 0.001 mm or less. The viewing, pointing, and digitizing components of these instruments should enable the operator to group multiple readings on any well-defined target or marked pass point within a maximum spread of 0.004 mm. Multiple readings are of more value if they are not consecutive; however, this reading scheme is often not practical. If multiple readings are made consecutively, the operator must move off the image and re-point between each reading. The instrument used should be capable of measuring a photo coordinate with an RMSE not greater than 0.003 mm. It is mandatory to end the measuring of a photograph with a reading on the first point measured (usually a fiducial mark) to assure that the instrument encoders have not drifted or skipped counts.

Table 2.6: Criteria for Evaluating Analytical Aerotriangulation

Analytical Aerotriangulation Procedures	*Criteria*
Photo Coordinate Measurements: Monocomparator or Analytical Stereoplotter	
Least Count of Stage Coordinate	0.001 mm
Interior Orientation: Transformation to Fiducial Coordinates	
Minimum Number of Fiducials	4
Maximum Residual (After Affine Transformation)	0.020 mm
Preliminary Sequential Strip Formation and Adjustment	
Stereomodel Relative Orientation	
Minimum Number of Points, y-parallax Residuals	6
RMSE	60.005 mm
Maximum	0.015 mm
Stereomodel Joins	
Minimum Number of Points	3
x,y Pass Point Coordinate Discrepancy	
RMSE	H/12,000 ft
Maximum	H/6,000 ft
z Pass Point Coordinate Discrepancy	
RMSE	H/10,000 ft
Maximum	H/5,000 ft
Polynomial Strip Adjustment	
X,Y Control Point Coordinate Residual	
RMSE	H/10,000 ft
Maximum	H/6,000 ft

Contd...

Table 2.6–*Contd...*

Analytical Aerotriangulation Procedures	*Criteria*
Z Control Point Coordinate Residual	
RMSE	H/7,000 ft
Maximum	H/5,000 ft
Simultaneous Bundle Adjustment	
RMSE of Photo Coordinate Residual	0.004 mm
Maximum Variance Factor Ratio	1.5

(b) Interior Orientation

Interior orientation refers to the geometric relationship between the image plane and the perspective center of the lens.

(1) The initial step is to transform the measured stage coordinates into the photo coordinate system defined by the calibrated fiducial coordinates. An affine transformation, which accounts for differential image scale and shear, is typically used to establish the photo coordinate system. The transformation parameters are determined by a least squares adjustment using at least the four midside or four corner fiducials. If both types of fiducials are present in the camera, eight fiducial points can be used to increase the redundancy of the adjustment.

(2) After the photo coordinate system is established, the image measurements must be corrected for systematic errors. This procedure is called photo coordinate refinement. Corrections are applied for principal point offset, radial lens distortion, tangential lens distortion, and atmospheric refraction. Procedures for making these corrections may be found in Wolf (1983) and Moffitt and Mikhail (1980). Correction for tangential lens distortion may be neglected if it is found to be insignificant in the camera calibration report. Photo coordinate refinement may be performed by the analytical stereoplotter software or the aerotriangulation software. Typically, the interior orientation and refinement parameters are considered known based on the calibration report. Then the photo coordinate refinement is performed before the photo coordinates are used for the aerotriangulation adjustment. If self-calibration aerotriangulation software is used, the camera interior orientation parameters are considered to be approximations, and they are adjusted as a parameter in the aerotriangulation solution.

(c) Preliminary Sequential Aerotriangulation

This process (Table 2.6) refers to the sequential assembly of independent stereomodels to form a strip unit and the polynomial strip adjustment into the ground coordinate system. The sequential procedure is a preliminary adjustment that develops initial approximations for the final simultaneous bundle adjustment.

The sequential procedure also serves as a quality control check of the photo and ground coordinate data. The guidelines listed in Table 2.3 are not rigorously enforced, but they are used to evaluate the building blocks of a larger strip or block configuration.

(1) Relative orientation of each stereo pair is performed by a least squares adjustment using the collinearity equations. The stereomodel is created in an arbitrary coordinate system, and the adjustment is unconstrained by ground coordinate values. Therefore, the photo coordinate residuals should be representative of the point transfer and measuring precision. The photo coordinate residuals should be examined to detect misidentified or poorly measured points. The minimum number of points that will uniquely determine a relative orientation is five, and the photo coordinate residuals will be zero. The six-point minimum recommended in Table 2.3 results if the standard nine pass points per photo configuration is used. With only one redundant pass point, the photo coordinate residuals will still be quite small unless a large blunder is present. Typically, more than the minimum six pass points are used. The RMSE and maximum residual values listed will more likely be reached when larger numbers of pass points are used.

(2) When stereomodels are joined to form a strip, the pass points shared between models will have two coordinate values, one value in the strip coordinate system and one value in the transformed model coordinate system. The coordinate differences or discrepancies between the two values can be examined to evaluate how well the models fit to one another. Horizontal coordinate discrepancies will typically be smaller than vertical discrepancies since the image ray intersection geometry is weaker in the vertical direction. As the stereomodels are transformed into the strip, one after the other, the pass point coordinate discrepancies should be uniform and no outliers should be observed. The coordinate discrepancy criterion is expressed as a fraction of the flying height above terrain because the magnitude of the discrepancy in ground units is dependent on the photo scale.

(3) Polynomial strip adjustment is a preliminary adjustment that produces initial ground coordinate values for all the pass points in a strip. The pass point coordinate values will be adjusted again by the final bundle adjustment. The polynomial correction curve is fit to the coordinate errors at the control point locations using a least squares adjustment. The residuals of the least squares curve fit can be examined to evaluate the adequacy of the polynomial adjustment. The residual criterion is expressed as a fraction of the flying height above terrain because the magnitude of the discrepancy in ground units is dependent on the photo scale. The evaluation of the polynomial residuals is the least critical check in the aerotriangulation process, and a great deal of latitude can be allowed in meeting these criteria. From project to project, large variations in the residuals may occur because of the number of stereomodels in the strip,

the polynomial function used, and the distribution of the control points. It is more important to check the X, Y, and Z error curves after the linear transformation of the strip into the ground coordinate system and before the polynomial correction. These error curves should be smooth Second- or Third-Order curves. Outliers from a smooth continuous curve are an indication of a blunder in the photogrammetric value or the ground survey value at a control point.

(d) Simultaneous Bundle Adjustment

Fully analytical aerotriangulation must be adjusted by a weighted least squares adjustment method. The adjustment software must form the collinearity condition equations for all the photo coordinate observations in the block and solve for all photo orientation and ground point coordinates in each iteration until the solution converges.

(1) The exterior coordinate system used for the adjustment should be a local rectangular coordinate system as defined in previous chapter. This coordinate system contains no earth curvature or map projection distortions. These effects may be judged to be negligible for small project areas and low flying heights, but they are significant factors for large project areas and high flying heights.

(2) The least squares adjustment results should be examined to check the consistency of the photo coordinate measurements and the ground control fit. Residuals on the photo coordinates should be examined to see that they are representative of the random error expected from the instrument used to measure them. Residuals should be randomly plus or minus and have a uniform magnitude. The residuals should be checked carefully for outliers and systematic trends. The standard deviation of unit weight computed from the weighted adjusted residuals should not be more than 1.5 times the reference standard deviation used to compute the weights for the adjustment. A large computed reference variance indicates inflated residuals and possible systematic errors affecting the adjustment. For example, if photo coordinates were judged to have an overall measurement standard deviation of 0.005 mm and this value was used to compute observation weights, the standard deviation of unit weight computed by the adjustment should not exceed 0.0075 mm.

(3) Accuracy of aerial analytical triangulation should be measured by the RMSE and the maximum error in each coordinate (X, Y, and Z) direction for the combined checkpoints. The maximum allowable error should be checked at the midpoint of the bridging distance between ground control points using checkpoints or test drop points surveyed for this purpose. Table 2.7 lists the accuracy criteria suggested for each class of mapping. These criteria are the final and most important check of the aerotriangulation results.

Table 2.7. Aerotriangulation Accuracy Criteria.

Map Class	Aerotriangulation Method	Allowable RMSE at Test Points[1]	
		Horizontal[2]	Vertical[2]
1	Fully Analytical	H/10,000	H/9,000
2	Fully Analytical	H/8,000	H/6,000
3	Fully Analytical or Semianalytical	H/6,000	H/4,500

1. The maximum allowable error is 3 RMSE.

2. One-sigma level

(e) Semianalytical Aerotriangulation

When semianalytical aerotriangulation is to be the final adjustment for pass point coordinates, final adjustment of strips or blocks must be performed using numerical polynomial strip adjustment or simultaneous stereomodel block adjustment. Analog or graphical adjustment methods must not be used. This criterion includes analytically computed independent stereomodels using monocomparator photo coordinates. Strips must be assembled by numerical coordinate transformations of successive stereomodels. Strips of stereomodels may be assembled using base-in/base-out capabilities on analytical or universal analog stereoplotters.

13.2 Stereoplotter Settings

Fully analytical aerotriangulation determines the six camera exterior orientation parameters for each photograph, camera position (XL, YL, and ZL), and angular orientation (the omega-phi-kappa system). By relating these parameters to the flight line between each two successive camera stations and scaling to the stereomodel, data are obtained for setting up the stereoscopic model in the stereoplotter. For a specific analog stereoplotter, orientation settings can be derived that will enable the operator to set up the stereomodel for compilation much faster than the manual empirical method, thereby saving expensive stereoplotter time.

Adjustment of Errors

While carrying out aerial triangulation over a large area, covered by a number of strips, and specially while using analytical or semi analytical approach, the ground control point requirement can further be reduced, if all the strips are triangulated together in one block. This process is termed as Block Triangulation, while triangulation of each strip individually is known as Strip Triangulation.

While the minimum ground control points required for Absolute Orientation is just three number of points, it is a usual practice to use much more number of ground control points for this purpose. The surplus points are used to check and control the accuracy. Care is taken to make sure that all the ground control points are uniformly and properly distributed throughout the area. After ground coordinates of all the points are computed, the residual errors at check points (surplus ground control points) is determined, and corrections are applied to all the points by a suitable mathematical model, in order to make the combined magnitude of residual errors to be minimum. This operation is known as Adjustment of errors. In most of the adopted practices, in Analytical Aerial Triangulation, this operation is carried out along with other photogrammetric computations, and entire computing operation is known as Block Adjustment.

Modern Workstations

Modern analytical Photogrammetric Instruments usually contain a microcomputer with entire measuring, recording and data storage system with computation, analysis and error adjustment programs, and in a system. This facilitates the entire operation at a single workstation and thus much faster in delivering the final output.

13.3 Ground Methods of Provision of Control

There are many methods of provision of horizontal and vertical control by ground methods. The principles of important ones are being mentioned below:

(a) Horizontal or Planimetric Control

(i) Triangulation

It is the process of measuring the angles of a chain or network of triangles formed by stations marked on the surface of the earth. The calculation involved is the trigonometrical proposition.

$$\frac{Sin\,A}{a} = \frac{Sin\,B}{b} = \frac{Sin\,C}{c}$$

Hence if any side of the triangle is known, the triangle can be solved. In decreasing order of the quality, precision and instruments used, triangulation is classified as:

- ✰ Primary or First Order
- ✰ Secondary or Second Order
- ✰ Tertiary or Third Order

(ii) Traverse

In this process the lengths and directions of azimuths of a series of connecting lines, from a point of known position to the point whose positions is to be determined, are measured. Like triangulation, traverse is also classified as Primary, Secondary and Tertiary.

(iii) Astronomical Fixings (Astro-fix)

Latitudes and Longitudes of a point are determined by astronomical observations. The method is very fast but the degree of accuracy with which the coordinates are determined is quite low and is not suitable for accurate work. However, it is used in exploratory and reconnaissance surveys.

(iv) Using Doppler or GPS

Latitudes, longitudes and heights can be fixed for a point using a Doppler or Global Positioning System using a artificial Geo-satellites. GPS are getting more popularity these days.

(b) Vertical Control

Levelling is the method of providing vertical control and can be subdivided into 3 main groups.

(i) Spirit Leveling

It is the process of determination of elevation of points (usually called Benchmarks) with respect to each other or with respect to a common datum by means of instruments using a spirit level or a precision pendulum. This is the most accurate method of providing vertical control.

(ii) Trigonometrical Levelling

In this method the elevation difference between two points is obtained by means of observed vertical angles combined with the length of the line joining the two points. This method is combined with triangulation and traverse procedures and is the commonly used method.

(iii) Barometric Levelling

The method is based on variation of the atmospheric pressure with height. Heights of points can be determined with great rapidity but the degree of accuracy is quite low and is unsuitable for accurate work. It is used for preliminary reconnaissance and exploratory surveys.

13.4 Requirement for Stereo Plotting

In order to fit the stereo model with the ground at certain scale, a few control points are required. For scaling at least two plan (horizontal) control points are needed but third is needed for a check. For levelling 3 points at suitable locations are required and a fourth point for a check. Thus, we require 4 height control point and 3 plan control points. If these points are plan and height control both, then we need 4 such points located at the corners of a model. A fifth point is always required to check for the deformation. The suitable location for this point is the centre of the model. A model once oriented absolutely in these points enable other detail in the model accurately surveyed in plan as well as in height.

13.5 Requirement for Rectification

Four or more planimetric control points are required at the corners of the photograph for its rectification.

13.6 Model Set-up with AT Results

While products produced from AAT can easily be used on both DPW and analytical instruments, DAT has limitations in this area. DAT products cannot easily be used on analytical systems. DPW processes do not regularly produce products such as pugged diapositives. Scanning is usually performed on original negatives or unpugged diapositives. "Digital pugging" is performed automatically or manually to identify tie and control points. Since pugged diapositives are not an output product, model set-up on analytical stereo plotters cannot be performed. However, advanced analytical stereoplotters can make use of DAT results directly. This is usually limited to installations where the analytical machine and DAT adjustment package are from the same manufacturers: For example, SD2000/3000 vs. Microsoft/Orientation Management (ORIMA) from LH Systems, Planicomp P_1/P_3 vs. PHODIS from Carl Zeiss. Multi-vendor compatibility is possible with certain orientation and data conversion programs, *e.g.*, AP32 from SoftMap and CONVERT from LH Systems. Another possibility is to sketch the point of interest when using the DPWs to be used for further identifications on the analytical plotter.

However, a couple of related aspects of DPWs are often overlooked as being rather trivial, but represent applications of automation. After triangulation it is effortless to load images and their orientation parameters; no diapositives have to be placed on plate carriers. Once IO is performed on an image, its parameters are included in the definition of the pixel's coordinates. This relieves the operator from reestablishing the IO in the case of re-measurement of previously used imagery. Indeed, in the Leica-Helave systems, it is possible to define stereomodels as pairs of already triangulated images.

13.7 Problems with Digital AT

DAT is developing quite rapidly and today it seems that this technique is more efficient than triangulation on analytical plotters. However, the advantage of automation is rapidly lost if the system is not reliable or the correction of mismatching takes too much time (Kölbl, 1996). Automatic point measurement (APM) is a key component of DAT. If the APM process can achieve a high success rate, little or no manual intervention is required. The success rate of APM is dependent on, a good initial block set-up, good quality photo, photo scale and the terrain characteristics. The following conditions may generate problems for APM.

- ☆ Large variation in height (image scale difference) within the triangulation block.
- ☆ Inferior photo quality due to scanning, exposure, development, or handling.
- ☆ Extensive photo shadow or sun angle effects.
- ☆ Densely forested areas and/or featureless terrain, *i.e.*, lakes.
- ☆ Occlusions, large relief displacement and movement of shadow (predominant in larger scales).

Digital Orthophotography

Digital photogrammetry, and particularly digital orthophotography, is making a significant impact in the mapping sciences. To paraphrase a famous photogrammetrist, Dr. F. Ackermann, softcopy photogrammetry is taking photogrammetry out of photogrammetry. What this means is that this tool is available to the masses. No longer is it a necessity to have a highly skilled photogrammetric technician create the orthophoto. Using the digital imagery and a companion digital elevation model (DEM), non-photogrammetrists, with little knowledge of the science of photogrammetry, can create adequate map products for many applications.

The one caveat of Dr. Ackermann's statement is that with this power comes the possibility of misuse. To put it in other terms, without the proper input data, the derived outputted map may not be accurate enough for its intended use. Garbage in, garbage out. To understand what is garbage still requires someone knowledgeable of the limitations with particular data sets.

1. What is a Digital Orthophoto?

An orthophotograph is an image, which has been processed such that the features on the image represent an orthographic projection. In other words, the picture will assume the same characteristics that are found in a conventional vector map. This is done through a differential rectification process whereby the effects of tilt and relief displacement are removed from the image. There is a big difference between a photograph and a map. The former is created through a perspective projection of the object space onto the photograph. Here all points enter the lens through the center of the lens, referred to as the nodal point, and is then projected onto the film radially from this nodal point. A map, on the other hand, represents an orthographic projection where each point on the earth is projected onto the map in a direction perpendicular to the map sheet.

The amount of correction, though, is limited. For example, in urban settings, or other areas where there are features with very sharp vertical features, it is impossible to create a truly orthographic projection for all features. Buildings viewed off-nadir will obscure features. Moreover, the sides of those buildings facing the center of the photograph will display the side of the building (Figure 2.118). This cannot be eliminated without some special processing capabilities. There are some techniques one can use to minimize the effect of building lean. For example, Merrick and Company was awarded a digital orthophoto project for Cook County, Illinois, which included Chicago. To reduce the effects of tall building, Merrick modified their normal acquisition of the aerial photography. One method employed was the use of 80 per cent endlap and sidelap in the photography, which means that only about the center 1½" of the photograph was used for the orthophoto. Moreover, they also employed a 12" focal length camera over the central business district of Chicago. To maintain the same scale of photography used with the normal 6" focal length camera, the longer focal length imagery was flown at a higher altitude. Finally, they also acquired spot or pinpoint photography over the taller buildings (Jacoby, 1999). All of these modifications add to the cost of the project.

2. Advantages of Digital Orthophotos

Despite these limitations, an orthophoto is a very useful mapping tool. It has the interpretative qualities inherent in an image and the geometric properties of a map. This means that the features on the image can be accurately measured, just

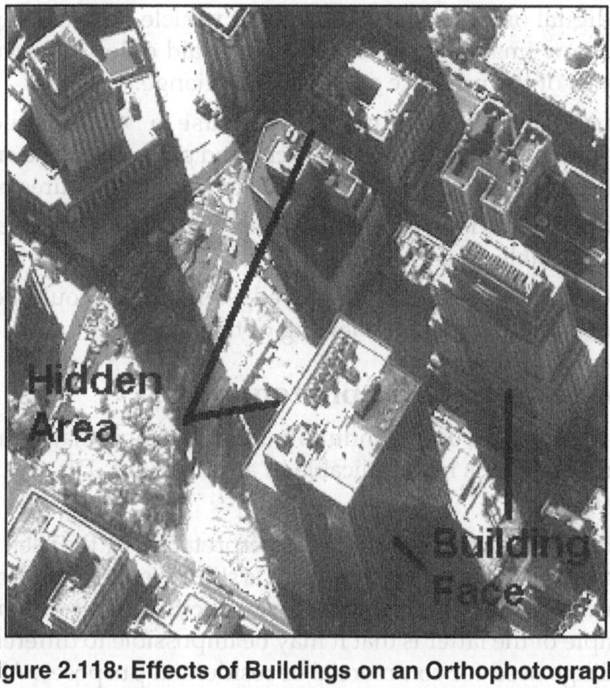

Figure 2.118: Effects of Buildings on an Orthophotograph.

like one might want to do with a conventional line map. Because of this, orthophotographs form an excellent base or control layer for a GIS. It is also relatively inexpensive, especially when one considers the costs incurred in conventional line mapping.

Associated with cost is the speed of map production. According to Greenfeld (2004), the U.S. Geological Survey (USGS) has experienced an improvement in map production of approximately 50 per cent. For revisions, the improvement is much higher, approaching 80 per cent. Other advantages of orthophotography include:

- ☆ The output image can be either a hard copy analog image or stored in digital form.
- ☆ The orthophoto is an image therefore the viewer sees what is depicted on the image whereas in a conventional vector map the information is not complete because it conveys what the cartographer felt the map should contain for its intended purpose. In other words, the map has undergone generalization.
- ☆ Digital orthophotos are GIS compatible since most software packages can incorporate the image into their display. In fact, it has been the growth of GIS that has contributed significantly to the use of digital orthophotography.
- ☆ The image can be overlaid with existing vector data. This is particularly useful when performing assessment of map accuracy.

- ☆ The digital orthophoto is an excellent vehicle for assessing change in an area. New imagery can be simply overlaid older orthophotography for analysis of the presence and extent of change.
- ☆ Orthophotos, being images, can use a wider spectrum of the electromagnetic spectrum for specialized purposes. For example, infrared or filtered imagery could be used to emphasize features on the ground.
- ☆ The orthophoto is an excellent medium from which mapping can be done in inaccessible areas.
- ☆ Although not frequently employed, stereomates could be created from an orthophoto for stereo viewing of the area.

3. Disadvantages of Orthophotography

It is important for the user community to understand that orthophotography is just a tool and as such is not applicable in all situations. Disadvantages include (Greenfeld, 2004):

- ☆ The data are an image that needs interpretation by the user. While features can be depicted on the image, there is still a wealth of information missing such as feature classification and specific feature identification. An example of the latter is that it may be impossible to differentiate between buildings used for commercial or residential purposes. All the view sees is a building.
- ☆ There are hidden areas where no data exist (Figure 2.118).
- ☆ Data shown on the image only represents that data above ground, and even this may sometimes be hidden.
- ☆ There are no spatial analysis capabilities. For example, a line could be identified within a vector map and an attribute table can be displayed. But, clicking on that same feature on a rasterized image will result in displaying pixel location and that is all.

4. Basic Components of an Orthoimage

A digital orthophoto is nothing more than an orthoimage stored in digital form. The image consists of an array of pixels that record the ground reflectance values for that pixel. The resolution of the image is dictated, in part, by the size of the pixels and as we increase the resolution to finer levels, the size of the image in the computer increases accordingly.

There are basically four data sources needed to create a digital orthophoto (FGDC, 1999). These are:

1. An unrectified raster image file acquired by either scanning an image or collected directly by a digital sensor,
2. A digital elevation model (DEM) or digital terrain model (DTM) of the same area covered by the imagery,
3. Ground control, and
4. Sensor calibration data.

The calibration data is required to compensate for the distortion characteristics that are inherent in any measurement system. One set of calibration data is the interior orientation parameters that help define the camera or sensor used to acquire the image. A second set of calibration values represent the characterization of the sensor that may have been used to convert the analog picture into a digital form. The ground control provides the absolute orientation of the image and allows us to georeference each of the pixels within the image. The DEM/DTM is used to compensate for the effects of relief displacement. This can be obtained from a

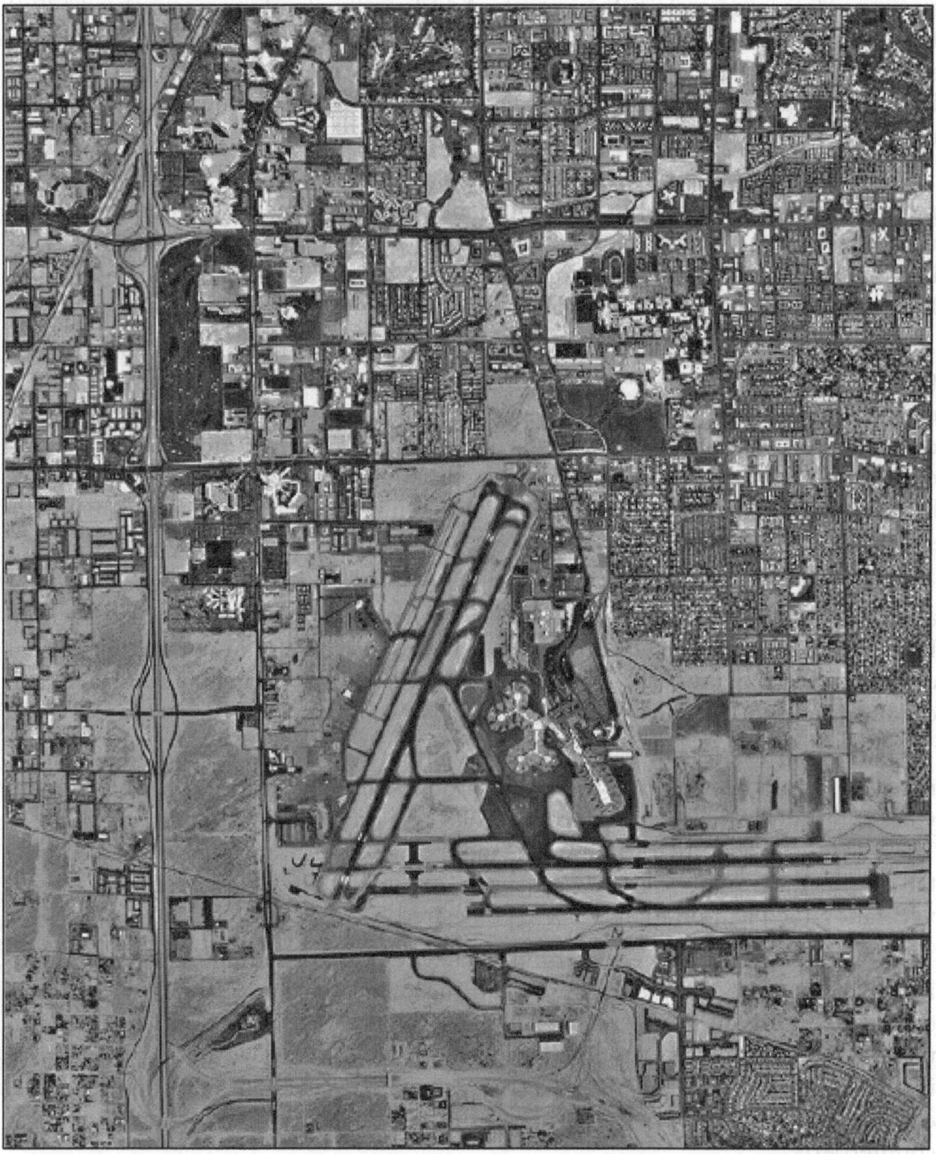

Figure 2.119: Example Digital Orthophoto.

number of different sources, but one must be careful that the density of the ground sampling be consistent with the area being mapped. Finally, one needs an unrectified image file-the picture itself. An example digital orthophoto is shown in Figure 2.119.

Imagery used in the creation of a digital orthophoto can be of different types: black-and white (B/W), color, color infrared (CIR), and other imagery captured in different regions of the electromagnetic spectrum (URISA, 2001). Black-and-white imagery consists of shades of gray extending from pure white to pure black. It is very versatile and yields excellent resolution if properly exposed. B/W film can accommodate large-scale enlargements. Additionally, it only requires about 1/3rd the storage space of color. The disadvantage of B/W imagery is that it may not be as helpful for analysis such as vegetation monitoring or when color or heat is important. If used for interpretation purposes, more training of the analyst is usually required.

Color film is often a medium that users prefer to work with because it yields a picture that closely resembles how humans view the scene. It does not require as much training for interpretation. Additionally, detail that may be lost in shadows in a black-and-white film, particularly light shadows, may still be visible in color. It is more expensive than B/W film and requires more storage space.

Color infrared or false color film is similar to color film except that it is sensitive to the green, red, and near-infrared regions of the electromagnetic spectrum. For example, vegetation will appear as red in CIR film, although one can change the colors of the different bands when the image is displayed in digital form. This type of film is particularly helpful in delineating differences in vegetation since reflectance between vegetation features is markedly different in this part of the electromagnetic spectrum.

While color and black-and-white films are the most common means of creating a digital orthophoto, other parts of the electromagnetic spectrum can also be used. Radar, as an example, is an active sensor that is very useful in obtaining a digital elevation model of the earth's surface. Moreover, it is helpful in obtaining an image of the surface under many different weather conditions.

An important consideration when obtaining any kind of imagery is when the imagery should be acquired. In other words, one must consider the season of the year. If terrain features are important then leaf-off imagery should be collected. The best time to acquire the imagery is the spring because snow has melted and the tall grasses that might be otherwise present are matted. If the purpose is to analyze vegetation then leaf-on imagery will be desired.

Imagery can be acquired in a number of ways. Conventionally, an aerial camera is used to acquire the picture and then the processed negative is scanned and converted to digital form. Imagery can also be obtained directly in digital form using a digital camera. Currently there are four vendors providing digital cameras in a large format and they are Leica Geosystems, Vexcel Imaging, Dimac Systems, and Z/I Imaging. These cameras are much more expensive than their film counterparts.

If the pictures are film based then they need to be converted to digital form and this is done using a scanner. There is a wide array of scanners available in the marketplace today, from less than $100 to tens of thousands of dollars. For orthophotography, a high-end scanner should be used. This equipment is more stable, has better geometric fidelity and yields a more robust digital orthophoto product by incorporating scanner calibration values.

Scanners come in different speeds. Normally, it takes 10-15 minutes to scan a complete 9" square format image. They can also handle different types of input media including cut film, glass diapositives, roll film, etc. Some scanners are highly automated in that it will take a film roll and automatically scan each frame. In a similar vein, most systems also provide different types of output formats. A TIFF file is the most common method of showing the output data. This data represents the raw input data. While systems do allow for compression of data, it is recommended that the data be stored in TIFF format and, if desired, a working copy stored in the same compression format can be used if the user wants. It takes no more time to scan the image so it is beneficial to store that raw data image.

Two important aspects of scanners play critical roles in the geometric relationship of the scanned image. First, the scanner calibration will define how the scanner behaves in its measurement process. There are manufacturing/service producers who claim that they have "distortion free" optics, but this is unrealistic. The second aspect is resolution. As it will be seen later, the selection of the scanner resolution will have important repercussions upon the orthophoto products.

Image data are commonly stored in files called tiles. When the tiles are brought together they should form a seamless map of the project area. For proper data management, an image catalog should be created and provided to the user (this may be transparent to the user). The catalog locates all of the tiles of orthophotography. Some systems use what is called an image pyramid. This consists of a series of images sampled at different ground resolutions, such as 1′, 2′, 5′, and 10′. The idea is to provide rapid image display by automatically loading only those images that are needed for the current views extent with the appropriate pixel resolution.

Applying a filter over the data during a resampling routine forms the pyramid. While a number of different types of filters are available, the Gaussian filter seems to be the best at preserving the content in the original image while reducing computation time. Some examples of aerial negative scale, ground coverage, flying height, map scale, contour interval, and size of TIF file (uncompressed) are given in Table 2.8.

5. Digital Orthophoto Problems

The creation of a digital orthophoto brings with it competing issues. These include accuracy, quality, cost, and the hardware/software display and manipulation capabilities. Image quality is dependent upon a number of production components such as (Triathon, 1998):

☆ Camera quality

- ☆ Diapositive/negative sharpness
- ☆ Photo to orthophoto map scale magnification
- ☆ Orthophoto diapositive density range or bits in the scanner scan pixel (radiometric resolution).
- ☆ Sample scan rate in micrometers2 or dots per inch (dpi) and the photo scale
- ☆ Rectification procedures
- ☆ Final pixel size in ground units (pixel ground resolution)
- ☆ Electronic auto-dodging or radiometric image smoothing after the rectification process.
- ☆ Selection of control points
- ☆ DEM data density
- ☆ Modulation Transfer Function
- ☆ Pixel output being proportion to density or to the "transmissivity" of the medium

Assuming that the correct inputs are used, the accuracy that can be achieved in orthophotography is comparable to that found in line maps. Accuracy of a digital orthophoto is a function of:

- ☆ Magnification
- ☆ Geometric accuracy of the scanner
- ☆ Quality of the DEM
- ☆ Control
- ☆ Focal length of the taking camera

One of the most abused aspects of digital data on the computer is the use of scale or magnification. Computers have the ability to zoom in or out very simply. This may give the user a false sense of the accuracy of the map product. As an example, field measurements may be taken of features with one meter positioning capabilities, such as with "resource-grade" global positioning system (GPS) receivers. But in the computer, these positions could be displayed at the millimeter level. Clearly, displaying data at this range is inappropriate for data collected at the "coarse" meter range. The same applies to orthophoto imagery. Remember that the farther the camera is away from the ground, there is a loss of detail in the features imaged on the photo. For example, a manhole might not be imaged on the photo because it is too small at the scale in which the photography was taken. Therefore, the size of the smallest feature that needs to be depicted on the orthophoto will be a major consideration when designing the scale of the photography.

Magnification also affects the image quality. The recommended magnification range is 8 or 9 times enlargement. Magnification of ten times or more will degrade the image quality because the distance between the silver crystals on the film become noticeable. Below five times enlargement does not noticeably improve the image

Table 2.8: Example Values for Coverage, Pixel Size, etc. for given Photographic Scales

Photogrammetry Mapping Scales and Digital Orthopho to Resolution						
Negative scale	1" = 250'	1" = 300'	1" = 417'	1" = 833'	1" = 666'	1" = 2,640'
Representative Fraction	1:3,000	1:3,600	1:5,000	1:10,000	1:20,000	1:31,680
Flying Height above mean terrain	1,500'	1,800'	2,500'	5,000'	10,000'	15,840'
Ground coverage	1,575' × 1,575'	1,890' × 1,890'	2,621' × 2,621'	5,280' × 5,280'	2 miles × 2 miles	3 miles × 3 miles
Coverage in PLS section	1/4 × 1/4 Section (40 Acres)		1/4 Section	1 Section	2 × 2 Section	3 × 3 Section
Map scale	1 = 40'	1 = 50'	1 = 100'	1 = 200'	1 = 400'	1 = 500'
Contour Interval	1'	1'	1'	2'	4'	10'
15 µm pixel size (Ground)	1.8"	2.1"	3.0"	5.9"	11.8"	18.7"
30 µm pixel size (Ground)	3.5"	4.3"	5.9"	11.8"	23.6"	37.4"
15 µm TIF Image size	98.0 mb	144.0 mb	144.0 mb	144.0 mb	144.0 mb	120.0 mb
30 µm TIF Image size	24.0 mb	36.0 mb	36.0 mb	36.0 mb	36.0 mb	30.0 mb

quality. Therefore, a range of 5-9 times enlargement is the optimum range, depending upon the area being mapped. This means that if the desired final orthophoto scale is 1" = 100' then the photo scale should not be less than 1" = 900'. Note that this would be for optimum terrain. Larger photo scales, such as 1" = 700', may be required to meet the needs of the client.

Radiometric resolution relates to the ability to discern small changes in the tonal change within an image. The Content Standard recommends that 8-bit binary data be used for black and white imagery and 24-bit, 3-byte data for color imagery. This gives the user 256 gray levels over the image (0-255). The value zero represents black and 255 is white. Radiometric corrections such as contrast stretching, analog dodging, noise filtering, destriping, and edge matching are frequently applied to the data before it is given to the user. The standard recommends that these processing techniques be used sparingly to minimize the amount of data loss.

Image quality is also affected by the resolution of the scanner. The scanner and the scanning process have inherent errors associated with them. It is important that high precision scanners be used in converting the image into a digital form. Additionally, it must be calibrated to ensure that the performance of the equipment is within the minimum specifications for the mapping. Many of the softcopy instruments used today have the capability of adding the scanner calibration parameters into the program to correct for the distortion scanning introduces.

It should be evident that the coarser the resolution (larger the pixel size), the more "steplike" lines and features become (recall that this is called pixelization). The important issue is the relationship between the size of the scan pixel to the scale of the photography and the desired output orthophoto scale. One suggestion is to scan the photo at about 240 dpi for each magnification range. This means that if the desired photo to final orthophoto magnification range is 5 times, then the photo should be scanned at 5 × 240 = 1200 dpi as a minimum. This represents a pixel size of approximately 20 μm (micrometers) at the photo scale. Taking the magnification recommendation to its limit of 9 times yields a sampling rate of 2160 dpi with a pixel size of roughly 12 micrometers. While a smaller pixel size may yield better resolution, it does not necessarily mean higher accuracy since accuracy is affected by a number of factors like the survey control, flying height, and focal length of the camera along with pixel size.

These results are consistent with other studies. For example, it has been pointed out that an approximate 15 μm resolution is needed to maintain photographic resolution of aerial film. Higher resolution does little to enhance interpretability of the image. In fact, 20-30 μm scan rates are commonly utilized in industry. These levels are both economical and meet the needs of most mapping applications.

Another issue affecting image quality is the pixel size expressed in ground units. This is frequently performed by resampling the pixel values to create a smoother image in terms of it tone. When this is done, the preference will be to sample to a coarser resolution, such as sampling at half a foot and resample to the foot level. As a rule of thumb, a 1.2 times or larger factor should be applied to the scanned pixel. For example, using this factor to a one-foot scan, the final orthophoto would have at least a 1.2' pixel size. Subsampling should only be applied within the limits defined by the Nyquist theorem (FGDC, 1999), which limits the resampling to a maximum of 2X. This limit is arrived at to avoid undesirable aliasing.

The accuracy of the orthophoto is dependent upon two primary factors: control and DEM accuracy. Survey control is required to fix the map to the ground. It is used to remove/reduce many of the random errors associated with the imagery, such as terrain relief, platform position/orientation, and faulty elevation data. Photogrammetrists often use aerotriangulation to provide control between the primary ground controls on a project. In some instances, control for the orthophoto is derived from existing maps of the area. Significant errors can be introduced into the process thereby degrading the orthophoto. For large-scale mapping, ground targets that will be imaged on the photo should be used. The control needs to meet the specifications for the mapping. With the global positioning system (GPS) and an inertial measurement unit (IMU), it is operationally feasible to perform the mapping without control. The combined GPS/IMU allows for direct sensor orientation (DSO). Since control is used to fix the exterior orientation, measurement of these parameters negates the necessity to obtain control, although it is often prudent to obtain control for quality control/quality assurance purposes and to make the photogrammetric solution more robust.

DEM accuracy is critical to the final quality of the orthophoto. The appropriate DEM must be selected to match the scale of the orthophoto, terrain conditions, focal length of the camera used to acquire the photography, and the magnification. The sampling interval for collecting the elevation data depends upon the terrain conditions. Where the ground is relatively flat, a coarser DEM can be used. On the other hand, if there is a lot of an elevation change (or surface roughness) in the area, a denser sampling rate is required. It is generally acknowledged that the density and accuracy of the DEM for orthophotography does not need to be as accurate as a DEM used for contouring or 3-D modelling. For large-scale mapping, it is important to also include break lines in the data collection. A break line is where the terrain changes direction in slope, such as the bottom or toe of a hill. These break lines control the modelling of the characteristics of the surface and fixes the placement of contour lines on the site. While density is important, the quality of the break lines is more significant. In fact, experience indicates that the sample rate can be very course provided that sufficient break lines exist in order to correctly capture the characteristics of the terrain surface (Michael and Krolikowski, 2001).

It is not uncommon to process the elevation data prior to importing it into the orthophoto rectification process. Generally, a rectangular grid is created which makes the estimation of elevation easier in the orthophoto production process. But, this interpolation process does degrade and smooth the DEM, which affects both the vertical and horizontal accuracy of the orthophoto.

The effect of a DEM on the accuracy of the orthophoto can be shown by the following example. Assume that the map is to meet National Map Accuracy. Then, the shift in the placement of well-defined points should not exceed 1/50" at the map scale. Therefore, for a map at a scale of 1" = 100' (1:1200), the maximum shift that should occur is 2' at the ground scale. This is found from the formula (1/50" × 100'/1") = 2'. Then, if the aerial camera has a 6" focal length lens, using basic trigonometry, a 22.5" error in the DEM would bring about a 2' error in the extreme edge of a 9" × 9" format aerial photo. Since only the neat model area is used in mapping, and using the dimension of 6.3" × 7.2" for the neat model, then the DEM can be in error by 30" at the extremity of the neat model and still fall within the 1/50" criteria. It can be seen that the closer to the center, the greater the tolerance. The formula for computing the errors in digital orthophotos can be expressed as:

$$e_{\text{Ortho}} = e_{\text{DEM}} \times \tan A$$

where,

e_{Ortho} = Error in digital orthophoto,

e_{DEM} = Error in the digital elevation model, and

A = Viewing angle in degrees outward from the center of the photo.

Cox (2002) gives a slightly different value for the effect of a DTM error on the horizontal error in an orthophoto. He presents the following rule of thumb that states that the "vertical error in the terrain model at the outer edge of the usable area of a photograph (flown with 30 per cent sidelap), will result in half the horizontal error in the orthophoto". Cox given an example of a 6′ DTM error on the edge of the model will result in a 3′ horizontal error.

Another factor to consider in DEM accuracy is the magnification ratio and the density of the DEM. Generally, the density of the DEM needs to be denser with smaller magnification ratios. As a rule of thumb, if the magnification is less than 3 times then the spacing for the DEM needs to be 4-8 mm at the final map scale. If the magnification is 3-8 times, then the spacing at the final map scale should be 8-16 mm. Over 8 times magnification allows a grid spacing of 12-24 mm at the final map scale. The operator needs to be aware that the density is greater when the terrain changes rapidly on the site and can be relaxed or spread farther when the terrain is flat.

DEM characteristics change with the terrain therefore it is impossible to outline minimum criteria that would be applicable for all surfaces. It is even possible to find a lot of variability within a single map sheet or tile. Because terrain variability can exist within a project, the map itself may meet accuracy specifications but local anomalies can exist where the map can fail stipulated testing. This is particularly true in areas where elevation changes are abrupt or where bridges, elevated highways and the like are present.

6. Problems with Digital Orthophotos

Problems with digital orthophotos that need to be looked at include:

- ☆ ***Image Completeness:*** The image area is not adequately covered by a DEM resulting in an inaccurate orthophoto. One of the biggest problems is cloud cover. Either the cloud itself or the shadow from a cloud may obscure ground detail. It is the users responsibility to ensure that if the image contains cloud cover that the percentage of obstruction is acceptable for the intended purpose.
- ☆ ***Image Stretch (Blurring,* also called *image smear):*** This is typically caused by anomalies within the DEM data resulting in a spike or large error. Excessive relief on the edge of the photography can also be the cause of this problem. The result is that ground image data is hidden from view. Smearing can occur when an interpolation program is used to assign brightness values to the hidden area using the surrounding visible image. There is no easy way of correcting for this problem except by using subjective visible inspection. It is up to the user to determine whether the amount of these smear artifacts affect the image data for their intended use.
- ☆ ***Image Distortions:*** For large-scale orthophotos, local distortions can exist as was discussed earlier. Figure 2.120(a) shows a distortion along a bridge deck due to reliance on a regular grid of elevations. These distortions can be reduced, but not eliminated, by either collecting break line information or densification of the DEM grid. Figure 2.120(b) shows how this distortion can be edited to create a more faithful rendition of the terrain.

Figure 2.120: Figure Shows Local Distortion Over a Bridge Overpass with the Distortion being Eliminated after Image Editing.

- ☆ *Double Image:* This is when the adjacent orthophotos are compared and the same feature is mapped on both photos when this should not occur. In other words, the maps should be mutually exclusive. The problem may be either improper orientation in the control or less accuracy in the DEM where ground elevations are given that are larger than the reality.
- ☆ *Missing Image:* The causes of this error are the same as the double image except that the DEM gives elevations lower than the real ground values. This error is hard to detect but is clearly evident when looking at linear features where sections may be missing.
- ☆ *Inaccurate Planimetry:* If the planimetric positions of the pixels are in error, look at the control by comparing the visible control on the orthophoto and the photogrammetric control used to control the project.
- ☆ *Image Replication:* Some clients have experienced problems with tone in their digital imagery in that it appears to vary depending on different computing environments. While a contractor may adjust the tone over the entire map, when it is ported to the client's computer environment then the tone quality may be noticeably different.
- ☆ *File Size:* More data results in larger data files. It is necessary to ensure that the computing environment can handle the image data. The size of the data file is a function of the resolution and size of the project area. For example, using a 6" ground resolution for the pixel size and 2,500′ × 2,500′ tiles, it will take about 100 MB per square mile to store the image data. This means that one CD can hold approximately 6 square miles, or about the size of a township.

7. Survey Control

Survey control is required to orient the stereo photos during the absolute orientation process. It is also critical in the creation of an orthophoto. It is ideal that sufficient control be placed at the edges of the project to ensure that the resulting map product meets with the specifications set forth by the client. Control consists of two aspects, horizontal or planimetric control and vertical or height control. These aspects can be combined when the surveyor used a 3-D method of data collection, such as found in the global positioning system (GPS). Since establishing control is a very expensive operation, photogrammetrists have sought ways of using the tools of the photography to provide supplemental control over the project. This process is called aerotriangulation.

Basically, aerotriangulation allows the photogrammetrist to require a sparser network of control actually placed in the field and then to "bridge" between these control points photogrammetrically by establishing secondary control between these ground control points. In the days of analog photogrammetry, ground control would be placed in every four or five models along a strip of photography. Exploiting the geometry of aerial photogrammetry, photo control would be established in the intermediate models and checked and adjusted when coming upon the model with the actual ground control. Here, the ground control would be considered as the true value and the photo control adjusted to it.

Recent advances in the theory of aerotriangulation have increased the reliability of the photo control, thereby allowing for more sparse patterns of ground control. This decreases the cost of the project. The use of airborne GPS, where the GPS receiver is tied together with the aerial camera, has allowed further reductions in the use of ground control. As was stated earlier, in theory, airborne GPS will allow the project to proceed without any ground control, but this is very dangerous since there are no independent checks on the results. Control is also very important in the establishment of the proper mapping datum surface/coordinate system.

Softcopy photogrammetry has the ability to increase the productivity of the aerotriangulation process as well. With correlation techniques, the system can be "trained" to look for objects on the image that are on subsequent photographs. Ideally, these points will be imaged on a number of different photos. Unlike the manual approach to aerotriangulation where the photogrammetrist needs to mark the points on the photos before measurement, correlation can be used. This allows the process to actually pick more photo control points per model than would be done in the manual mode. The result is increased accuracy in the aerotriangulation process.

8. Quality Assurance

Whenever a map product is produced it is critical that the data undergo quality assurance (QA) review. Like any QA program, the assessment process should be well documented both in terms of the testing process and the results from the evaluation. The process should be standardized to ensure a standardized output. When there are departures from the established process, document what was done and why. Some of the items that a QA program should look at are:

- ☆ Foreign matter and scratches in the imagery
- ☆ Mosaic matching

- ☆ Look at the entire image tile by tile by panning through the image in a systematic manner
- ☆ If current digital orthophotography is available, compare the new image to the old.
- ☆ Overlay any vector data that may be available and check for matching
- ☆ Print out a hardcopy on different systems and media to see how the output behaves.

A QA program is essential when accepting data from either a contractor or in-house development. It provides a level of confidence for the map product. An assessment of accuracy is an important part of any QA/QC process. The problem with measuring the accuracy of digital orthophotography lies in the inherent limitations of digital imagery. For example, it could be very difficult to take an individual pixel and relate that to a "well defined" feature. Additionally, it is also difficult to determine precise distance within a precision of about 2 pixels.

9. Applications

The use of digital orthophotography is continuing to rise and innovative applications are continually being explored. The burgeoning growth of GIS finds increased utilization of digital orthophotography. Using digital orthophotography yields some unique issues that users need to address. These include:

- ☆ Lack of matching existing vector data with the image
- ☆ Displacement of vertical structures
- ☆ Lack of stereo viewing that helps in the identification of features within the image.

These issues force the user to question the data. For example, if vector data doesn't match the image data then adjustments must be made and this could lead to inconsistencies throughout the image.

10. Viewing Digital Orthophotography

Most modern GIS software packages allow the user to view digital orthophotography. For example, within ArcView, one needs to just activate the appropriate extension in which the data are stored (Figure 2.121). Then create a new view and load the image. The image is a sample digital orthophoto of the Rolla, Missouri, available through the US Geological Survey.

11. Conclusion

Digital orthophotography has dramatically changed the nature of mapping. It has almost become an essential part of a GIS since it gives the user a spatial tool with excellent interpretative characteristics along with the geometric properties one expects from a good quality map. With the developments of image processing and the incorporation of photogrammetric theory into current software suites, most individuals who are familiar with basic mapping concepts can easily generate a digital orthophotograph. While this has almost become a "black box" technology,

Figure 2.121: Sample Digital Orthophoto as a Part of ArcView's Display.

there are a number of issues that need to be addressed. This lesson has looked at different kinds of orthophotography programs, the concepts of creating the orthographic projection, and the problems involved in the generation of the orthophoto product.

Satellite Photogrammetry

Stereo images can be obtained from satellites also in the same way as they are collected from an aircraft, and the study is called as satellite photogrammetry or satellitegrammetry. In case of stereo photography from satellites the basic mathematical and geometrical aspects remain same as in aerial photogrammetry. However in case of satellite photogrammetry, the flying height is of order of some hundreds of kilometers in comparison with, a few kilometers, in aerial photogrammetry. As a result of this, the earth's curvature plays more than usual role in geometry. The ratio of earth's relief with flying height also becomes too small in satellite photogrammetry, thus seriously effecting the perception of relief in stereo images, obtained from satellites.

Satellite photogrammetry imaging-systems designed for photogrammetric geodesy can be characterized as stereoscopic imaging systems capable of producing data from which a photogrammetrical can determine 3-dimensional coordinates, and related topographic information. The users' requirements must play a key role in the design of an earth, moon or planetary mapping system. Before attempting to specify a system, which would collect data that would permit useful topographic mapping from space of the earth or extraterrestrial bodies, it is necessary to examine some of the essential factors involved.

1. A topographic map contains three kinds of information.
2. Content - the cultural and natural features represented on the map
3. Horizontal location - the reference graticule, grid, datum.
4. Elevation - spot heights, contour lines, profiles and elevations.

A satellite photogrammetry imaging-system designed for mapping must be able to provide all three kinds of data. Map content is determined, in part, by photographic resolution and scale, or more directly from resolved distance on the ground (resolution). It is difficult to establish a linear relation between map scale and resolution required, because some features, like roads and railroads, must be shown on a map regardless of its scale. Hence, these features are nearly independent of map scale; and therefore resolution required is not linearly related to scale. On the moon and planets, however, cultural features have, so far, not been detected so that a linear relation seems more justified. Considering the photograph as a base (photo mosaic or orthophotograph) for the map, a useful criterion can be produced.

The smallest feature, which can be depicted on a map, is assumed to have a least dimension of 0.25 mm (0.010 inch). In order for an object to be photographically identifiable it must be imaged by about 5 resolution elements. It follows; then, that the resolution required for photography can be estimated as:

Ground Resolution = 0.2 × 0.25 mm × map scale number or $R_g = 5 \times 10^{-5} S_m$

where,

R_g = Ground resolution required (metres) for the sensor

S_m = Map scale number

The second kind of map information is horizontal location in an absolute coordinate system. For mapping the earth, this is generally provided by reference to ground control points. But such control does not exist on the other planets. However, orbital-tracking data can provide spacecraft position to a high order of accuracy. When this is coupled with precise data on attitude, and time of exposure, an independent means of determining absolute horizontal location is available. This removes the need for ground control. Within a stereoscopic model, or a photogrammetric triangulation, the internal (relative) positional accuracy is approximately equal to the product of the scale of the photograph and the error in measuring, or identifying a point on the picture:

$$\sigma_p = S_p \, \sigma_m$$

where,

σ_p = Standard error in relative position of a ground point

σ_m = Precision of measuring an image point

S_p = Photo-scale number

An imaging system designed for photogrammetric surveys must be concerned with two basic categories. The first is the metric characteristics of the system and spacecraft, which are as follows.

1. Coordinate (X, Y, Z) on the orbit as a function of time. If the Global Positioning System is used to determine the orbit, an accuracy of ± 10 m may be expected.
2. Time (tt) of exposure correlated with the coordinates. One millisecond precision (0.001 sec) was achieved with Apollo 15, 16 and 17.
3. Attitude – Orientation angles (ω, φ, κ) can be obtained from stellar sensors such as those on Apollo 15, 16, and 17 with accuracy of the order of 5" to 10".
4. Altimetric distance – Radar and laser distance measuring devices that are aimed down the optical axis of the mapping camera produce distance from the spacecraft to the ground with a precision below 1 m.
5. Calibration – Determination of the radial and decentering distortion should reduce distortion errors to less than 10 micrometers in the image plane of the sensor.

The second category is the geometrical configuration of the sensor. This takes into account the stereoscopic coverage of the surface, considering both forward and side-lap of the imagery. Also, the mode of operation whether it be vertical or convergent and its overall effect on precision in extracting relative height measurements (σh) from the stereoscopic pairs is important.

1. Elements of Satellite Photogrammetry

In satellite photogrammetry, the perspective centres of the imaging systems may be assumed to be on the satellite's orbit. It follows, then, that the satellite must be observed to determine the imaging system's position in the orbit as a function of time. Examples of equipment used for observing satellites are the tracking stations of Deep Space Network and the Spaceflight Tracking and Data Network. These stations are distributed around the globe.

These tracking stations determine direction, distance (range) and/or radial velocity (range rate) of the object being tracked. These data are converted, by a process called orbit determination, to a set of six constants, called "orbital elements", that specify the orbit. The orbital elements most commonly used are so-called "Eulerian elements".

The satellite's coordinates are then usually the radius vector *r*, whose components are the radial distances from geocenter, and the true anomaly *f*. For satellite photogrammetry, the Eulerian elements are usually replaced by another set: the location and velocity of the satellite at a specific time. The satellite's coordinates are then its rectangular Cartesian coordinates.

Thus the coordinates (X, Y, Z) of the perspective centres of the camera (imaging system) along an orbital and are not independent quantities, but instead are functions of the six orbital elements and time. A list of coordinates and corresponding times is called an Ephemeris.

The other important quantities in satellite photogrammetry are the three angles of orientation (attitude) of the camera at the time of exposure. In summary, the tracking data are converted to an ephemeris of (X, Y, Z) coordinates as a function of time t. The steller camera provides three orientation angles based on the star's positions. These are the six elements required to solve a photogrammetric problem. They may be used as first approximations in strip and block adjustments.

2. Stereo Images from Satellites

Cameras as well as scanners have been used in satellites to obtain stereo images. During early stages of manned and unmanned satellite missions, small 70 mm cameras were used to take photographs, at very small scale and low resolution. Later, attempts were made to obtain better stereo images from satellites using large format cameras and scanners, which are described, in the following sections.

2.1 Stereo Images from Large Format Camera

The Large Format Camera (LFC), a special purpose camera with high resolution and high geometric fidelity, was flown in NASA Space Shuttle mission in October 1984. The LFC in Space shuttle mission is a precision cartographic camera, with image format 230 × 460 mm and focal length 305 mm and advanced image motion compensation mechanism to account for the shuttle ground velocity of 27000 km per hour. The photographs were taken from varying flying height of 235 to 375 km above earth. Stereoscopic coverage was taken with variable overlap (20 to 80 percent) in the same way as in aerial photography. The photographs retained sharp details even after enlargement and were found to have high cartographic value. The original photos were at scale of 0.8 million to 1.2 million.

On another mission, a 230 × 230 mm format photogrammetric camera with 305 mm focal length lens was also used with a colour infrared film. The ground resolution was found to be of order of 25 meters. Skylab, a manned satellite, carried an Earth Terrain Camera (ETC, S-190B Experiment) had a 457 mm focal length lens and acquired photographs with a ground coverage of 109 by 109 km. The photographs were collected in normal color, black and white and IR colour at a scale of 1:950,000 on 11.4 cm square film. The overlap was 60 percent, which is well suited for photogrammetric purposes.

The satellite images taken from Russian Satellite during mid-eighties, from a camera with focal length of one meter and ground resolution of about two meters are now available for the most part of world.

The photographs taken from satellites at altitude of few hundred kilometers on a photographic film are seriously effected by atmospheric haze. Therefore either color infrared, or panchromatic films are used with minus blue filter to cut the blue haze.

2.2 Across Track Scanners for Stereo Imaging

The common optical mechanical scanners, using a mirror, sweeping the small area (pixel) being viewed, across the flight path are known as Across Track Scanners (and also as whiskbroom scanners). The Multi Spectral Scanner (MSS) used in Landsat series of satellite comes under category. The orbit of Landsat and other similar satellites is such that, at equator adjoining images have about 14 percent overlap. This overlap increases towards Polar Regions to as much as 85 percent. In such areas of high overlaps, the two images taken from adjoining satellite path could be used as stereo images. The base to height ratio varies from 0.174 at the equator to 0.031 at Polar Regions. In such images obtained by across track scanners, the relief displacement is in cross track scan direction, (outwards from the center of the scan line in each scan line). However, due to limited ground resolution, and unsuitable base to height ratio, the relief perception is not suitable for cartographic application.

2.3 Along Track Scanners with Steering Capability

The along track scanners are also known as pushbroom scanners. They consist of linear arrays of numerous Charge Coupled Device (CCD). High Resolution Visible (HRV) imaging system used in SPOT satellite series and PAN (a panchromatic camera) in IRS-1C satellite belong to these type. The satellites are provided with steering capability to tilt the sensor with an angle towards right or left. Thus, off nadir view could be obtained, covering the same area from a different orbit. The amount of tilt is variable to plan for a suitable look angle from available orbits to cover the area of interest. In these scanners also, the relief displacement is normal to the ground track. The stereo images are found to be suitable for medium scale mapping. Many researchers have found that SPOT image (PAN) with 10 meter resolution in stereoscopic mode, can be used for topographic mapping at 1:50,000 scale with 20 meter contour interval.

2.3.1 Geometry of PAN Camera in IRS-1C Satellite

This uses reflective optics along with 4096-element CCD linear array (7 micron × 7 micron) for imaging. A special arrangement, comprising of an isosceles prism reflector, is used for covering full swath of 70 km. spatial resolution at nadir is 5.8m. The swath steering range is ñ 26°, the step size of ñ 0.09 and the repeatability for stereo coverage is 0.1°. The data is collected in panchromatic mode having spectral band 0.50 μm to 0.75 μm. The space segment specifications are given in Table 2.9. PAN data is encoded in 6-bits.

Table 2.9: Space Segment Specifications of IRS-1C

Orbit type	Polar Sun synchronous
Altitude	817 km
Inclination	98.69°
Distance between adjacent traces	117.5 km
Repetivity for LISS-3	24 days
Repetivity for WIFS	5 days
Off-nadir coverage ñ 26° for PAN	398 km
Stereo viewing capability	5 days

2.3.2 Geometry of HRV Camera in SPOT Satellite Series

SPOT pushbroom scanning system does not employ a scanning mirror, rather, it employs a linear array of CCDs arranged side by side along a line perpendicular to the satellite orbit track. The HRV contains four CCD subarrays of 6000 elements each, acquiring data in panchromatic mode to record 10m resolution data. Three 3000 element subarrays are employed in the multispectral mode at 20 m resolution. HRV optical system has a plane mirror, which can be rotated to either side by, ground command, through an angle of ñ 27° (in 45 steps of 0.6° each). This allows each instrument to image any point within a strip extending 475 km to either side of the satellite ground track. At latitude of 45°, there are six possible occasions during the 26-day orbit cycle on which successive day stereo coverage may be obtained. At the equator, only two stereo viewing opportunities on successive days are possible. The base height ratio also varies with latitude from approximately 0.50 or 45° to approximately 0.75 at the equator. If stereoscopic coverage need not be acquired on successive day basis, the range of possible viewing geometry greatly increases. The specifications of the orbit of the satellite is given in Table 2.10.

2.4. Along Track Scanner with Fore and Aft looking

Some pushbroom scanners employed fore and aft looking mode with the help of external mirror to produce stereo images. This external mirror attachments provides forward looking and an aft looking channel in addition to the normal downward looking (nadir looking) channels. In Multispectral Electro-optical Imaging Scanner (MEIS II), which was developed for the Canada Centre for Remote Sensing, acquired data in eight spectral bands ranging from 0.39 μm to 1.1 μm, with an IFOV of 0.7 mrad and a total field of view of 40°. This was the first airborne pushbroom scanner to be used operationally.

Table 2.10: Orbital Characteristics of SPOT Series Satellites

Altitude	832 km
Orbital period (min)	101
Inclination (degrees)	98.7
Equatorial crossing time	10.30
Sensors	HRV
Repetivity	26 days
Stereo viewing capability	5 days

2.5. RADAR Stereo Images

RADAR is acronym to Radio Detection And Ranging where microwave pulses are transmitted to illuminate the terrain and back-scattered pulses from the terrain are received to produce images of the terrain. The operating principle and other details are discussed separately in other sections. Stereo images can be acquired using Radar also in the same way as the across track scanners are used *i.e.*, images are collected from two adjacent paths. However, because the Radar side lighting effect will be reversed on the two images of the stereo pair, stereoscopic viewing is little difficult using this technique. Radar can acquire stereo images by varying the flying height also, if it is airborne, or by having two antennae with two different look angles (Table 2.11). There are so many other parameters are to be considered in reproducing stereo models from these type of Radar stereo pairs other than from photogrmammetric point of view.

Table 2.11: Salient Futures of SAR Systems

Mission and Year	*Frequency*	*Polarisation*	*Look Angle (°)*	*Swath (km)*	*Resolution (m)*
SIR-C 1994 (USA/Germany)	L, C and X-Bands	HH and VV HV and VH VV	15-55	15-90	30
ERS-2 1994 (Europe)	C-Band	VV	23.5	80	25
ALMAZ-2 1996 (Russia)					
RADARSAT 1995 (Canada)	C-Band	HH	20-40	100	25
			20-40	150	35
			37-49	45	10
			49-59	300/500	100
			49-59	75	30

Contd...

Table 2.11–*Contd...*

Mission and Year	*Frequency*	*Polarisation*	*Look Angle (°)*	*Swath (km)*	*Resolution (m)*
EOS 1998 (USA)	L, C and X-Bands	VV, HH HV and VH HH and VV	15-55	30-120 700 (Scan)	30 15
ENVISAT 1998 (Europe)	C-Band	VV, HH	20-50	100-400	30

Measuring parallax and feature heights and the related study, from these images is called Radargrammetry. Among Space borne Radar systems (Table 2.11), SIR-B and SIR-C, the space shuttle imaging Radars and RADRASAT, a Canadian satellite presently in the orbit, collected stereo images. SIR-B Radar experimental mission flown during October 1984, collected stereo images in L-band, HH polarization, with varying look angles from 150 to 60°. The azimuth resolution was 25 m and the range resolution varied from 14 m at a look angle of 60° to 46 m at a look angle of 150. SIR-C collected images in mutli-frequency, multi-polarization modes with varying look angles ranging from 15 to 55°. Radarsat SAR is a C-band system operating with HH polarization. The system can be operated in a variety of beam selection modes providing various swath widths, resolutions and look angles.

The Modular Optoelectronic Multispectral Scanner (MOMS) was developed in Germany and works like pushbroom scanner with linear array of CCD. The size of actual ground swath covered varies with pointing angle employed. At the 27° maximum value, the swath width for each instrument is 80 km. When the two instruments are pointed so as to cover adjacent image fields at nadir, the total swath width is 117 km and the two fields overlap by 3 km. While each HRV instrument is capable of collecting panchromatic and multispectral data simultaneously, resulting in four data streams, only two data streams can be transmitted at one time. Thus either panchromatic or multispectral data can be transmitted over a 117 km wide swath, but not both simultaneously.

Radargrammetry

1. Basic Geometric Concepts

Radargrammetry initially requires a mathematical model relating an object point in some three-dimensional world coordinate system to the time and range coordinates, t and r, measurable from the side-looking radar record. Radargrammetry begins with the SLR image, not with the raw electronic signals acquired by the antenna. A second radargrammetric issue is the so-called sensor model that relates the radar recording to time t and slant range r. Each specific SLR system will have a unique model.

2. Definitions

Let us assume first that we deal with a radar image as presented on film. The coordinates along the SLR film are denoted by *x*, the *y*-coordinates are defined across the film. The origin of the x coordinate may be defined arbitrarily; for example, at the beginning of the flight line. The coordinates relate to the time at which an object point has been imaged. The *x*-coordinates axis is called the along-track or azimuthal axis. The *y*-axis is the cross-tract or range direction. The *x*-coordinates are proportional to the time of flight, the y coordinate relates to range between the antenna and an object or target.

3. Viewing Geometry

Viewing geometry is a team used to describe the basic imaging geometry of a sensor without consideration of imaging errors due to platform motion, system defects, the atmosphere, and so on. Thus, only an idealized SLR image is considered. The electromagnetic pulse illuminates the terrain and creates an image line. It is evident that ground objects will be imaged into locations on the image line as a function of their distance from the antenna. This permits us to define the radar "projection" and relate it to the geometry of a map or camera photograph.

For the present purpose, a map at scale 1:1 can be assumed to be an orthogonal projection of the object or terrain. This is defined by the fact that the lines connecting image and object are all straight, parallel, and normal to the projection plane. Similarly, conventional camera photography presents a central perspective projection; that is, the projection lines connecting image and object points are all straight and all pass through one point, namely, the projection center. The SLR projection (at scale 1:1) is basically different. The projection lines in this case are concentric circles around the antenna. The basic property of the SLR projection, which sometimes also is referred to as range projection, leads to a number of differences between SLR imagery and an orthogonal projection of the imaged object.

3.1 Image of a Distance

Consider first the image of a distance E_g, extending from point *A* to point *B* in the object space. The distance E_g would be represented in its correct dimension; but in a slant range projection, we would obtain E_s. Having given a slant range presentation, the scale number *f* was defined as a constant for the imagery. If, however, the scale number of a slant range presentation is defined as the ratio between a distance in the image and the corresponding distance on the ground, then we obtain from the distance e_g and e_s at the image scale ($e_s = E_s/f$, $e_f = E_g/f$).

3.2 Image of a Vertical Structure – Relief Displacement

Let us now consider a point, A, that does not lie in the datum or reference plane above which the flying height, H, is measured but is situated on top of a vertical structure of some height.

It is obvious that, in both the ground range and slant range presentations, relief displacement, pg and ps is introduced for any point not in the reference or datum plane. We can conclude that the auxiliary angle α is nearly equal to the depression angle θ. Therefore, we obtain, in ground range presentation, and with slant ranges.

3.3 Image of a Vertical Structure – Lay Over

Another interesting fact to note is the direction of the relief displacement. From aerial photography, interpreters are used to the fact that vertical structures, like houses, fall away from the nadir point. In radar the situation is the opposite: relief displacement is towards the nadir line, N. If AB is a solid vertical structure, then A, B will be its image in a ground range presentation. This is inverted when compared to photography: the top of the structure is closer to the nadir than the bottom, and we obtain the so-called radar lay over.

3.4 Image of a Vertical Structure – Shadows

Finally, there are shadows in SLR images. We can see that no reflections will be received at the antenna during the time that signals should be returning from the area between points A and S but no objects are hit between these two points. Because the radiation is blocked by the vertical structure, no reflections are received from the ground points between A and S. On the image, area AS will be black, called a shadow of the vertical structure. The length of this shadow area is given sg in the ground range and as ss in the slant range presentations. We find

$$S_g = h/\tan\theta$$

and

$$S_s = h/\tan\theta$$

From these formulas we easily can see that shadows get longer if the structure is at smaller depression angles (if the aircraft flies at a lower altitude, H, or the distance to the object is larger) or if the height, h, increases.

3.5 Image of a Slope

Let us now consider the images of slopes. Because the relief displacement is directed towards the nadir line, slopes directed towards the antenna will appear "foreshortened", "laid over", or, as a limiting case between foreshortening and lay over, a slope might appear as a line in the image. This last possibility exists only for slopes that are parallel to the flight line and occurs if the slope angle equals the elevation angle = 90< –θ.

Three possibilities exist for slopes facing away from the flight line. First, they may be in the radar shadow; second, the radiation may only just strike the slope: or finally, it may be fully imaged. Slopes facing away from the flight line always will show a longer image than in an orthogonal map projection.

3.6 Squint

Radar pulses do not propagate along a plane vertical to the antenna's longitudinal axis but along a conical surface of which the antenna is the axis. This always will occur if the real aperture antenna is end-fed, or if a synthetic aperture is created using nonzero Doppler frequencies. Radar engineers often denote any imaging mode by "squint", in which imaging is systematically not done in a plane normal to the flight direction. We therefore may find the concept of squinted-mode imaging used for imaging with an antenna that is swung around a vertical axis.

This can be accomplished only with real apertures; with synthetic aperture the "beam" always is a cone with the velocity vector is as its axis. The cone degenerates to a plane when the zero Doppler frequency is used for image formation.

Digital Photogrammetry

Digital photogrammetry or Soft Copy photogrammetry is the latest development in the field of Photogrammetry. This has a lot of similarity with the already existing analytical Photogrammetry. Until last two or three years analytical and digital Photogrammetry terms were used synonymously, since the internal processes use the same mathematical models used for solving unknowns are same in both. Also in both the systems most of the end products are same. However, there exists a considerable difference in input, viewing system, and level of automation in some areas, when we compare these two branches of photogrammetry.

1. What is Digital or Softcopy Photogrammetry?

Also known as digital photogrammetry, softcopy uses manual and automatic techniques for capturing, processing, and analyzing digital imagery-beginning either by scanning analog film or by direct digital image capture. The core concepts and mathematical functions remain the same as in traditional photogrammetric techniques developed over the past 100 years; the main difference is that softcopy uses digital imagery. The capability of handling image data of various sources with different geometric properties in a single workstation makes the system powerfully versatile and productive.

2. Historical Perspective

Analog photogrammetry was the first stage of development for functional mapping systems. It employed optical-mechanical components and analog stereo photographic film images recorded on film or glass diapositives. The solution was achieved by means of physically altering the positions and rotations of the stages holding the film media relative to each other. Early systems produced only hardcopy maps while the advent of computers made possible modified systems capable of limited digital data recording.

Analytical photogrammetry represented the second developmental stage. Similar hardware and film components were employed, but the solution was achieved by mathematical methods using a computer to calculate solutions and record the compiled data. In the current softcopy system, analog film has been replaced by digital imagery. The costly optical-mechanical components have been completely replaced by computers and monitors. The solution is obtained mathematically using basic photogrammetric concepts and image processing techniques.

3. Softcopy Evolution

With the advent of increasingly affordable scanners, computers, and high-capacity storage, softcopy system development started in the late 1970s and was introduced to the photogrammetric community in the late 1980s. Early systems lacked high image quality and speed, but these problems were easily overcome by later computer innovations. At present, the transition from analytical to softcopy is nearly complete.

Results from numerous research and development efforts by a host of vendors have led to significant efficiencies through increasing automation of aerial triangulation, stereo model orientation, and DTM extraction. With the availability of digital cameras and satellite imagery the transition to softcopy systems has been inevitable.

The digital photogrammetry/soft copy photogrammetry involves using digital or digitized images imagery with digital viewing systems, such as a digital stereoplotter. Softcopy is changing the world of mapping. The production of orthophotography, aerial triangulation and DEM (digital elevation model) extraction are being used in a production environment by a growing number of companies.

Taken to its theoretical limit, digital photogrammetry will automate the mapping process. This should significantly increase productivity and reliability. But, since this is a technology at its infancy, there is still a lot of development necessary before the full benefits of softcopy are realized. One of the reasons is that the process mimics the conventional process currently being used in mapping. For example, photography, for the most part, still has to be obtained using conventional analog cameras.

Being a digital mapping system, one of the first tasks of the photogrammetrist is to create the digital data. Recently, four different manufacturers have introduced large-format digital photogrammetric cameras. Because it is difficult to manufacture an effective 23 × 23 cm format array of CCDs to capture the image in one array, two different methods of data collection have been devised. Leica Geosystems offers the ADS40 Airborne Digital Sensor (Figure 2.122). This camera utilizes the three-

Figure 2.122: Leica Geosystems ADS40 Digital Camera.

Figure 2.123: Z/I Imaging DMC Digital Camera.

Figure 2.124: UltraCam Digital Camera from Vexcel.

line-scanner principle to provide multiple views of the terrain in both the panchromatic and multispectral bands. Z/I Imaging has introduced the Digital Mapping Camera (DMC) system (Figure 2.123). Their approach is different than Leica's in that the camera consists of 4 high-resolution $7k \times 4k$ panchromatic and 4 multispectral $3k \times 2k$ cameras that expose their sensors simultaneously. The images are then "stitched" together. This approach by Z/I has also been used by Vexcel in their UltraCam digital camera (Figure 2.124).

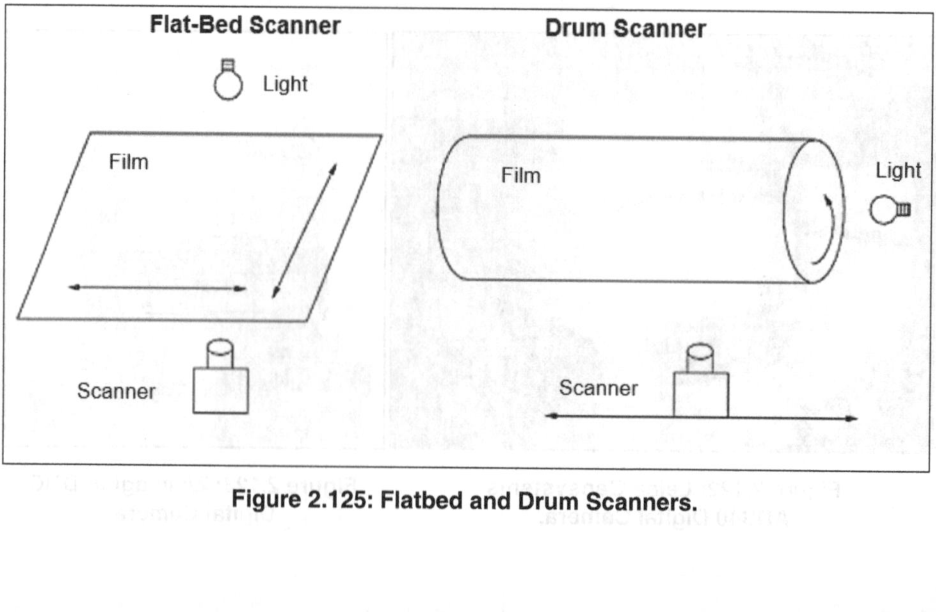

Figure 2.125: Flatbed and Drum Scanners.

If conventional film is used to acquire the image it needs to be scanned. This can be done using two different types of scanners: flatbed or drum scanners (see Figure 2.125). The flat-bed scanner is generally used for precision scanning. It has greater accuracy and better stability.

When a photo is scanned the resultant image is composed of an array or raster of picture elements called pixels or cells. The size of these cells gives us the resolution of the image. Figure 2.126 shows the effects of creating a raster image containing a straight line such as a road. On the left size, the cells are overlaid the straight line. Then, let each pixel containing the line be classified as a road. The right hand side depicts the result of this process. You will note that the road has a step-like structure. This is referred to as pixelization. If the pixels are smaller than the effects are lessened, but the storage of the data increases.

How the pixels are created depends on the scanner being used. With the flatbed scanner, the photo is placed in a plate carrier and a light is illuminated from above. Below is the actual scanner, which measures the amount of energy, received after the light has passed through the film. A digital number (DN) that represents the illumination in an area is then assigned that pixel. With the flatbed scanners, the film moves first along a row before moving to the next row.

The drum scanner operates in a similar fashion, except that the drum rotates. Once the sensor has captures a line of data, it moves longitudinally to the next line. Increasing the resolution of a digital image results in a corresponding increase in the file size. A neat model is a part of the overlap area between two successive photos that are usable for mapping. The edges of the photo are usually not usable because distortions can become larger the farther away from the center of the photo. In this case, a model of 7.2 inches by 6.6 inches is used. The latter value is in the direction of flight while the 7.2 inches is the width of the coverage. The values in table are valid for uncompressed black and white imagery. Color photography would be three times larger.

Digital photogrammetry permits vastly extended automation possibilities. The first digital photogrammetric workstations were exhibited in 1988 at the ISPRS Kyoto Congress. In addition to the capabilities offered by analytical plotters, digital workstations now permit the use of newly developed automated technology such as:

- ☆ Automated or automatic aerial triangulation.
- ☆ The derivation of digital elevation models by image matching techniques.
- ☆ Added display and analysis techniques.
- ☆ Integration into GIS systems.

A digital stereo workstation consists of the following parts:

- ☆ A central processing unit with sufficient performance of hundreds to thousand MHz.

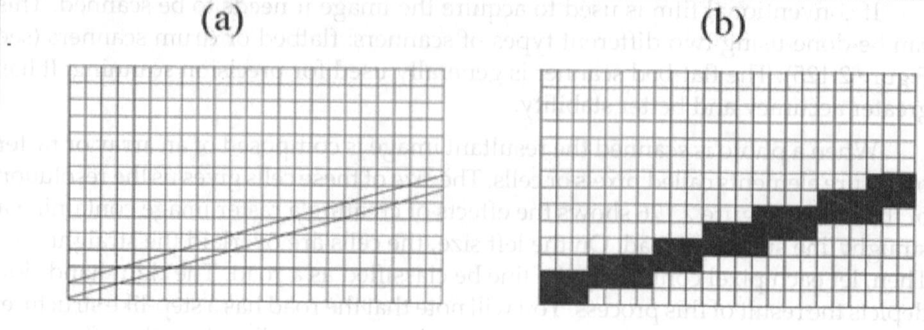

Scan Sample Rate (mm)	7.5	15	22.5	30	45	60	90	120
Dots per inch	3386	1693	1128	846	564	423	282	211
File Size in megabytes	496	124	55	31	14	7.7	3.4	1.9

Figure 2.126: Converting Vector Data into Raster Data.

- ☆ A 32-bit operating system.
- ☆ A large enough memory with at least 64Mb RAM, preferably 256 Mb or 515 Mb RAM.
- ☆ A large enough storage system to store a sufficient number of images at least in the tens of GB.
- ☆ A graphics system to permit stereo viewing and measurement.
- ☆ A graphic user interface.

The prime manufacturers of digital workstations are:

- ☆ LH-Systems with SOCET-SET
- ☆ Z/I Imaging with Image Station SSK and with Image Station 2001 or Z4 (see Figure 2.127).
- ☆ Virtuozo.

4. Principles of Digital Photogrammetry

Fundamental to the modern treatment of photogrammetry are the tools of matrix algebra and of least square adjustment. With these tools, the problems of spatial networks and their coordinate system conversions can be efficiently treated.

4.1 Coordinate Transformations between Image and Terrain

Basic to photogrammetric restitution is the conversion of two-dimensional image coordinates into three-dimensional object coordinates and vice versa. The basic relations are shown in Figure 2.128.

4.2 Image Coordinates and Local Cartesian Object Coordinates

The relations can be expressed as three-dimensional vectors between the following points:

Figure 2.127: The Z/I Imaging Image Station 2001.

- ☆ *O*, the origin of the Cartesian object coordinate system, *x*, *y*, *z*.
- ☆ P_i, the object point with its coordinates, x_i, y_i, z_i, in that system, expressed by the vector *x_i*.
- ☆ *O_*, the exposure station with its coordinates *x_o*, *y_i*, *z_o*, in that system, expressed by the vector *x_o_*.
- ☆ With these two vectors the vector OP_i forms a spatial triangle. O_P_i can be expressed by the image coordinates, *x_*, *y_* and *f* or the vector *p_i_*. Prerequisite for the perspective transformation is that the origin of the image coordinate system is linked to the projection centre of the exposure station, *O_*. Furthermore, the image coordinate system requires a spatial rotation expressed by a three-dimensional rotation matrix, *R*, and a scale change, *__i*, between image measurements and the object coordinates. The vectors in the triangle OO_P_i then can be added: *x*.

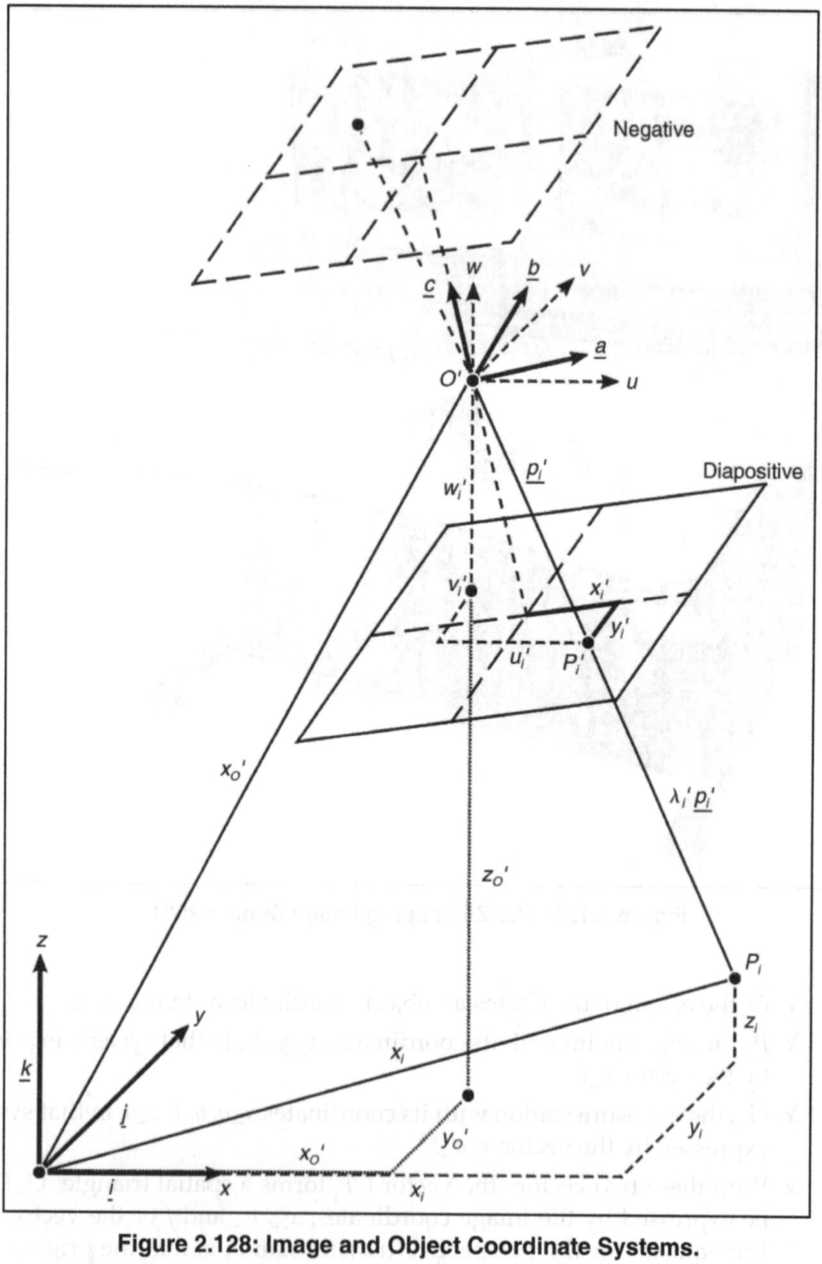

Figure 2.128: Image and Object Coordinate Systems.

5. Softcopy Advantages

Clearly softcopy techniques and systems have quietly but surely become the mainstream photogrammetric system of choice due to their inherently greater flexibility and efficiency (Table 2.12). The advantages are numerous.

Table 2.12: Functional and Structural differences between Softcopy and Analytical Photogrammetry

Functional and Structural Differences	*Softcopy*	*Analytical*
The Aerial Triangulation (AT) Process	The geometric quality of the stored image never changes; however the radiometric properties can be changed by a click of a button to accommodate the preferences of different operators and the requirements of different applications. Once the aerial triangulation process is complete, the georeferenced stereo-models can be recalled instantly.	Properties of diapositives are set once in the photo lab and cannot be altered. A stereo-model reset using an Analytical instrument can require a half hour effort to re-compute the orientations and achieve a stereo solution.
The Automatic Extraction of	The entire process is achieved on a single workstation and is mostly automated, with results in a far superior solution due to the added redundancy of Many connection points between models. The process takes a few minutes per image.	Lengthy and requires costly and bulky opticalmechanical equipment. Diapositives are physically drilled at Predefined locations to establish connections between neighboring photos and flight lines. These locations are measured along with the established ground control points, and the computations are performed using a computer based adjustment package. The process takes 45 minutes to one hour per photo.
Elevation Data	Now widely used in various Applications such as orthophoto generation and terrain analysis.	Not possible.

5.1 Economics

The cost of a complete softcopy system is a fraction of an analytical system. Software modules can be added as needed. Nearly all photo lab operations are eliminated and, when paired with a digital camera, scanning is also eliminated. Considerable additional cost reduction can be achieved due to automated processes in AT and elevation data extraction as well as in overall data collection time.

5.2 Image Quality

Various sources of digital images widen the user's scope to collect and interpret the required data more efficiently and accurately. Eliminated are data integrity and stability concerns caused by shrinkage and other degradations. Dynamic radiometric corrections allow the user to change the image characteristics to show details otherwise hidden in the image. In the case of direct digital camera recording, data is stored at a bandwidth of up to 16 bits/pixel without grain noise. This allows for better matching accuracy and success in automated processes.

5.3 Quick Response to Evolving Situations

Using today's technology in airborne GPS control and inertial measurement data, AT time can be reduced or eliminated, and mapping can often be delivered to clients in days instead of months.

5.4 Simplified Workflow

Once the model orientations are established by automatic means, stereo-imagery can be displayed instantly on the screen. This enables compilers to spend more time collecting data and less time solving orientation problems.

5.5 Versatility

The same softcopy system can be employed for different tasks and products. Add-on software modules can be easily integrated without the need for additional hardware components.

6. Future Trends

Advances in airborne GPS and inertial systems will further reduce the need for aerial triangulation and ground control. All intermediate processes including fi lm processing, scanning, and photo lab operations will be completely eliminated as the digital camera is adopted by more mainstream mapping organizations.

The time required for compilation and processing of photogrammetric mapping data will be further reduced by the final transition from analog and analytical approaches to a complete softcopy workflow (Table 2.12). Already in process elements such as automatic AT, automatic elevation data extraction, and orthophoto generation, time has been reduced to a small fraction of that required by previous methods. Further developments are underway on all these fronts.

The locus of skills needed in softcopy photogrammetry will continue to shift from the nuts-and-bolts dexterity required for operating analytical stereoplotters to the macrowriting, computer networking, and web-based operations skills complementary to the computer-based workflow. In a fully digital environment, these skills have become critical to the success of mapping organizations.

Without great fanfare softcopy techniques and systems have quietly matured to become the photogrammetric system of choice. However, the full potential of softcopy is yet to come. The quiet softcopy revolution will continue to dramatically enhance the abilities of aerial mappers to respond rapidly and efficiently to an increasing variety of client-driven geospatial needs.

Photogrammetric Equipment

The exploitation of hard copy imagery, *i.e.*, negatives, diapositives, and paper prints, required the development of special purpose instruments. Each task, such as stereoplotting, rectification, image measurement, point transfer, and triangulation would have instruments devoted to its execution. Many variations of these instruments were introduced to optimize particular aspects of the task. For example, stereoplotting instruments might be configured especially for oblique models, small formats, profiling and cross sectioning, unusual focal lengths or distortion patterns, close range imaging or small scale compilation. Dimensional stability and precision demanded that such instruments be expertly manufactured, rugged, and therefore expensive. In recent decades imagery has come to be more often collected, stored, manipulated, and presented in digital format. Very often a general purpose computer operating on digital imagery can perform the functions that used to require four or five specialized instruments using hardcopy imagery. This evolutionary change has generally yielded a gain in productivity.

One virtue of analog optical and mechanical instruments is that, by viewing the apparatus, the user could gain insight into the geometric solution of the problem. Watching a computer execute its commands does not convey the same insight into the algorithms that are being executed. A challenge for modern photogrammetrists is therefore to find other ways to convey the fundamental principles of photogrammetric process, most likely by creative simulations, animations, and computer graphics.

Most analytical photogrammetric procedures require photo coordinates as measured quantities. This, in turn, requires accurate, reliable and efficient devices for measuring points on stereo images. The accuracy depends on the application. Typical accuracies range between three and ten micrometers. Consequently, the measuring devices must meet an absolute, repeatable accuracy of a few micrometers over the entire range of the photographs that is over an area of 230 mm × 230 mm. In this section we discuss the basic functionality and working principles of analytical plotters and digital photogrammetric workstations.

1. Stereoscopic Model

The power of photogrammetry lies in the fact that stereo photographs are taken of the same scene from two different locations (see Figure 2.129). The distance between the two exposure stations is called the air base (B). H' is the flying height above the ground. Then, when viewing each photo individually, with the left eye viewing the left photo and the right eye the right photo, the photogrammetrist can see the image in three dimensions. This is a requirement for measuring height of the desired point. The two photographs are referred to as a model or stereoscopic model.

1.1 Stereoscopy

Stereoscopy is based on stereoscopic or binocular vision. Stereoscopy is the name given to the phenomenon of an apparent three dimensional "model" created by viewing two photographs of the same object, one photograph with each eye, simultaneously. When the eyes are focused on an object, the optical axes of the two eyes converge on that point intersecting at an angle called-parallactic angle. Nearer the object–greater the parallactic angle and vice a versa. This stereomodel is a subjective phenomenon and is not physically real, but is formed by the brain using the normal process associated with binocular vision Depth cues are given by object sizes, perspective, occlusion, movement parallax and change in eye focus.

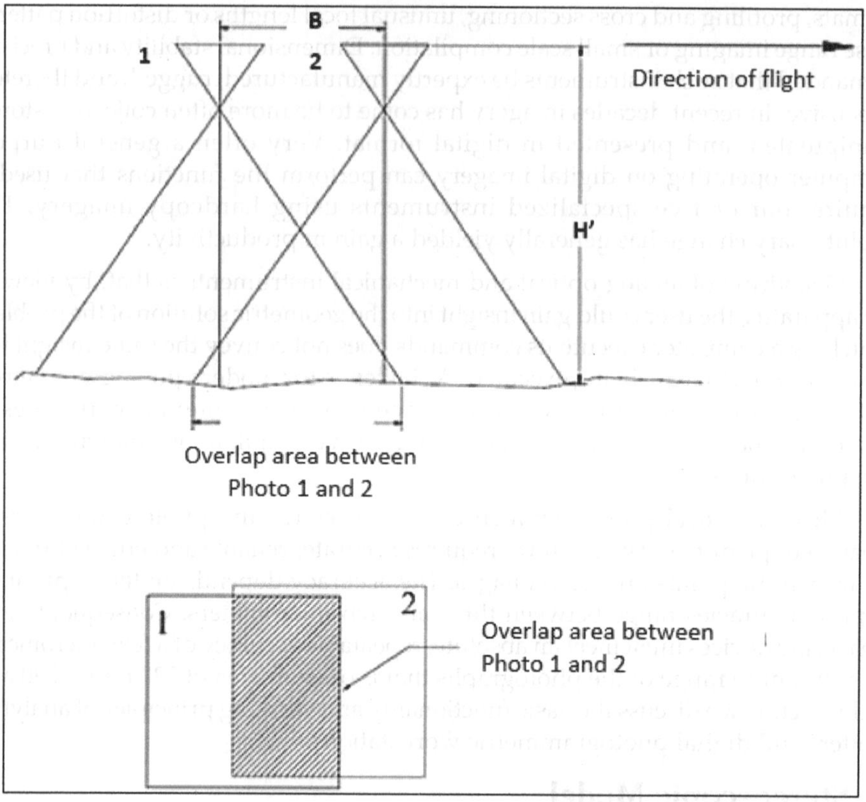

Figure 2.129: Overlapping Aerial Photography.

1.1.1 Depth Perception

Human beings can distinguish depth instinctively. However, there are many aids to depth perception, for instance, closer objects partly cover distant objects or distant objects appear smaller than similar objects nearby (Figure 2.130). These aids apply to monocular vision. For short distances binocular vision is more important and is of interest to Photogrammetrists, for it is binocular vision which enables us to obtain a spatial impression of a MODEL formed by two photographs of an object (or objects) taken from different viewpoints.

Normally, our eyes give us two slightly different views, which are fused physiologically by the brain, and result in a sensation of seeing a model having three dimensions. This three-dimensional effect, due to binocular vision, is very limited however, decreasing rapidly beyond a viewing distance of one metre. Thus it may be concluded that binocular vision is primarily an aid in controlling and directing the movements of one's limbs. A small percentage of the people do not have the facility of binocular vision and no amount of training will give it to them. Unfortunately, there is no known physical aid to provide stereoscopic sight to such person who does not possess it naturally, but training can help those having weak fusion.

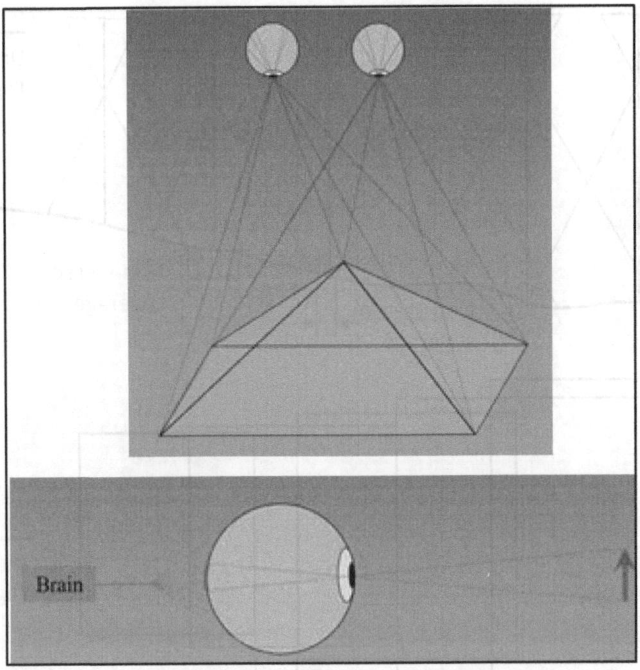

Figure 2.130: Depth Perception.

How far should the air base be? To be able to see stereo, all of the terrain needs to be viewed stereoscopically. Therefore, at least 50 per cent of each photo needs to image the same terrain features. But 50 per cent would mean that there is no room for error when flying. Thus, 60 per cent overlap is usually the minimum requirement, but this amount of overlap may need to be modified. In Figure 2.131, the effects of sloping terrain on stereoscopic imagery are shown assuming that the pilot it maintaining a constant flying height above the datum. Eventually stereoscopic viewing is lost. Moreover, the amount of detail imaged on the photo also decreases thereby requiring more flight lines closer together. While this situation is kind of unique and easily solved for by proper planning, it does show that variations in the terrain, along with fluctuations in the aircraft altitude, can affect stereoscopic coverage. Other situations may also result in a loss of stereoscopic coverage and they include excessive tilt of the camera axis at the instant of exposure and failure to maintain a constant flying height.

One of the unique and advantageous outcomes of stereophotogrammetry is that the vertical scale is exaggerated when viewed by the observer. This allows for accurate determination of height by the photogrammetrist. While increasing the flying height is advantageous for economical mapping, it will decrease the amount of vertical exaggeration. Again, if we look at our snapshots taken during our last vacation, far away objects appear flat on the image. By increasing the air base, vertical exaggeration is increased. But, there is a physical limitation here in that if the base increases too much, stereoscopic coverage is lost.

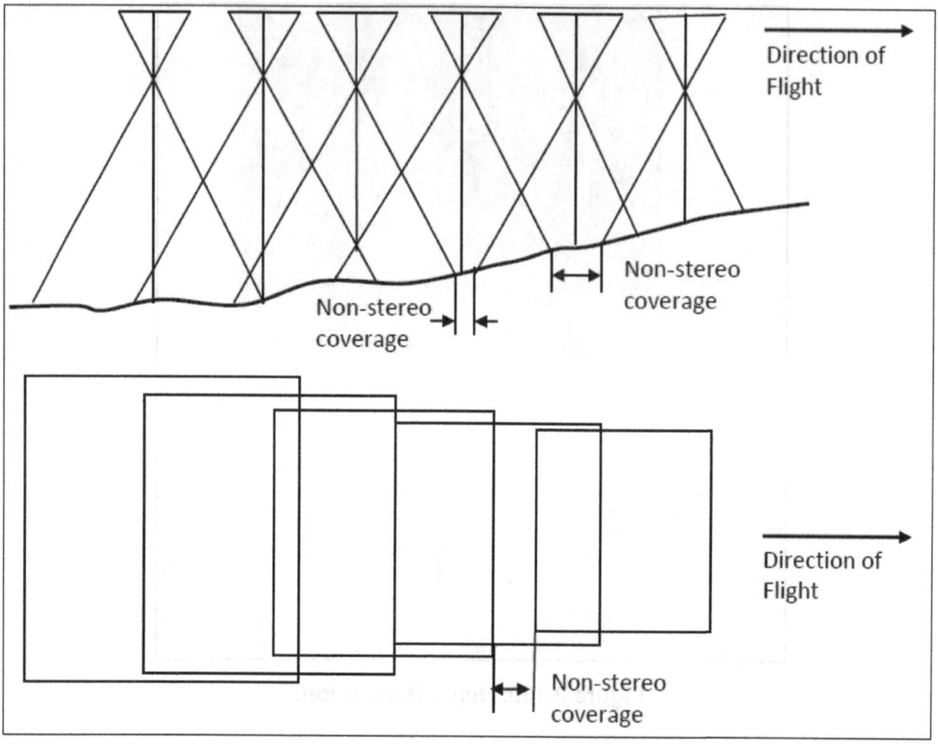

Figure 2.131: Effect of Sloping Terrain on Aerial Photography taken at a Constant Heightabove the Datum.

Parallax is the phenomenon required for stereoscopic viewing. Parallax is basically the difference in the location of a point on two subsequent photographs. In other words, the object will not appear in the exact location on two photos. This makes a lot of sense when we consider that the aircraft has moved during the two exposures. There are two components. x-parallax is the difference in the position in the x-direction or direction of flight. This component is critical in the determination of height. In fact, there is a direct correlation between the amount of x-parallax in two images and the elevation of a point. The second component is y-parallax and this is in a direction perpendicular to the flight direction. If y-parallax exists, stereoscopic viewing is not possible. Tilt in the photography is the biggest culprit in introducing y-parallax. Therefore, a critical function of the photogrammetrist is to eliminate the y-parallax within the model.

The advantages of stereoscopy, or stereoscopic vision, include the obvious – the ability to view a scene in three dimensions. It also helps in the identification of features that may exist in the scene. A feature may appear flat when viewed monoscopically but when depth is added, features can stand out from their surroundings making them easier to identify. Stereoscopy is also critical when an individual desires to extract height or contours from imagery when creating a topographic map.

1.1.2 Accommodation and Convergence

If we have a pair of stereoscopic photographs in front of us, on paper, glass plates or projected with projectors and they are oriented in such a way that epipolar lines are situated in the way described before we can observe them in different ways. In order to evaluate the different ways of observation, we have to use the terms accommodation and convergence.

Accommodation refers to focussing of eye-lens to see objects sharply at different distances. An un-accommodated eye is considered to be focussed at infinity. Convergence refers to the directing of lines of sight (*i.e.*, the optical axes) of the two eyes to the same point. The optical axis of the eye can be changed in direction by rotating the eye in its socket. The angle the eye base subtends at the point is called angle of convergence or parallactic angle.

Normal reading distance is 250 mm, *i.e.*, while reading we accommodate and converge the eyes at this distance. As the eye-base is one an average about 65 mm (2.5 inches) for human eye, the angle of convergence them is approximately 16 degrees. (The line joining the nodes of the eyes is called eye-base or the interocular or interpupillary distance (Figure 2.132). The relation between the accommodation distance (d) and angle of convergence (in radians) is given by

$$= \frac{E}{d}$$

E being the interpupillary distance

Normally accommodation and convergence are automatically linked up. If we look at a point at a certain distance, accommodation and convergence are set for that distance. We can disconnect this link but not without much strain on eyes. A lot of practice is required for accommodation at a distance other than the distance of convergence.

There are three ways of observation of stereoscopic photographs:

(a) Observation with Crossed Eye Axes

This involves looking with the right eye at the left photograph and with the left eye at the right photograph. The convergence and accommodation are at two different distances, and this type of observation is, therefore, very tiring. Large photographs can be used conveniently by this method, but due to strain on the eye, this method is not used in practice.

(b) Observation with Parallel Eye Axes

This method is possible without any optical aids, but is tiring as well as the eyes are converged on infinity, yet accommodating at approximately 250 mm. It is less tiresome if positive lenses are placed between the eyes and the photographs so that the photos are placed at the focal length of the lenses. The accommodation then corresponds with the convergence and the eyes are viewing naturally. The 'pocket-stereoscope' was developed on this principle.

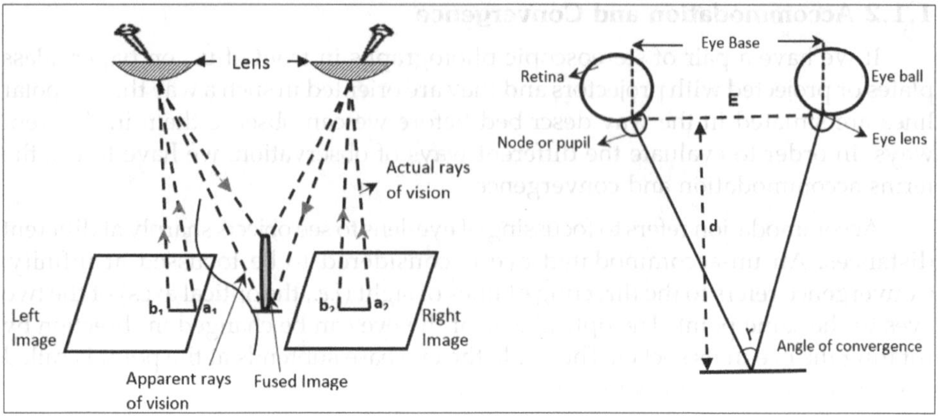

Figure 2.132: Depth Perception.

(c) Observation with Convergent Eye-Axes

When the accommodation and convergence are at the same distance the viewing is least tiring and this is the normal method of viewing. But in order to view the photos stereoscopically they must be superimposed, such that the point A and the corresponding point A′ on the other photo lie at the point of convergence. The images have to be separated so that left eye sees only the left hand photographs and the right eye only the right hand photograph. The resulting stereoscopic perception is similar to that of normal 3 dimensional perception. The separation may be achieved by colour filters or by polarized filters.

There is an interesting phenomenon in Stereoscopy. In viewing terrain in aerial photography a reversal of the relief is sometimes obtained by the eyes. Such a phenomenon is known as pseudoscopic illusion or Pseudoscopy. Such an impression can be obtained by viewing the photos with crossed eye axes. Sometimes, viewing with the shadows in case of excessive relief (*e.g.*, hills) away from the observer can also result in pseudoscopy. So, in the initial stages, to avoid pseudoscopic view, it is desirable to view the photographs with shadows of objects falling towards the observer.

1.1.3 Separation by Colour Filters

The photos are either projected or printed in two different colours. By placing a filter of the same colour over each eye corresponding picture is observed by one eye only. In practice this problem is difficult to solve completely. The human eye is sensitive for light with wavelength from 400 to 720 millimicrons (mμ). The vertex lies at about 560 mμ.

A possibility for separation of the two superimposed images would be to use filters of which one cuts off all wave length over 560 mμ (its colour would be blue-green) and the other all under 560 mμ (orange-red).

The image projected in orange-red can be observed with an orange-red glass in front of the eye. With the blue-green image it is just opposite. This means, we have on one of our retina at bluish image from one projector, and on the other a red one from the other projector. We seem to be able to fuse these different images to one stereoscopic whiteblack image.

In the case of anaglyphs printed on paper the condition is different from that described above. The two images are printed in red and blue. The eye covered with the red filter sees both the red image is indistinguishable and only the blue image is visible as varying shades of gray. Similarly the eye covered with the blue filter seems the red image only. If the spectacles are reversed, we see the LH photograph with the right eye and vice versa. A pseudoscopic image will result.

1.1.4 Separation by Polarized Filters

Light has the characteristics of a wave motion in which the waves vibrate in all possible planes perpendicular to the direction of prorogation. These are called transverse waves. It is possible to analyze the transverse waves into separate components along two axes perpendicular to each other and to the direction of propagation by means of filters.

For stereoscopic vision the filters are placed so that polarized light rays forming the left image are at right angles to the light rays forming the right image. There are several advantages in using polarized light:

- ✰ Light loss is about 50 per cent only in both projections,
- ✰ There is not colour contrast between the two picture, and
- ✰ It is possible to use colour photography on this principle.

However, there is one big disadvantage in using the method, which has so far prevented its use in photogrammetry. With the type of plotting instrument, which uses this system, it is important, that the screen on which the image is projected be diffuse, so that it can be viewed equally well from all directions but a diffuse surface acts as a depolarizer and so no stereoscopic image would be apparent.

1.2 Subjective Spatial Model

The subjective spatial model observed with a stereoscope when photographs having overlap are viewed is termed as Stereo-model. If one observes the ground from an aeroplane one does not see a spatial model. The eye base is so small (65 mm average) compared with the flying height of the aeroplane that the two-retine image is virtually the same. Therefore, there is no true comparison between the natural view and the stereoscopic view of a model.

It may be assumed that we see natural relief if we observe an object with a normal base-height ratio. In the light of what has been said before, *i.e.*, that binocular vision was mainly an aid in controlling the movements of the limbs, we could say that a normal baseheight ratio is near about 65/250, *i.e.*, about 1/4 to 1:1 or even 1:0.6, this could lead to the conclusion that the stereoscopic image formed by aerial photographs is always different and distorted. However, there are other factors, which influence the subjective model.

Assume that the photographs are taken with a vertical optical axis and that they are observed flat on a table, oriented according to the epipolar rays.

(a) The first difference is that the eye-base has been changed from say 800 mm to 65 mm. This change only alters the scale of the model and the two views remain similar in every other respect.

(b) The second difference is that the photographs are observed at a distance, which is not equal to the principal distance. This, not only, alters the magnification of the model but simultaneously alters the ratio between the *x*, *y* scales against the z scale. We get an affine flattened model if this distance is smaller than the principal distance and exaggerated if it is greater than the principal distance. This corresponds with what one finds in practice.

(c) The third difference is that our eyes are moved away from the vertical through the principal points. This produces deformations difficult to construct or visualize in a diagram.

(d) The fourth difference is that one of the photographs is moved during observation, so that the corresponding points are seem vertically. This shift is equal to the stereoscopic parallax (P), and makes the rays from the corresponding points to the respective observation. However, this parallelism renders the construction of the spatial image impossible as it means that the spatial model should be formed at infinity. In practice the image does not appear at infinity but an indeterminate distance varying from 250 mm to 1 meter according to the personal idiosyncrasies of the operator.

(e) Lastly the shape of the object, the shadows, the natural association of the observed data and relative distance all influence the process of depth perception.

2. Stereoscopes

The simplest approach to viewing stereo is with a stereoscope (see Figure 2.133). There are three common types of stereoscopes: mirror stereoscope, pocket stereoscope and Scanning stereoscope.

- ☆ The mirror stereoscope effectively widens the eye separation so that the full overlap area of a pair of aerial photographs may be viewed conveniently.
- ☆ Both the pocket and mirror stereoscopes must be physically moved over the photographs to view all parts of the available overlap.
- ☆ The scanning stereoscope also allows the full overlap area of an aerial pair to be viewed, but has a mechanical rail or optical lens system which allows scanning across the overlap area.

The purpose of the stereoscope is to force the left eye to view the left image and the right eye to view the right image. The advantage of the mirror stereoscope is that the two photos can be laid out side by side for viewing. With the pocket stereoscope, the photographs have to be overlaid onto one another by a small amount for stereo viewing. This requires the user to flip the photographs to see the

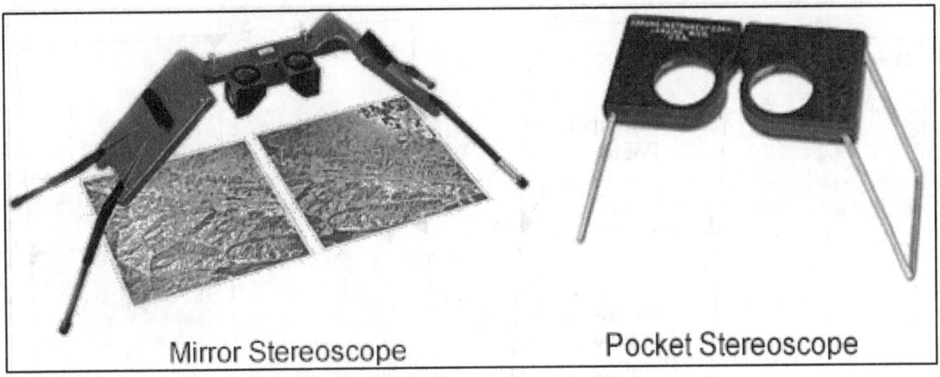

Figure 2.133: Example Stereoscopes.

area that is hidden from view. The stereoscope is aligned such that the axis between the two eyes is parallel to the direction of flight.

To illustrate this, let's look at a photograph taken from a metric camera. Located on the edges of the photography are a series of index marks, called fiducial marks (see Figure 2.134). These fiducial marks can be located midway along each side (called midside fiducials), in the corners (called corner fiducials), or in both locations. Let's assume we have midside fiducials. Now, draw lines from opposite fiducials (Figure 2.134). The point of intersection of these two lines is called the principal point.

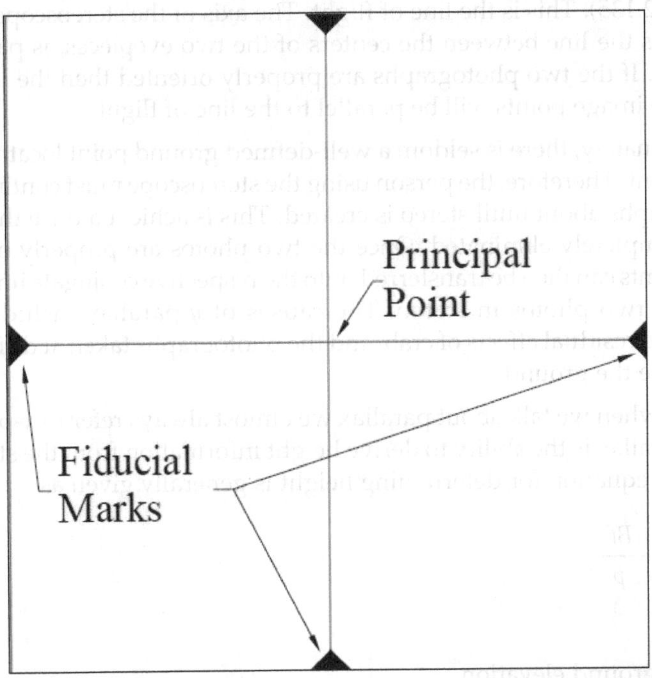

Figure 2.134: Aerial Photograph with Fiducial Axes and Principal Point Location.

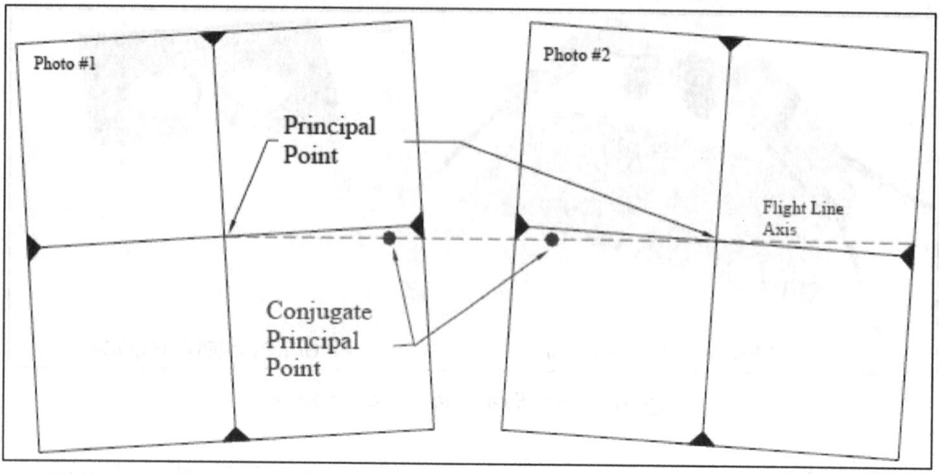

Figure 2.135: Flight Line Axis System.

Each photograph has its own principal point. Since we have 60 per cent overlap between successive photographs (as a minimum), this means that the principal point on the left photograph is imaged on the right photograph and the principal point on the right photograph is imaged on the left. These are commonly referred to as conjugate principal points. The photographs are properly aligned when the line through both principal points and both conjugate principal points is a straight line (Figure 2.135). This is the line of flight. The axis of the stereoscope, which can be defined as the line between the centers of the two eyepieces, is parallel to the line of flight. If the two photographs are properly oriented then the line between all conjugate image points will be parallel to the line of flight.

Unfortunately, there is seldom a well-defined ground point located under the principal point. Therefore, the person using the stereoscope must continually move the photographs about until stereo is created. This is achieved once the y-parallax has been completely eliminated. Once the two photos are properly oriented, the principal points can then be transferred onto the respective conjugate imagery while viewing the two photos in stereo. The causes of y-parallax include tilt in the photography, residual effects of crab, and the photographs taken at different flying heights above the ground.

Hence, when we talk about parallax we almost always refer to x-parallax. The power of parallax is the ability to derive height information from the stereo image. The parallax equation for determining height is generally given as

$$h = H - \frac{Bf}{p}$$

where,

h is the ground elevation,

H is the flying height above the datum (such as sea level),

B is the air base or ground distance measured between the two exposures,

f is the focal length of the camera, and

p is the parallax.

Example: *A pair of photographs were taken with a camera having a focal length of 152.3 mm at a height of 4,050′ above sea level. The air base was found to be 1,280′. If the parallax at a point was measured as 3.61″, what is the elevation of the point?*

Solution: First, convert the focal length from millimeters to inches

$$153.3\left[\frac{1''}{25.4\,mm}\right] = 6.035''$$

Next, use the parallax equation to find the elevation of the point.

$$h = H - \frac{Bf}{p} = 4{,}050' - \frac{1{,}280'(6.035'')}{3.61''} = 1{,}910'$$

If a second point on the photograph was found to have a parallax measurement of 3.78", its elevation becomes

$$h = 4{,}050' - \frac{1{,}280'(6.035'')}{3.78''} = 2{,}006'$$

Thus we see that as the parallax increases, so does the elevation of the point on the ground. There are other approaches to computing elevation. For example, one can use the differences in parallax to determine the height of a point, provided that the elevation of one point is known. This equation is given as:

$$h_A = h_B + \frac{\Delta P(H - h_B)}{p_a}$$

Where the subscript *A* indicates point *A* on the ground and subscript *B* is point B on the ground.

The difference in parallax is designated as $\Delta_p = p_a - p_b$. From the previous example, assume we know the elevation of the first point (h_B = 1,910′). Then, the elevation of the second point can be found as

$$\Delta_p = 3.78'' - 3.61'' = 0.17''$$

$$h_A = 1{,}910' + \frac{0.17''(4{,}050' - 1{,}910')}{3.78''} = 2{,}006'$$

3. Stereo Plotters

Stereoplotter provides rigorous, accurate solution for object point location (X, Y, Z) from images appearing in overlapping pair of photos. The three main component systems in all stereoplotters are the projection system, the viewing system, and the measuring and tracing system. Stereoplotters are most often grouped according to the type of projection system used in the instrument.

(a) Direct-optical Projection Stereoplotters

Direct-optical projection stereoplotters project a transparent positive of the original negative optically through a lens to recreate the bundle of image rays recorded by the cameras. This type of stereoplotter was the precursor of modern mechanical projection and analytical stereoplotters.

Briefly, these stereoplotters use a fixed focal length lens for projection so that the principal distance and the projection distance of the instrument are fixed. Thus, the enlargement ratio from positive to stereomodel is also fixed. For a typical "Kelsh-type" instrument, the enlargement ratio of the stereomodel is five times the contact positive scale. The stereomodel is viewed directly on a reflecting platen in the model space. The floating mark is on the platen, and as the mark is moved through the model, its position is traced directly onto a map manuscript below the mark. The vertical movement of the platen is read out at model scale as the elevation of the floating mark. Direct-optical projection stereoplotters are classified as the least accurate instrument type. They have been replaced in practice by mechanical projection and analytical stereoplotters.

(b) Mechanical Projection Stereoplotter

Mechanical projection stereoplotters are analog instruments that recreate the image ray using a metal space rod. The space rod pivots in a fixed gimbal joint representing the lens nodal point. Two space rods from adjacent diapositives intersect in the model space to define the terrain point. Since no lens is included in the projection system, the principal distance and projection distance along the space rod can be varied. Usually a small range of camera focal lengths can be accommodated, and model scale can be varied within the mechanical limits of the instrument.

(1) The viewing system of a mechanical projection stereoplotter is an optical train system of lenses and prisms similar in principle to a mirror stereoscope. The operator perceives the stereomodel by looking directly at the positive image. Image magnification is included to increase the precision of the floating mark placement in the stereomodel.

(2) The floating mark is included in the optical train. Its position in the model space corresponds to the intersection of the space rods representing the two conjugate image rays. The model carriage where the space rods intersect may be free moving. The model carriage may include a pencil mount, and the carriage is hand-driven to trace out a map manuscript. The model carriage may be connected to a mechanical drive that can be digitally encoded. Thus, the movement of the floating mark is converted to digital *x-y-z* model coordinates in three dimensions.

(3) The accuracy of mechanical projection stereoplotters will typically be in the medium to high range. Plotters designed for map compilation are in

the medium range. These plotters typically have free-moving model carriages that are hand-driven during map compilation. The Kern PG2, Galileo G6, and Wild B8 are some examples of medium-accuracy mechanical projection stereoplotters used for map compilation. Mechanical projection stereoplotters in the high-accuracy range warrant more precise measurement of the stereomodel space. In these instruments, the model carriage is mechanically driven and controlled by operator hand wheels and foot disk or similar control devices. These stereoplotters are also more massive and stable instruments than the medium-accuracy compilation instruments, and typically they are universal instruments capable of analog aerotriangulation. The Wild A8, Zeiss C8, and Jena Stereometrograph are some examples of these instrument types.

(c) Analytical Stereoplotters

Analytical stereoplotters use a mathematical image ray projection based on the collinearity equation model. The mechanical component of the instrument consists of a precise computer controlled stereocomparator. Since the photo stages must move only in the x and y image directions, the measurement system can be built to produce a highly accurate and precise image measurement. The x and y photocoordinates are encoded, and all interior and exterior orientation parameters are included in the mathematical projection model. Except for the positive format size that will fit on the photo stage, the analytical stereoplotter has no physical constraints on the camera focal length or model scale that can be accommodated.

(1) The viewing system is an optical train system typically equipped with zoom optics. The measuring mark included in the viewing system may be changeable in style, size, and color. The illumination system should have an adjustable intensity for each eye.

(2) The measuring system consists of an input device for the operator to move the model point in three dimensions. The input device is encoded, and the digital measure of the model point movement is sent to the computer. The software then drives the stages to the proper location accounting for interior and exterior orientation parameters. These operations occur in real time so that the operator, looking in the eyepieces, sees the fused image of the floating mark moving in three dimensions relative to the stereomodel surface. The operator's input device for model position may be a hand-driven free-moving digitizer cursor on instruments designed primarily for compilation, or it may be a hand wheel/foot disk control or similar device on instruments supporting fine pointing for aerotriangulation.

(3) Analytical stereoplotters are more accurate than analogic stereoplotters because the interior orientation parameters of the camera are included in the projection software. Therefore, any systematic error in the photography can be corrected in the photocoordinates before the photogrammetric projection is performed. Correcting for differential film deformations, lens distortions, and atmospheric refraction justifies

measuring the photocoordinates to accuracies of 0.003 mm and smaller in analytical stereoplotters. To achieve this accuracy, the analytical stereoplotter must have the capability to perform a stage calibration using measurements of reference grid lines etched on the photo stage.

(4) Analytical stereoplotters capable of precise map compilation and aerotriangulation measurements include (but are not limited to) the Zeiss Planicomp P series, the Wild (now Leica) AC and BC series, the Kern (now Leica) DSR series, the Intergraph Intermap, and the Galileo Digicart (Figures 2.136-2.138).

(d) Digital Stereoplotters

The next generation of stereoplotters will be digital (or soft copy) stereoplotters. These instruments will display a digital image on a workstation screen in place of a film or glass diapositive. The instrument will operate as an analytical stereoplotter except that the digital image will be viewed and measured. The accuracy of digital stereoplotters is governed by the pixel size of the digital image. The pixel size directly influences the resolution of the photocoordinate measurement. A digital stereoplotter can be classified according to the photocoordinate observation error

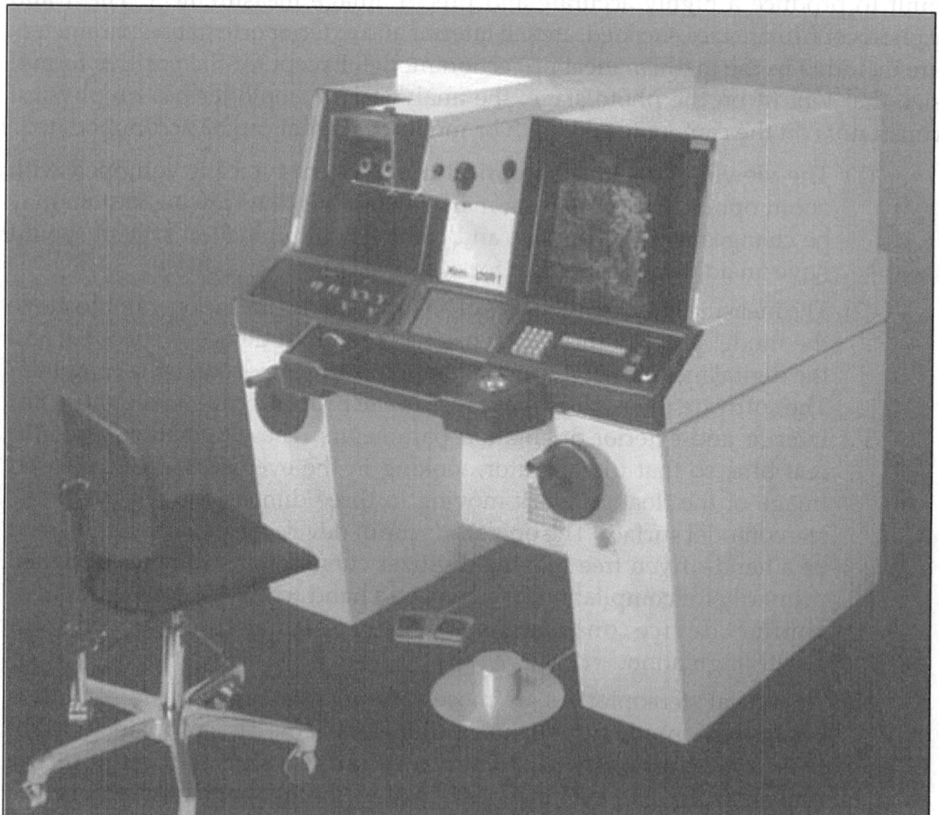

Figure 2.136: Kern DSR1 Analytical Plotter.

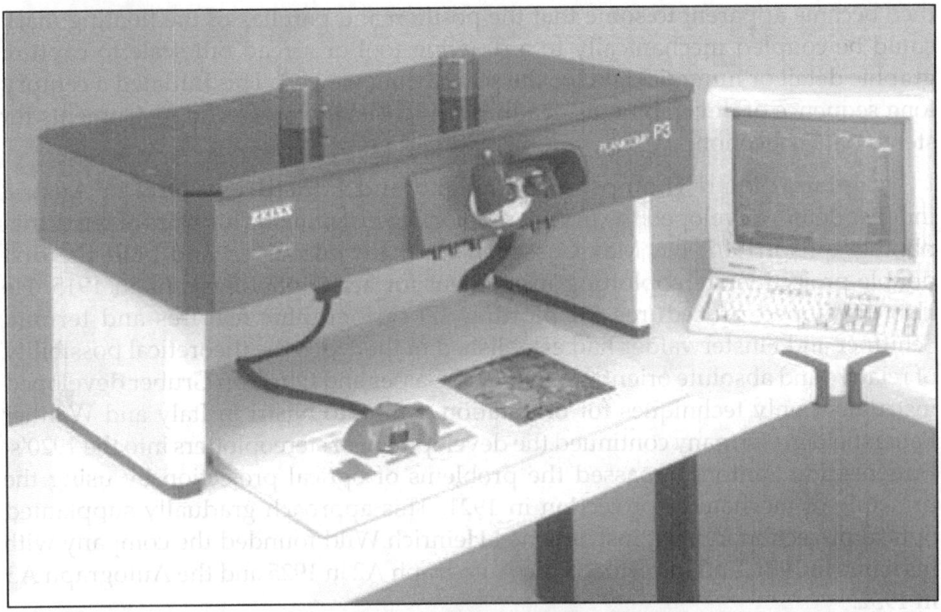

Figure 2.137: Zeiss P3 Analytical Plotter.

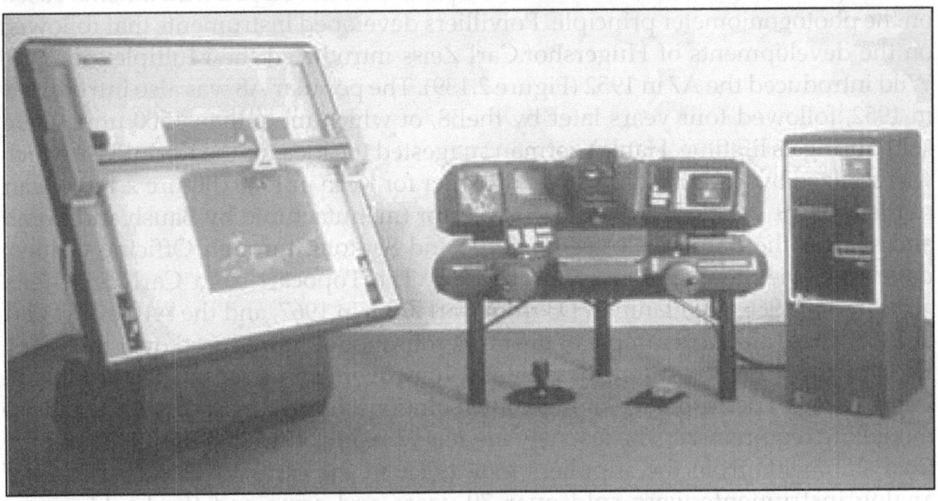

Figure 2.138: Wild AC1 Analytical Plotter.

at image scale. Then it should be comparable to an analog or analytical stereoplotter having the same observation error, and the standards and guidelines in this manual should be equally as applicable.

3.1 Types of Stereo Plotters

3.1.1 Analog Stereo Plotters

Stereoscopes and stereo viewers were staples of Victorian drawing rooms. In 1892 Stolze developed the concept of the floating measuring mark. The possibility then became apparent to some that the position and parallax of the floating mark could be coupled mechanically to a drawing tool or a read out scale to capture graphic detail or numeric data for the scene being viewed. This initiated a century long sequence of developments, leading to increasingly powerful instruments for stereo data collection.

Captain Vivian Thompson in England and Eduard von Orel in Austria independently developed the first stereoplotting instruments for pairs of terrestrial photographs in1907. Dr. Max Gasser in Germany conceived and built the first double-projection stereoplotting instrument for aerial photo-graphs in 1915. He also developed procedures for plotting 3D cartographic features and terrain. Schiffner and Finsterwalder had established in the 1890s the theoretical possibility of relative and absolute orientation. In 1915 Gasser and Otto von Gruber developed instrument only techniques for orientation. Umberto Nistri in Italy and Walther Bauersfield in Germany continued the development of stereoplotters into the 1920's. Ermenegildo Santoni bypassed the problems of optical projection by using the principle of mechanical projection in 1921. This approach gradually supplanted optical projection in most instruments. Heinrich Wild founded the company with his name in 1921, and introduced the Autograph A2 in 1925 and the Autograph A3 in 1932.

Hugershof, Fourcade, and E.H.Thompson each developed instruments based on the photogoniometer principle. Poivilliers developed instruments that followed on the developments of Hugershof.Carl Zeiss introduced the Multiplex in 1933. Wild introduced the A7 in 1952 (Figure 2.139). The popular A8 was also introduced in 1952, followed four years later by theB8, of which more than 1500 units were sold during its lifetime. Hank Yzerman suggested the idea for the KerinPG-2 which was realized by Rudolf Haller, chief designer for Kern in 1960 (Figure 2.140). Bean of the U.S.G.S. developed the ER-55 projector (manufactured by Baush and Lomb and sold as the Balplex plotter) in 1951, and Santoni, through Officine Galileo, developed the Stereocartograph V in 1963. The Topocart from Carl Zeiss Jena appeared in 1966, the Planimat D3 from Carl Zeiss in 1967, and the A10 from Wild in 1968. These are but a sample of the numerous models introduced over the years. Each of them incorporated incremental improvements over earlier models to improve price/performance, or ingenious solutions to various aspects of the precise modelling requirement. The last new models were introduced in the 1980s and the final shipments from the suppliers took place in the early 1990s. Thousands of analog instruments were sold over 70 years and remained the backbone of production photogrammetry until the 1980s.

Over the decades of development of these analog stereoplotting instruments, their capabilities evolved from the simple plotting of cartographic features, into strip triangulation, electronic coordinate readout and recording, independent model triangulation, CAD data entry, image superimposition, and constrained movement for profile collection.

The complex, precisely manufactured solutions summarized above were necessarily expensive. Over the years, cheaper alternatives were developed, from

Figure 2.139: Wild A7 Stereoplotter.

simple paper print instruments, often based on stereoscopes or sketchmaster principles, to intricate mechanical solutions designed to solve the miniaturization problem in an approximate but sufficiently accurate manner, such as the Zeiss Stereotope, Santoni Cartographic Stereomicrometer, or Cartographic Engineering CP1.

3.1.2 Analytical Plotters

3.1.2.1 Background

The analytical plotter was invented in 1957 by Helava. The innovative concept was met with reservation because computers at that time were not readily available, expensive, and not very reliable. It took nearly 20 years before the major

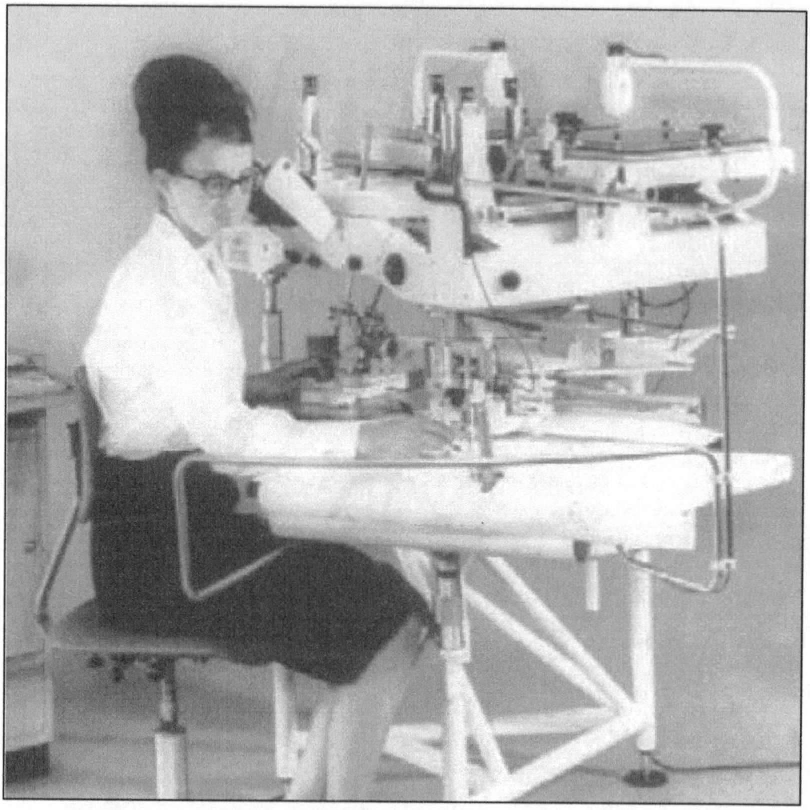

Figure 2.140: Kern PG-2 Stereoplotter.

manufacturers of photogrammetric equipment embarked on the idea and began to develop analytical plotters.

At the occasion of the ISPRS congress in 1976, analytical plotters were displayed for the first time to photogrammetrists from all over the world. Figure 2.140 shows a typical analytical plotter. Slowly, analytical plotters were bought to replace analog stereoplotters. By 1980, approximately 5,500 stereoplotters were in use worldwide, but only a few hundred analytical plotters. Today, this number increased to approximately 1,500. Leica and Zeiss are the main manufacturers with a variety of systems. However, production of instruments has stopped in the early 1990s.

3.1.2.2 System Overview

Figure 2.141 and 2.142 depicts the basic components of an analytical plotter. These components comprise he stereo viewer, the user interface, electronics and real-time processor, and host computer.

Stereo Viewer: The viewing system resembles closely a stereo comparator, particularly the binocular system with high quality optics, zoom lenses, and image rotation. Also, the measuring mark and the illumination system are refined versions of stereocomparator components. The typical viewer contains a binocular system,

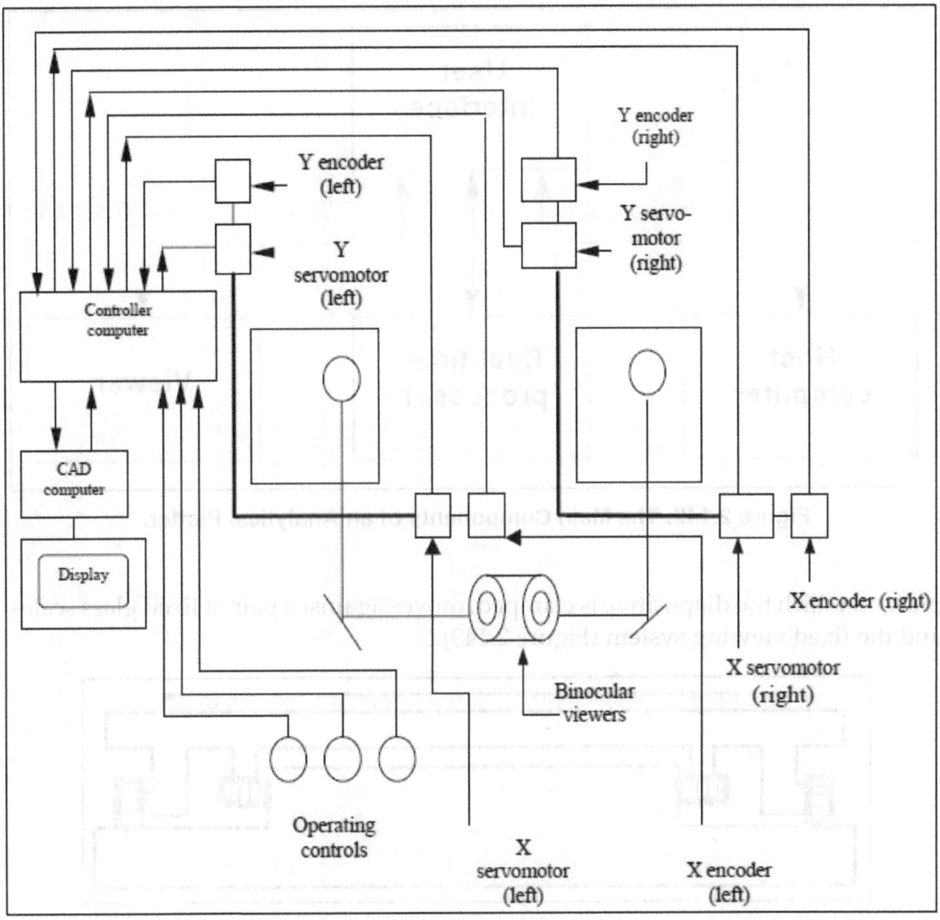

Figure 2.141: Schematic Diagram of Components and Operation of an Analytical Plotter. (*Source*: Wolf, 2000).

the stages, and the knobs for adjusting the magnification, illumination and image rotation. The size of the stages must allow for measuring aerial photographs. Some instruments offer larger stage sizes, for example 18 × 9 in. to accommodate panoramic imagery.

An important part of the stereo viewer is the measuring and recording system. As discussed in the previous section, the translation of the stages, the measuring and recording is all combined by employing either linear encoders or spindles.

Translation System: In order to move the measuring mark from one point to another either the viewing system must move with respect to a stationary measuring system, or the measuring system, including photograph, moves against a fixed viewing system. Most x-y-comparators have a moving stage system. The carrier

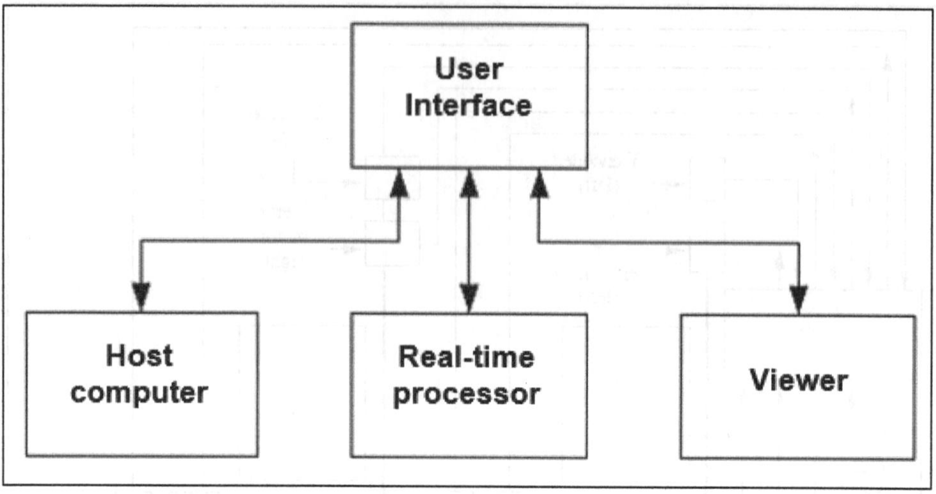

Figure 2.142: The Main Components of an Analytical Plotter.

plate on which the diapositive is clamped, moves against a pair of fixed glass scales and the fixed viewing system (Figure 2.143).

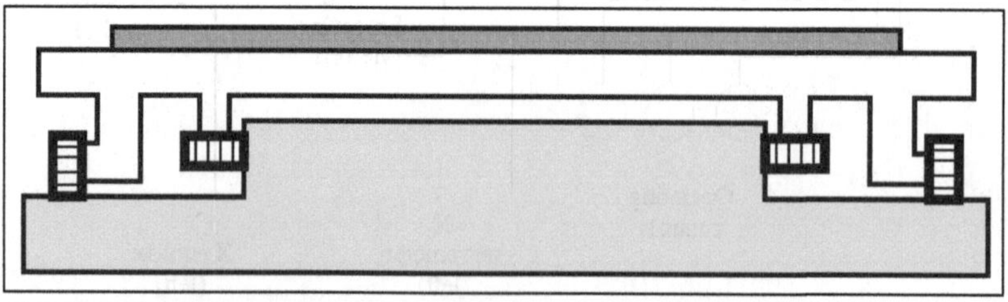

Figure 2.143: End View of Typical Translation Way.

In most cases, the linear translation is accomplished by purely mechanical means. Figure 2.144 depicts some typical translation guides. Various forms of bearings are used to reduce friction and wear and tear. An interesting solution are air bearings. The air is pumped through small orifices located on the facing side of one of two flat surfaces. This results in a thin uniform layer of air separating the two surfaces, providing smooth motion.

The force to produce motion is most often produced by threaded spindles or precision lead screws. Coarse positioning is most conveniently accomplished by a free moving cursor. After clamping the stages, a pair of handwheels allows for precise positioning.

Measuring and Recording System: If the translation system uses precision lead screws then the measuring is readily accomplished by counting the number of rotations of the screw. For example, a single rotation would produce a relative translation equal to the pitch of the screw. If the pitch is uniform, a fractional part of the rotation can be related to a fractional part of the pitch. Full revolutions are

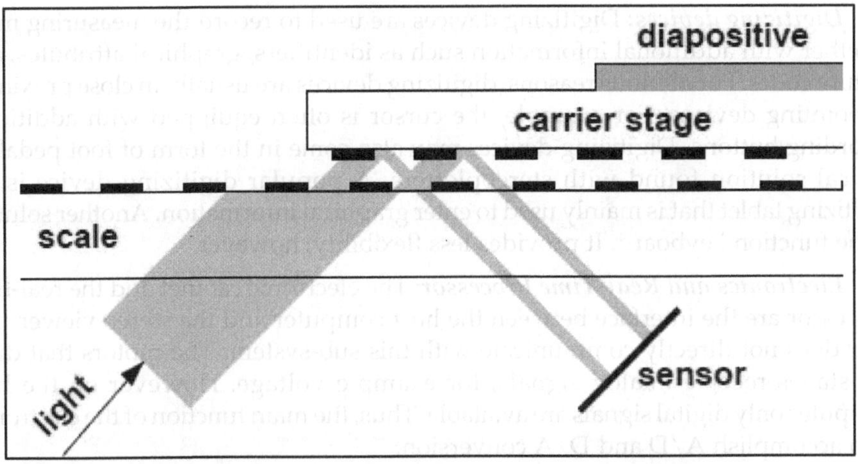

Figure 2.144: Working Principle of Linear Encoders.

counted on a coarse scale while the fractional part is usually interpreted on a separate, more accurate scale.

To record the measurements automatically, an analog to digital (A/D) conversion is necessary because the x-y-readings are analog in nature. Today, A/D converters are based on solid state electronics. They are very reliable, accurate and inexpensive.

Figure 2.144 illustrates one of several concepts for the A/D conversion process, using linear encoders. The grating of the glass scales is 40 μm. Light from the source L transmits through the glass scale and is reflected at the lower surface of the plate carrier.

A photo diode senses the reflected light by converting it into a current that can be measured. Depending on the relative position of plate carrier and scale, more or less light is reflected. As can be seen from Figure 2.144 there are two extreme positions where either no light or all light is reflected. Between these two extreme positions the amount of reflected light depends linearly on the movement of the plate carrier. Thus, the precise position is found by linear interpolation.

User Interface: With user interface we refer to the communication devices an operator has available to work on an analytical plotter. These devices can be associated to the following functional groups:

Viewer control buttons: It permit to change magnification, illumination and image rotation.

Pointing devices: These are necessary to drive the measuring mark to specific locations, *e.g.*, fiducial marks, control points or features to be digitized. Pointing devices include handwheels, footdisk, mouse, trackball, and cursor. A typical configuration consists of a special cursor with an additional button to simulate z-movement. Handwheels and footdisk are usually offered as an option to provide the familiar environment of a stereoplotter.

Digitizing devices: Digitizing devices are used to record the measuring mark together with additional information such as identifiers, graphical attributes, and feature codes. For obvious reasons, digitizing devices are usually in close proximity to pointing devices. For example, the cursor is often equipped with additional recording buttons. Digitizing devices may also come in the form of foot pedals, a typical solution found with stereoplotters. A popular digitizing device is the digitizing tablet that is mainly used to enter graphical information. Another solution is the function keyboard. It provides less flexibility, however.

Electronics and Real-Time Processor: The electronic cabinet and the real-time processor are the interface between the host computer and the stereo viewer. The user does not directly communicate with this sub-system. The motors that drive the stages receive analog signals, for example voltage. However on the host computer only digital signals are available. Thus, the main function of the electronics is to accomplish A/D and D/A conversion.

The real-time processor is a natural consequence of the distributed computing concept. Its main task is to control the user interface and to perform the computing of stage coordinates from model coordinates in real-time. This involves executing the collinearity equations and inverse interior orientation at a rate of 50 to 100 times per second.

Host Computer: The separation of real-time computations from more general computational tasks makes the analytical plotter a device independent peripheral with which the host communicates via standard interface and communication. The task of the host computer is to assist the operator in performing photogrammetric procedures such as the orientation of astereomodel and its digitization.

The rapid performance increase of personal computers (PC) and their relatively low price makes them the natural choice for the host computer. Other hosts typically used are UNIX workstations.

Auxiliary Devices: Depending on the type of instruments, auxiliary devices may be optionally available to increase the functionality. On such device is the superpositioning system. Here, the current digitizing status is displayed on a small, high resolution monitor. The display is interjected into the optical path so that the operator sees the digitized map superimposed on the stereomodel. This is very helpful for quickly checking the completeness and the correctness of graphical information.

3.1.2.3 Basic Functionality

Analytical plotters work in two modes: stereocomparator mode and model mode. We first discuss the model mode because that is the standard operational mode.

Model Mode

Suppose we have set up a model. That is, the diapositives of a stereopair are placed on the tages and are oriented. The task is now to move the measuring mark to locations of interest, for example to features we need to digitize. How do the stages move to the conjugate location?

The measuring mark, together with the binoculars, remain fixed. As a consequence, the stages must move to go from one point to another. New positions are indicated by the pointing devices, for example by moving the cursor in the direction of the new point. The cursor position is constantly read by the real-time processor. The analog signal is converted to a 3-D location. One can think of moving the cursor in the 3-D model space. The 3-D model position is immediately converted to stage coordinates. This is accomplished by first computing photo-coordinates with the collinearity equations, followed by computing stage coordinates with the inverse interior orientation. We have symbolically

X, Y, Z = derived from movement of pointing device

$x', y' = f(\text{ext.or}', X, Y, Z, c')$

$x'', y'' = f(\text{ext.or}'', X, Y, Z, c'')$

$x_m', y_m' = f(\text{int.or}', x', y')$

$x_m'', y_m'' = f(\text{int.or}'', x'', y'')$

These equations symbolize the classical real-time loop of analytical plotters. The real-time processor is constantly reading the user interface. Changes in the pointing devices are converted to model coordinates *X, Y, Z* which, in turn, are transformed to stage coordinates *xm, ym* that are then submitted to the stage motors. This loop is repeated at least 50 times per second to provide smooth motion. It is important to realize that the pointing devices do not directly move the stages. Alternatively, model coordinates can also be provided by the host computer.

Comparator Mode

Clearly, the model mode requires the parameters of both, exterior and interior orientation. These parameters are only known after successful interior and relative orientation. Prior to this situation, the analytical plotter operates in the comparator mode. The same principle as explained above applies. The real-time processor still reads the position of the pointing devices. Instead of using the orientation parameters, approximations are used. For example, the 5 parameters of relative orientation are set to zero, and the same assumptions are made as discussed earlier in relative orientation. Since only rough estimates for the orientation parameters are used, conjugate locations are only approximate. The precise determination of conjugate points is obtained by clearing the parallaxes, exactly in the same way as with stereocomparators. Again, the pointing devices do not drive the stages directly.

3.1.2.4 Typical Workflow

In this section we describe a typical workflow, beginning with the definition of parameters, performing the orientations, and entering applications. Note that the communication is exclusively through the host computer, preferably by using a graphical user interface (GUI), such as Microsoft Windows.

Definition of System Parameters

After the installation of an analytical plotter certain system parameters must be defined. Some of these parameters are very much system dependent, particularly those related to the user interface. A good example is the sensitivity of pointing devices. One revolution of a handwheel corresponds to a linear movement in the model (actually to a translation of the stages). This value can be changed.

Other system parameters include the definition of units, such as angular units, or the definition of constants, such as earth radius. Some of the parameters are used as default values, that is, they can be changed when performing procedures involving them.

Definition of Auxiliary Data

Here we include information that is necessary to conduct the orientation procedures. For the interior orientation camera parameters are needed. This involves the calibrated focal length, the coordinates of the principal point, the coordinates of the fiducial marks, and the radial distortion. Different software varies in the degree of comfort and flexibility of entering data. For example, in most camera calibration protocols the coordinates of the fiducial marks are not explicitly available. They must be computed from distances measured between them. In that case, the host software should allow for entering distances, otherwise the user is required to compute coordinates.

For the absolute orientation control points are necessary. It is preferable to enter the control points prior to performing the absolute orientation. Also, it should be possible to import a ground control file if it already exists, say from computing surveying measurements. Camera data and control points should be independent from project data because several projects may use the same information.

Definition of Project Parameters

Project related information usually includes the project name and other descriptive data. At this level it is also convenient to define the number of parallax points, and the termination criteria for the orientation procedures, such as maximum number of iterations, or minimum changes of parameters between successive iterations. More detailed information is required when defining the model parameters. The camera calibration data must be associated to the photography on the left and right stages. An option should exist to assign different camera names. Also, the ground control file name must be entered.

Interior Orientation

The interior orientation begins with placing the diapositives on the stages. Sometimes, the accessibility to the stages is limited, especially when they are parked at certain positions. In that case, the system should move the stages into a position of best accessibility. After having set all the necessary viewer control buttons, few parameters and options must be defined. This includes entering the camera file names and the choice of transformation to be used for the interior orientation. The system is now ready for measuring the fiducial marks. Based on the information in the camera file, approximate stage coordinates are computed for the stages to drive to. The fine positioning is performed with one of the pointing devices.

With every measurement improved positions of the next fiducial mark can be computed. For example, the first measurement allows to determine a better translation vector. After the second measurement, an improved value for the rotation angle is computed. In that fashion, the stages drive closer to the true position of every new fiducial mark. After the set of fiducial marks as specified in the calibration protocol is measured, the transformation parameters are computed and displayed, together with statistical results, such as residuals and standard deviation. Needless to say that throughout the interior orientation the system is in comparator mode. Upon acceptance, the interior orientation parameters are downloaded to the real-time processor.

Relative Orientation

The relative orientation requires first a successful interior orientation. Prior to the measuring phase, certain parameters must be defined, for example the number of parallax points and the type of orientation (*e.g.*, independent or dependent relative orientation).

The analytical plotter is still in comparator mode. The stages are now directed to approximate locations of conjugate points, which are regularly distributed across the model. The approximate positions are computed according to the consideration discussed in the previous section. Now, the operator selects a suitable point for clearing the parallaxes. This is accomplished by locking one stage and moving the other one only until a point is parallax free.

After six points are measured, the parameters of relative orientation are computed and results are displayed. If the computation is successful, the parameters are downloaded to the RT processor and a model is established. At that time, the analytical plotter switches to the model mode. Now, the operator moves in an oriented model. To measure additional points, the system changes automatically to comparator mode to force the operator to clear the parallaxes. It is good practice to include the control points in the measurements and computations of the relative orientation. Also, it is advisable to measure twelve or more points.

Absolute Orientation

The absolute orientation requires a successful interior and relative orientation. In case the control points are measured during the relative orientation, the system immediately computes the absolute orientation. As soon as the minimum control information is measured, the system computes approximate locations for additional control points and positions the stages accordingly.

3.1.2.5 Advantages of Analytical Plotters

The following table summarizes some of the advantages of analytical plotters over computer-assisted or standard stereoplotters. With computer-assisted plotters we mean a stereoplotter with encoders attached to the machine coordinate system so that model coordinates can be recorded automatically. A computer processes then the data and determines orientation parameters, for example. Those parameters must be turned in manually, however.

3.2 Stereoplotter Operations

All stereoplotters, whether analog or analytical, must be set up for measuring or mapping in three consecutive steps: interior orientation, relative orientation, and absolute orientation. The accuracy of Stereoplotter depends on these orientation processes. The process of setting up a stereoplotter for mapping requires three stages.

(a) Interior Orientation

The first is called interior (or inner) orientation. The purpose of interior orientation is to recreate the geometry of the taking camera. Interior orientation involves placing the photographs in proper relation to the perspective center of the stereoplotter by matching the fiducial marks to corresponding marks on the photography holders and by setting the principal distances of the stereoplotter to correspond to the focal length of the camera (adjusted for overall film shrinkage). Think of the process this way. Take two cameras out in the field and take a picture of the same area with both separated by a defined distance (air base). The axes of both cameras are parallel to one another. Lets further assume that a slide is taken much like a Polaroid picture (in other words, the film does not have to be removed for processing). Then, take the two cameras and mount them onto a stereoscopic viewing device. When the light hits the film and is projected through the lens, it would negate the distortion of the image that comes when the light travels from the terrain through the lens and onto the film. But it is impractical to have a bunch of cameras like this. Therefore, the projectors where we will place our film from the camera need to be adjusted so that they mimic the distortion of the taking camera. Interior orientation involves preparing the diapositives, accommodating for the lens distortion, centering the diapositives, and setting out the proper principal distance (focal length).

(b) Relative Orientation

Relative orientation involves reproducing in the stereoplotter the relative angular relationship that existed between the camera orientations in space when the photographs were taken. This is an iterative process and should result in a stereoscopic model easily viewed, in every part, without "*y*-parallax"– the separation of the two images so they do not fuse into a stereoscopic model. When this step is complete, there exists in the stereoplotter a stereoscopic model for which threedimensional coordinates may be measured at any point; but it may not be exactly the desired scale, and it may not be level-water surfaces may be tipped.

Once the interior orientation is performed, the stereoplotter operator needs to complete relative orientation. Here, the same relative angular relationship that existed between the two successive exposures are recreated within the stereoplotter. This is done by methodically eliminating the *y*-parallax that is present within the stereomodel. This can also be described as determining the differences in the pitch, yaw and crab between successive exposures and putting these differences in the stereoplotter. Once the relative orientation is completed, the operator is able to view the model in stereo. Recall that earlier it was pointed out that to view images stereoscopically that the *y*-parallax must be eliminated.

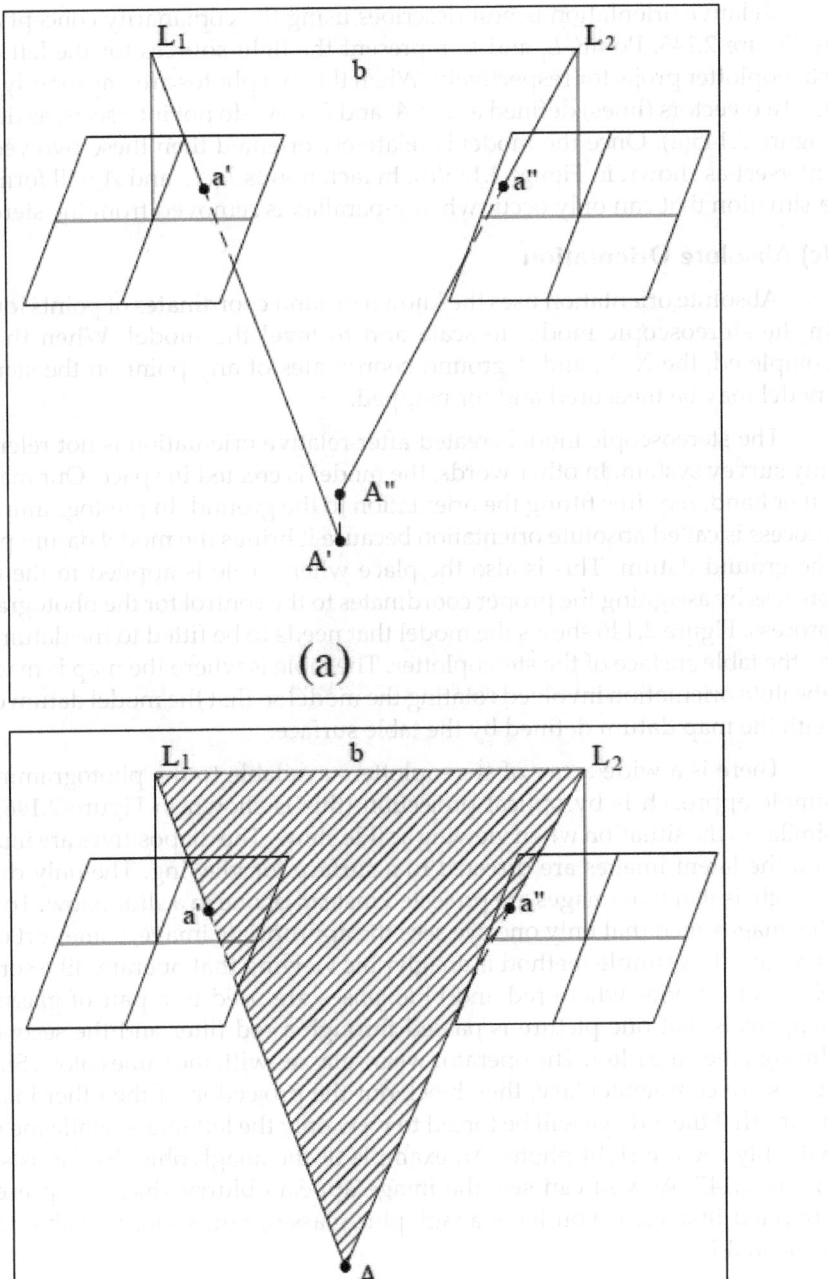

Figure 2.145: Coplanarity Concept.

Relative orientation is best describes using the coplanarity concept as shown in Figure 2.145. Points L_1 and L_2 represent the light sources for the left and right stereoplotter projector respectively. When the two photos are improperly oriented, the two vectors (lines) defined as $L_1a'A'$ and $L_2a''A''$ do not intersects, as depicted in Figure 2.145(a). Once the model is relatively oriented then these two vectors will intersect as shown in Figure 2.145(b). In fact, points L_1, L_2 and A will form a plane, a situation that can only occur when y-parallax is removed from the stereomodel.

(c) Absolute Orientation

Absolute orientation uses the known ground coordinates of points identifiable in the stereoscopic model to scale and to level the model. When this step is completed, the *X*, *Y*, and *Z* ground coordinates of any point on the stereoscopic model may be measured and/or mapped.

The stereoscopic model created after relative orientation is not referenced to any survey system. In other words, the model is created in space. Our map, on the other hand, requires fitting the orientation to the ground. In photogrammetry, this process is called absolute orientation because it brings the model datum parallel to the ground datum. This is also the place where scale is applied to the mapping process by assigning the proper coordinates to the control for the photogrammetric process. Figure 2.146 shows the model that needs to be fitted to the datum defined by the table surface of the stereoplotter. The table is where the map is made. Thus, absolute orientation involved rotating the model so that the model datum coincides with the map datum defined by the table surface.

There is a wide array of stereoplotters available to the photogrammetrist. A simple approach is by optical projection (this is shown in Figure 2.146). This is similar to the situation when viewing a slide show. The diapositives are illuminated and the latent images are directed to a surface for viewing. The only difference, though, is that two images are projected instead of one in a slide show. To separate the images such that only one eye sees the appropriate image, some sort of viewer is required. A simple method is anaglyphic viewing that operates like some of the older 3-D movies where red and blue lenses are used in a pair of glasses. What happens is that one picture is passed through a red filter and the second passes through the blue filter. The operator wears glasses with the same colors. Since these colors are complementary, they block out the projection of the other image. This means that the left eye will be forced to view only the left image while the right eye will only see the right photo. An example of the anaglyphic display is shown in Figure 2.147. As you can see, the image appears blurry since it represents two projected images. If you have anaglyphic glasses, you should be able to see the scene in 3-D.

Other systems use Polaroid filters to achieve the same result. Here the filters are oriented out of phase therefore blocking the opposite image from passing to the eye. Again, the image is passed through the Polaroid filters and the operator wears like-oriented Polaroid glasses. A third approach is to use shutters that alternate the picture display. Like the anaglyphic and Polaroid filters, the operator

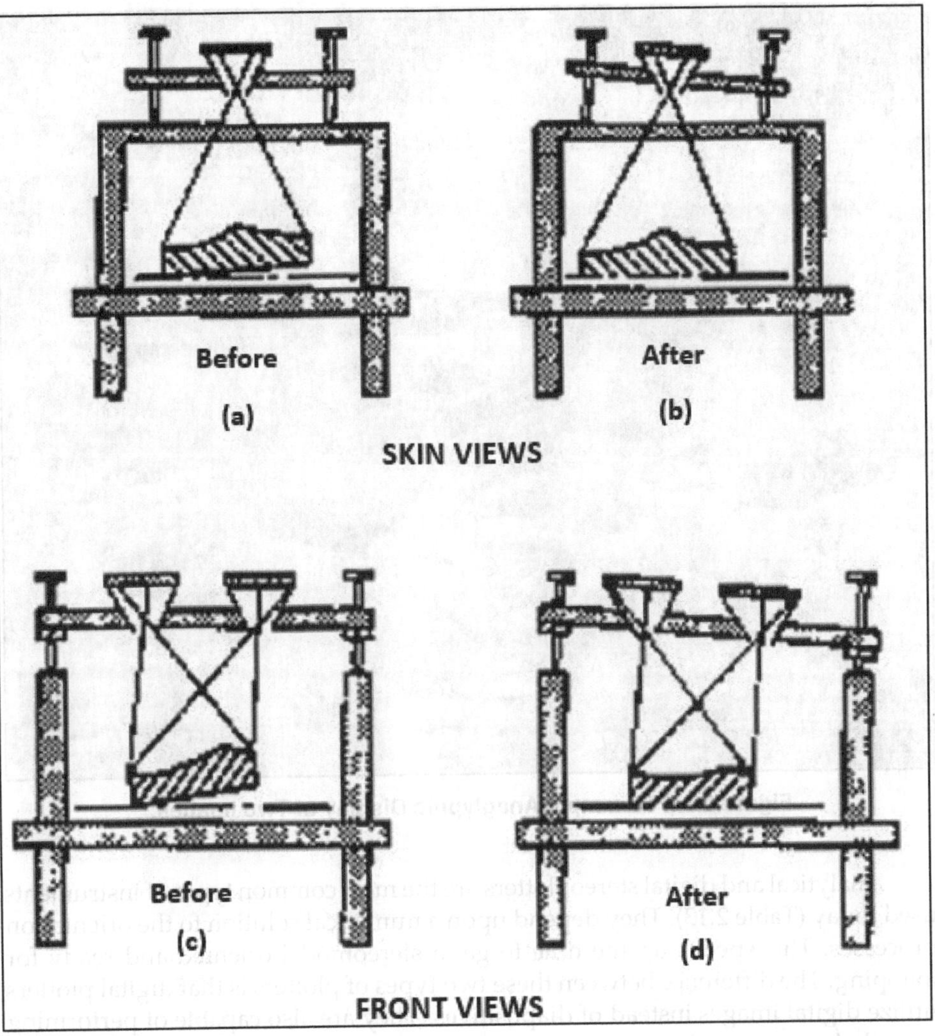

Figure 2.146: Schematic Diagram Showing the Concept of Leveling a Model.

must wear glasses with shutters to force the right eye to see only the right image and left eye the left image.

With mechanical stereoplotters the model is constructed using mechanical linkages on the instrument instead of projection of the image onto a surface. Here the viewing system is performed through an optical train. Both mechanical and optical projection plotters are referred to as analog plotters. These instruments require the operator to introduce angular changes to the projectors for the relative and absolute orientation. Analog plotters are still occasionally used but are now interfaced with computers to help expedite the orientation process. The orientation, though, are still done manually.

Figure 2.147: Example Anaglyphic Display of Two Images.

Analytical and digital stereoplotters are the most common types of instruments used today (Table 2.13). They depend upon a numerical solution to the orientation processes. This speeds up the time to get a stereomodel oriented and ready for mapping. The difference between these two types of plotters is that digital plotters utilize digital images instead of diapositives. They are also capable of performing extended operations that are not available to analytical plotters, namely the creation of digital orthophotos, automatic aerotriangulation and image correlation. While digital plotters offer more flexibility, they do have some loss in resolution of the imagery because they are scanned. Figure 2.148 is an example of an analytical plotter, the Zeiss P-3. Figure 2.149 shows an example of a digital photogrammetric workstation (SOCET SET).

Whether one is using an old analog stereoplotter or a newer analytical stereoplotter or digital workstation, the process of orientation is the same. The advantage of the latter two instruments is that they are capable of mathematically modelling the parallaxes within the model. These mathematical models make the mapping process easier to perform and increase productivity.

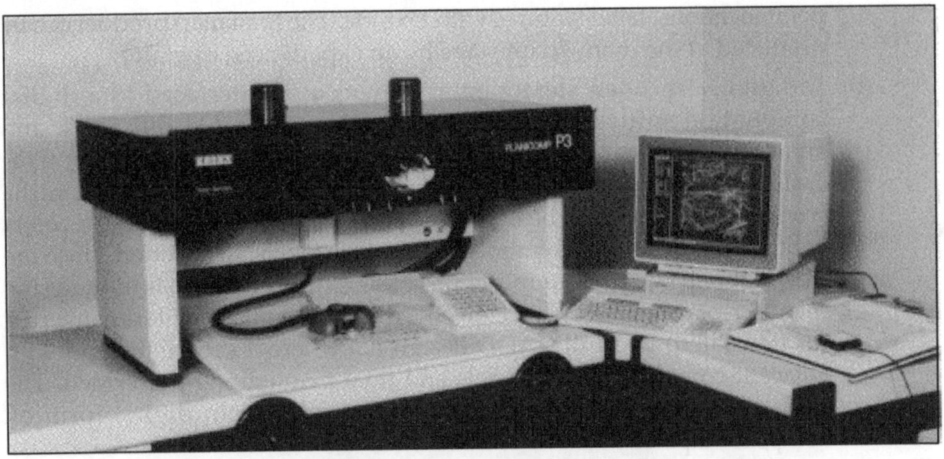

Figure 2.148: Zeiss P3 Planicomp Analytical Stereoplotter.

Figure 2.149: LH Systems SOCET SET Digital Photogrammetric Workstation.

3.3 Stereoplotter Accuracies

Stereoplotter accuracies are best expressed in terms of observation error at diapositive scale. In this way, instruments can be compared based on the fundamental measurement of image position on the positive. Measurement error on the positive can be projected to the model space so that expected horizontal and vertical error in the map compilation can be estimated. Stereoplotter accuracy affects the maximum allowable enlargement from photo scale to map scale and the minimum CI that can be plotted from given photo scale.

3.4 Stereoplotter Output Devices

Stereoplotters are connected to a variety of devices for plotting or recording line work and digital data. Although the direct tracing of map manuscripts on a stereoplotter is a form of output, it is expected that modern stereoplotters will be encoded and interfaced to a graphical or digital output device.

(a) Stereoplotters may be computer assisted. The movement of the measuring mark of these stereoplotters is digitally encoded. The digital signal is sent directly to a computer-controlled coordinatograph. The operator can include feature codes in the digital signal that the computer can interpret to connect points with the proper line weight and type, plot symbols at point locations, label features with text, etc. The final map product can be produced at the manuscript stage by an experienced operator.

(b) A digitally encoded stereoplotter may also be interfaced to a digital compilation software system in which the compiled line work and annotations appear on a workstation screen. Similar to the computer-assisted stereoplotter and coordinatograph, the compilation on the workstation screen can be displayed with proper line weights (or colors and layers), line types, symbols, and feature labels. In addition, since the map exists in a digital computer file, basic map editing can be done as the manuscript progresses. Typically, the digital compilation file is converted (or translated) to a full CADD design file where it can be merged with adjacent compiled stereomodels and final map editing performed. The deliverable product from these systems is often in digital form on computer tape or disk.

Table 2.13: Comparison Analytical Plotters/Stereoplottes

Feature	*Analytical Plotter*	*Computer Assisted Stereoplotter*	*Conventional Stereoplotter*
Accuracy			
Instrument	2 μm	≥ 10 μm	≥ 10 μm
Image refinement	Yeas	No	No
Drive to			
FM control points	Yes	No	No
Profile	Yes	Yes	Yes
DEM grid	Yes	No	No
Photography			
Projection system	Any	Only central	Only central
Size	≤ 18 × 9 in.	≤ 9 × 9 in	≤ 9 × 9 in
Orientations			
Computer assistance	High	Medium	Medium
Time	10 minutes	30 minutes	1 hour
Storing parameters	Yes	Yes	No
Range of or. parameters	Unlimited	ω, ≤ 5°	ω, ≤ 5°
Map compilation			
CAD	Many	Few	None
Time	20 per cent	30 per cent	100 per cent

4. Digital Photogrammetric Workstations

Probably the single most significant product of digital photogrammetry is the *digital photogrammetric workstation (DPW)*, also called a *softcopy workstation*. The role of DPWs in digital photogrammetry is equivalent to that of analytical plotters in analytical photogrammetry.

The development of DPWs is greatly influenced by computer technology. Considering the dynamic nature of this field, it is not surprising that digital photogrammetric workstations undergo constant changes, particularly in terms of performance, comfort level, components, costs, and vendors. It would be nearly impossible to provide a comprehensive list of the current products, which are commercially available much less describe them in some detail. Rather, the common aspects, such as architecture and functionality is emphasized.

4.1 Background

Great strides have been made in digital photogrammetry during the past few years due to the availability of new hardware and software, such as powerful image processing workstations and vastly increased storage capacity. Research and development efforts resulted in operational products that are increasingly being used by government organizations and private companies to solve practical photogrammetric problems. We are witnessing the transition from conventional to digital photogrammetry. DPWs play a key role in this transition.

Digital Photogrammetric Workstation and Digital Photogrammetry Environment

Figure 2.150 depicts a schematic diagram of a digital photogrammetry environment. On the input side we have a digital camera or a scanner with which existing aerial photographs are digitized. At the heart of the processing side is the DPW. The output side may comprise a filmrecorder to produce hardcopies in raster format and a plotter for providing hardcopies in vector format. Some authors include the scanner and filmrecorder as components of the softcopy workstation. The view presented here is that a DPW is a separate, unique part of a digital photogrammetric system.

Digital images are obtained directly by using electronic cameras, or indirectly by scanning existing photographs. The accuracy of digital photogrammetry products depends largely on the accuracy of electronic cameras or on scanners, and on the algorithms used. In contrast to analytical plotters (and even more so to analog stereoplotters), the hardware of DPWs has no noticeable effect on the accuracy. Figures 2.151 and 2.152 show typical digital photogrammetric workstations. At first sight they look much like ordinary graphics workstations. The major differences are the stereo display, 3-D measuring system, and increased storage capacity to hold all digital images of an entire project. The station shown in Figure 2.151 features two separate monitors. In this fashion, the stereo monitor is entirely dedicated to display imagery only. Additional information, such as the graphical user interface, is displayed on the second monitor. As an option to the 3-D pointing device (trackball), the system can be equipped with handwheels to more closely simulate the operation on a classical instrument.

The main characteristic of Intergraph's ImageStation Z is the 28-inch panoramic monitor that provides a large field of view for stereo display (see Figure 2.151, label 1). Liquid crystal glasses (label 3) ensure high-quality stereo viewing. The infrared emitter on top of the monitor (label 4) provides synchronization of the

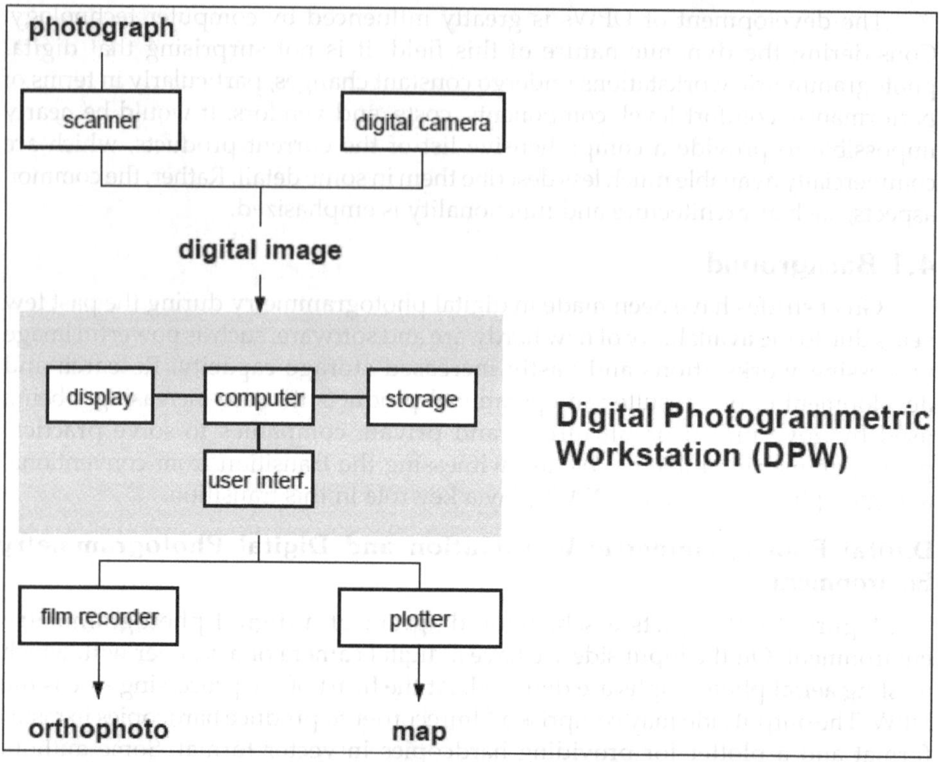

Figure 2.150: Schematic Diagram of Digital Photogrammetry Environment with the Digital Photogrammetric Workstation (Softcopy workstation) as the Major Component.

Figure 2.151: Typical Digital Photogrammetric Workstation. The system shown here offers optional handwheels to emulate operation on classical photogrammetric plotters. Courtesy LH Systems, Inc., San Diego, CA.

glasses and allows group viewing. The 3-D pointing device (label 6) allows freehand digitizing and the 10 buttons facilitate easy menu selection.

4.2 Basic System Components

Figure 2.153 depicts the basic system components of a digital photogrammetric workstation.

CPU the central processing unit should be reasonably fast considering the amount of computations to be performed. Many processes lend themselves to parallel processing. Parallel processing machines are available at reasonable prices. However, programming that takes advantage of them is still a rare commodity and prevents a more wide spread use of the workstations.

Figure 2.52: Digital Photogrammetric Workstation. Shown is Intergraph's Image Station Z. Main characteristic is the large stereo display of the 28-inch panoramic monitor. Courtesy Intergraph Corporation, Huntsville, AL.

OS the operating system should be 32 bit based and suitable for real-time processing. UNIX satisfies these needs; in fact, UNIX based workstations were the systems of choice for DPWs until the emergence of Windows 95 and NT that make PCs a serious competitor of UNIX based workstations.

Main memory due to the large amount of data to be processed, sufficient memory should be available. Typical DPW configurations have 64 MB, or more, of RAM.

Storage system must accommodate the efficient storage of several images. It usually consists of a fast access storage device, *e.g.*, hard disks, and mass storage media with slower access times.

Graphic system the graphics display system is another crucial component of the DPW. The purpose of the display processor is to fetch data, such as raster (images) or vector data (GIS), process and store it in the display memory and update the monitor. The display system also handles the mouse input and the cursor.

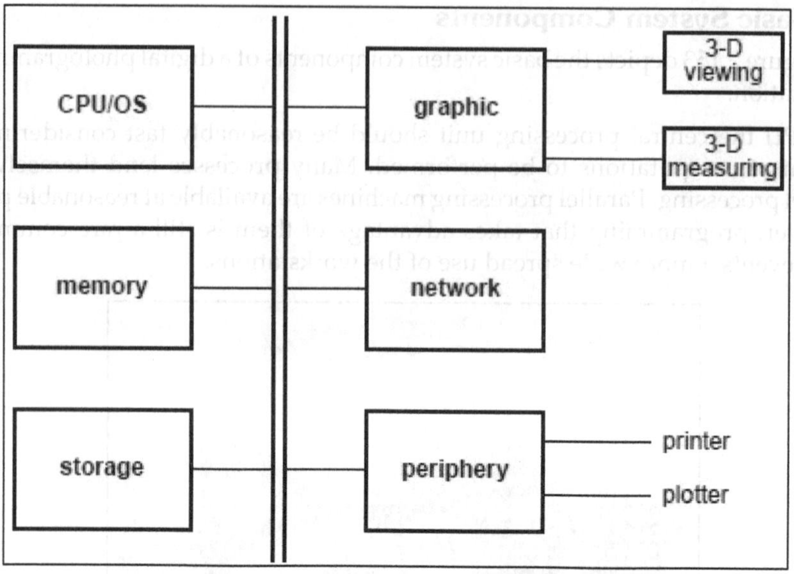

Figure 2.153: Basic System Components of a Digital Photogrammetric Workstation.

3-D viewing system is a distinct component of a DPWs usually not found in other workstations. It should allow viewing a photogrammetric model comfortably and possibly in color. For a human operator to see stereoscopically, the left and right image must be separated. Sec. 6.2.3 discusses the principles of stereo viewing.

3-D measuring device is used for stereo measurements by the operator. The solution may range from a combination of a 2-D mouse and trackball to an elaborate device with several programmable function buttons.

Network a modern DPW hardly works in isolation. It is often connected to the scanning system and to other workstations, such as a geographic information system. The client/server concept provides an adequate solution in this scenario of multiple workstations and shared resources (*e.g.*, printers, plotters).

User interface may consist of hardware components such as keyboard, mouse, and auxiliary devices like handwheels and footwheels (to emulate an analytical plotter environment). A crucial component is the graphical user interface (GUI).

4.3 Basic System Functionality

The basic system functionality can be divided into the following categories

1. *Archiving:* store and access images, including image compression and decompression.
2. *Processing:* basic image processing tasks, such as enhancement and resampling.
3. *Display and Roam:* display images or sub-images, zoom in and out, roam within a model or an entire project.

4. *3-D Measurement:* interactively measure points and features to sub-pixel accuracy.
5. *Superpositioning:* measured data or existing digital maps must be superimposed on the displayed images.

This section focus on the storage system, on the display and measuring system, and on roaming.

4.3.1 Storage System

A medium size project in photogrammetric mapping contains hundreds of photographs. It is not uncommon to deal with thousands of photographs in large projects. Assuming digital images with 16 K × 16 K resolution (pixel size approx. 13 μm), a storage capacity of 256 MB per uncompressed black and white image is required. Consider a compression rate of three and we arrive at the typical number of 80 MB per image. To store a medium size project on-line causes heavy demands on storage.

Photogrammetry is not the only imaging application with high demands on storage, however. In medical imaging, for example, imaging libraries in the terabyte size are typical. Other examples of high storage demand applications include weather tracking and monitoring, compound document management, and interactive video. These applications have a much higher market volume than photogrammetry; therefore, it is appealing for companies to further develop storage technologies. The storage requirements in digital photogrammetry can be met through a carefully selected combination of available storage technologies. The options include:

Hard disks: are an obvious choice, because of fast access and high performance capabilities. However, the high cost of disk space1 would make it economically infeasible to store entire projects on disk drives. Therefore, hard disk drives are typically used for interactive and real-time applications, such as roaming or displaying spatially related images.

Optical disks: have slower access times and lower data transfer rates but at lower cost (*e.g.*, \$10 to \$15 per GB, depending on technology). The classical CD ROM and CD-R (writable) with a capacity of approximately 0.65 GB can hold only one stereomodel. A major effort is being devoted to increasing this capacity by an order of magnitude and make the medium a rewritable one. Until such systems become commercially available (including accepted standards), CDs are used mostly as a distribution media.

Magnetic tape: offers the lowest media cost per GB (up to two orders of magnitude less than hard disk drives). Because of its slow performance (due to sequential access type), magnetic tapes are primarily used as backup devices. Recent advances in tape technology, however, make this device a viable option for on-line imaging application. Juke boxes with Exabyte or DLT (digital linear tape) cartridges (capacity of 20 to 40 GB per media) lend themselves into on-line image libraries with capacities of hundreds of gigabytes.

When designing a hierarchical storage system, factors such as storage capacity, access time, and transfer rates must be considered. Moreover, the way data is accessed, for example, randomly or sequentially, is important. Imagery requires inherently random access: think of roaming within a stereomodel. This seems to preclude the use of magnetic tapes for on-line applications. Clearly, one would not want to roam within a model stored on tape. However, if entire models are loaded from tape to hard disk, the access mode is not important, only the sustained transfer rate.

4.3.2 Viewing and Measuring System

An important aspect of any photogrammetric measuring system, be it analog or digital, is the viewing component. Viewing and measuring is typically performed stereoscopically, although certain operations do not require stereo capability.

Humans can discern 7 to 8 lp/mm at a normal viewing distance of 25 cm. To exploit the resolution of aerial films, say 70 lp/mm, it must be viewed under magnification. The oculars of analytical plotters have zoom optics that allow viewing the model at different magnifications. Obviously, the larger the magnification the smaller the field of view. Table 2.14 lists zoom values and the size of the corresponding film area that appears in the oculars. Feature extraction (compilation) is usually performed with a magnification of 8 to 10 times. With higher magnification, the graininess of the film reduces the quality of stereoviewing. It is also worth pointing out that stereoscopic viewing requires a minimum field of view.

Table 2.14: Magnification and Size of Field of View of Analytical Plotters.

Magnification	*Field of View (mm)*		
	BC 1	*C120*	*P 1*
5X		29	40
6X	32		
10X		21	
15X		14	
20X	9	10	10

Let us now compare the viewing capabilities of analytical plotters with that of DPWs. First, we realize that this function is performed by the graphics subsystem, that is, by the monitor(s). To continue with the previous example of a film with 70 lp/mm resolution, viewed 10 × magnified, we read from Table 2.14 that the corresponding area on the film has a diameter of 20 mm. To preserve the high film resolution it ought to be digitized with a pixelsize of approximately 6 μm [1000/(2 × 70)]. It follows that the monitor should display more than $3K \times 3K$ pixels. Monitors with this sort of resolution do not exist or are prohibitively expensive, particularly when considering color imagery and true color rendition (24+ bit planes).

If we relax the high resolution requirements and assume that images are digitized with a pixelsize of 15 μm, then a monitor with the popular resolution of

1280 × 1024 would display an area that is quite comparable to that of analytical plotters. Magnification, known under the more popular terms *zooming in/out*, is achieved by changing the ratio of number of image pixels displayed to the number of monitor pixels.

To zoom in, more monitor pixels are used than image pixels. As a consequence, the size of the image viewed decreases and stereoscopic viewing may be affected. The analogy to the floating point mark of analytical plotters is the three dimensional cursor that is created by using a pattern of pixels, such as a cross or a circle. The cursor must be generated by bitplane(s) that are not used for displaying the image. The cursor moves in increments of pixels, which may appear jerky compared to the smooth motion of analytical plotters. One advantage of cursors, however, is that they can be represented in any desirable shape and color.

The accuracy of interactive measurements depends on how well you can identify a feature, on the resolution, and on the cursor size. Ultimately, the pixelsize sets the lower limit. Assuming that the maximum error is 2 pixels, the standard deviation is approximately 0.5 pixel. A better sub-pixel accuracy can be obtained in

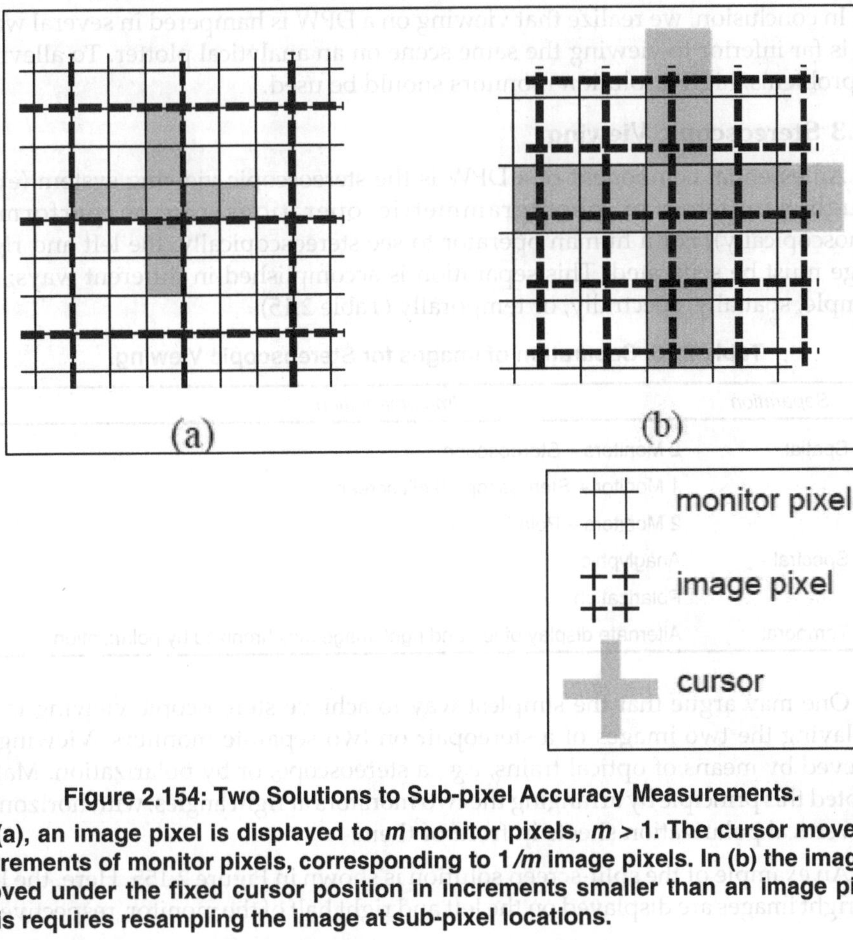

Figure 2.154: Two Solutions to Sub-pixel Accuracy Measurements.

In (a), an image pixel is displayed to *m* monitor pixels, *m* > 1. The cursor moves in increments of monitor pixels, corresponding to 1 /*m* image pixels. In (b) the image is moved under the fixed cursor position in increments smaller than an image pixel. This requires resampling the image at sub-pixel locations.

two ways. A straight-forward solution is to use more monitor pixels than image pixels. Figure 2.154(a) exemplifies the situation. Suppose we use 3 × 3 monitor pixels to display one image pixel. The standard deviation of a measurement is now 0.15 image pixels. As pointed out earlier, using more monitor pixels for displaying an image pixel reduces the size of the field of view. In the example above, only an area of 6 mm would be seen-hardly enough to support stereopsis.

To circumvent the problem of reducing the field of view, an alternative approach to sub-pixel measuring accuracy is often preferred. Here, the cursor is fixed in the monitor's center and the image is moved instead. Now the image does not need to move in increments of pixels. Resampling at sub-pixel locations allows smaller movements.

This solution requires resampling in real-time to assure smooth movement. Yet another aspect is the illumination of the viewing system; so crucial when it comes to interpreting imagery. The brightness of the screen drops to 25 per cent when polarization techniques are used. Moreover, the phosphor latency causes ghost images. All these factors reduce the image quality.

In conclusion, we realize that viewing on a DPW is hampered in several ways and is far inferior to viewing the same scene on an analytical plotter. To alleviate the problems, high resolution monitors should be used.

4.3.3 Stereoscopic Viewing

An essential component of a DPW is the stereoscopic viewing system (even though a number of photogrammetric operations can be performed monoscopically). For a human operator to see stereoscopically, the left and right image must be separated. This separation is accomplished in different ways; for example, spatially, spectrally, or temporally (Table 2.15).

Table 2.15: Separation of Images for Stereoscopic Viewing.

Separation	*Implementation*
Spatial	2 Monitors + Stereoscope
	1 Monitor + Stereoscope (split screen)
	2 Monitors + Polarization
Spectral	Anaglyphic
	Polarization
Temporal	Alternate display of left and right image synchronized by polarization

One may argue that the simplest way to achieve stereoscopic viewing is by displaying the two images of a stereopair on two separate monitors. Viewing is achieved by means of optical trains, *e.g.*, a stereoscope, or by polarization. Matra adopted this principle by arranging the two monitors at right angles, with horizontal and vertical polarization sheets in front of them.

An example of the split-screen solution is shown in Figure 2.155. Here, the left and right images are displayed on the left and right half of the monitor, respectively.

Figure 2.155: Example of a Split-Screen Viewing System. Shown is the DVP digital photogrammetric workstation. Courtesy of DVP Geomatics, Inc., Quebec.

A stereoscope, mounted in front of the monitor, provides viewing. Obviously, this solution permits only one person to view the model. A possible disadvantage is the resolution, because only half of the screen resolution is available for displaying the model.

The most popular realization of spectral separation is by anaglyphs. The restriction to monochromatic imagery and the reduced resolution outweigh the advantage of simplicity and low cost. Most systems today use temporal separation in conjunction with polarized light. The left and right image is displayed in quick succession on the same screen. In order to achieve a flicker-free display, the images must be refreshed at a rate of 60 Hz per image, requiring a 120 Hz monitor.

Two solutions are available for viewing the stereo model. As illustrated in Figure 2.156 (a), a polarization screen is mounted in front of the display unit. It polarizes the light emitted from the display in synchronization with the monitor. An operator wearing polarized glasses will only see the left image with the left eye as the polarization blocks any visual input to the right eye. During the next display cycle, the situation is reversed and the left eye is prevented from seeing the right image.

The second solution, depicted in Figure 2.156 (b), is more popular and less expensive to realize. It is based on active eyewear containing alternating shutters, realized, for example, by *liquid crystal displays* (LCD). The synchronization with the screen is achieved by an infrared emitter, usually mounted on top of the monitor. Understandably, the goggles are heavier and more expensive compared to the simple polarizing glasses of the first solution. On the other hand, the polarizing screen and the monitor are a tightly coupled unit, offering less flexibility in the selection of monitors.

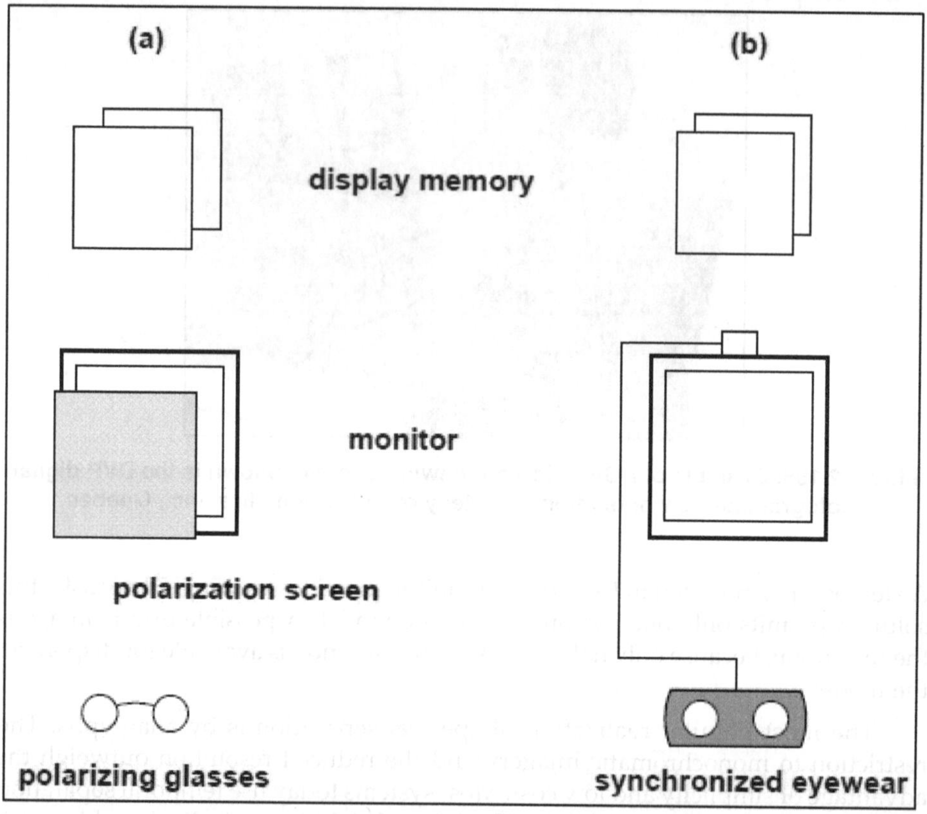

Figure 2.156: Schematic Diagram of the Temporal Separation of the Left and Right Image of a Stereopair for Stereoscopic Viewing.

In (a), a polarizing screen is mounted in front of the display. Another solution is sketched in (b). The screen is viewed through eyewear with alternating shutters. See text for detailed explanations.

4.3.4 Roaming

Roaming refers to moving the 3-D pointing device. This can be accomplished in two ways. In the simpler solution, the cursor moves on the screen according to the movements of the pointing device (*e.g.*, mouse) by the operator. The preferred solution, however, is to keep the cursor locked in the screen center, which requires redisplaying the images. This is similar to the operation of analytical plotters where the floating point mark is always in the center of the field of view.

The following discussion refers to the second solution. Suppose we have a stereo DPW with a 1280 × 1024 resolution, true color monitor, and imagery digitized to 15 μm pixelsize (or approximately 16*K* × 16*K* pixels). Let us now freely roam within a stereomodel, much as we would do it on an analytical plotter and analyze the consequences in terms of transfer rates and memory size.

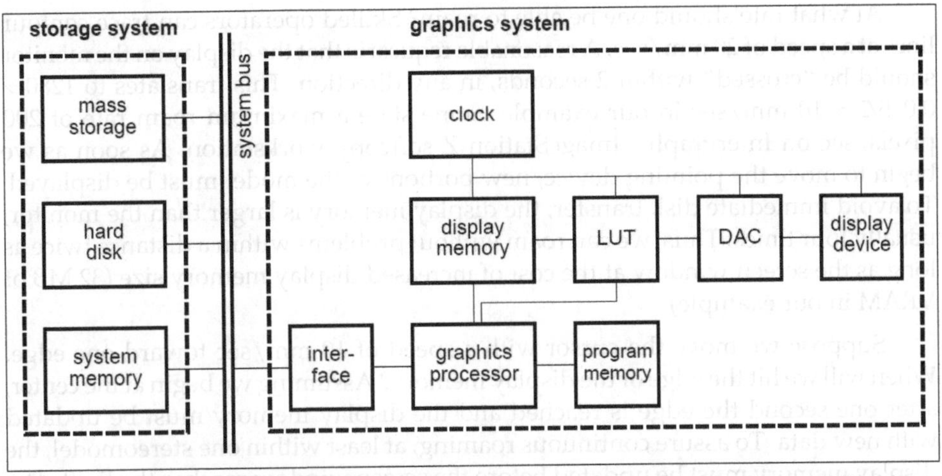

Figure 2.157: Schematic Diagram of Storage System, Graphic System and Display.

Figure 2.157 schematically depicts the storage and graphic systems. The essential components of the graphic system include the graphics processor, the display memory, the digital-to-analog converter (DAC), and the display device (CRT monitor in our case).

The display memory contains the portion of the image that is displayed on the monitor. Usually, the display memory is larger than the screen resolution to allow roaming in real-time. As soon as we roam out of the display memory, new image data must be fetched from disk and transmitted to the graphics system.

Graphic systems come in the form of high-performance graphics boards, such as Realizm or Vitec boards. These state-of-the-art graphics systems are as complex as the system CPU. The interaction of the graphics system with the entire DPW, *e.g.*, requesting new image data, is a critical measure of system performance. Factors such as storage organization, bandwidths, and additional processing cause delays in the stereo display. Let us further reflect on these issues. With an image compression rate of three, approximately 240 MB are required to store one color image. Consequently, a 24 GB mass storage system could store 100 images on-line. By the same token, a hard disk with 2.4 GB capacity could hold 10 compressed color images.

Since we request true color display, approximately 2×4 MB are required to hold the two images of the stereomodel6. As discussed in the previous section, the left and right image must be displayed alternately at a frequency of 120 Hz to obtain an acceptable mode. The bandwidth of the display memory amounts to $1280 \times 1024 \times 3 \times 120 = 472$ MB/sec. Only high speed, dual port memory, such as VRAM (video RAM) satisfies such high transfer rates. For less demanding operations, such as storing programs or fonts, less expensive memory is used in high performance graphic workstations.

At what rate should one be able to roam? Skilled operators can trace contour lines at a speed of 20 mm/sec. A reasonable request is that the display on the monitor should be "crossed" within 2 seconds, in any direction. This translates to 1280 × 0.015/2 ≈ 10 mm/sec in our example. Some state a maximum roam rate of 200 pixels/sec on Intergraph's ImageStation Z softcopy workstation. As soon as we begin to move the pointing device, new portions of the model must be displayed. To avoid immediate disk transfer, the display memory is larger than the monitor, usually four times. Thus, we can roam without problems within a distance twice as long as the screen window at the cost of increased display memory size (32 MB of VRAM in our example).

Suppose we move the cursor with a speed of 10 mm/sec toward one edge. When will we hit the edge of the display memory? Assuming we begin at the center, after one second the edge is reached and the display memory must be updated with new data. To assure continuous roaming, at least within one stereomodel, the display memory must be updated before the screen window reaches the limit. The new position of the window is predicted by analyzing the roaming trajectory. A look-ahead algorithm determines the most likely positions and triggers the loading of image data through the hierarchy of the storage system.

Referring again to our example, we have one second to completely update the display memory. Given its size of 32 MB, data must be transferred at a rate of 32 MB/sec from hard disk via system bus to the display memory. The bottle necks are the interfaces, particularly the hard disk interface. Today's systems do not offer such bandwidths, except perhaps SCSI-2 devices. A PCI interface (peripheral component interface) on the graphics system will easily accommodate the required bandwidth.

A possible solution around the hard disk bottleneck is to dedicate system memory for storing an even larger portion of the stereomodel, serving as sort of a relay station between hard disk and display memory. This caching technique, widely used by operating systems to increase the efficiency of data transfer disk to memory, offers additional flexibility to the roaming prediction scheme. It is quite unlikely that we will move the pointing device with a constant velocity across the entire model (features to be digitized are usually confined to rather small areas). That is, the content of the system memory does not change rapidly.

Figure 2.158 depicts the different windows related to the size of a digital image. In our example, the size of the display window is 19.2 mm × 15.4 mm, the display memory size is 4× larger, and the dedicated system memory again could be 4× larger. Finally, the hard disk holds more than one stereopair.

4.4 Analytical Plotters vs. DPWs

During the discussion of the basic system functionality and the orientation procedure, the advantages and disadvantages of DPWs became apparent. Probably the most severe shortcoming of today's DPWs is the viewing and roaming quality, which is far inferior to that of analytical plotters. Consider moving the floating point mark in an oriented model on the analytical plotter. Regardless on how quickly

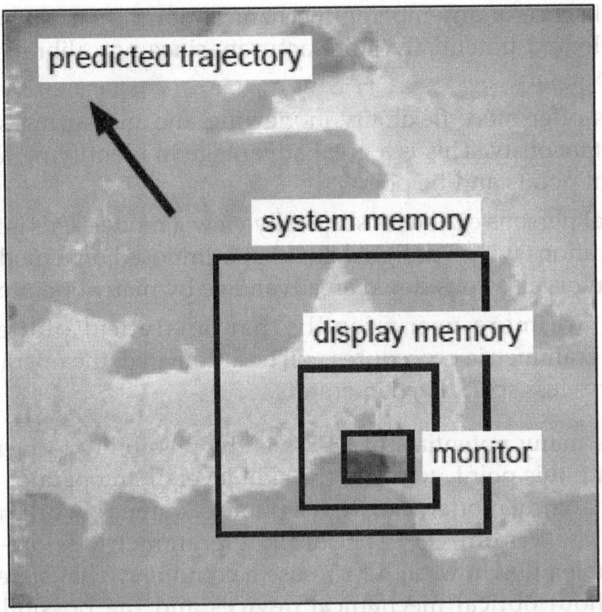

Figure 2.158: Schematic Diagram of the different Windows Related to the Size of an Image. Real-time roaming is possible within the display memory. System memory holds a larger portion of the image. The location is predicted by analyzing the trajectory of recent cursor movements.

you move, the model is always there, viewed at superb quality. In contrast, DPWs behave poorly when an attempt is made to move the cursor from one model boundary to another.

Several factors influence the viewing quality. For one, the monitor resolution sets a limit to the size of the field of view. Small fields of view reduce the capability to stereoscopically view a model. This makes image interpretation more difficult. Flickering, particularly noticeable on large monitors, is still a nuisance, despite of 120 Hz refresh rates. A third factor is the reduction in illumination due to polarizing and alternate image displays. Finally, the ease and simplicity of optical image manipulation, such as rotation, cannot be matched on DPWs. Resampling is a time consuming process and may even reduce the image quality.

The advantages of DPWsoutweigh these shortcomings by far, however. The benefits are well documented. The following are a few noteworthy factors:

- ☆ Image processing capabilities are available at the operator's fingertips. Enlargements, reductions, contrast enhancements, and dodging do not require a photo lab anymore-DPWs have a built-in photo lab.
- ☆ Traditional photogrammetric equipment, such as point transfer devices and comparators, are no longer required-their functionality is assumed by DPWs. Digital photogrammetric workstations are much more universal than analytical plotters.

- ☆ The absence of any moving mechanical-optical parts make DPW more reliable and potentially more accurate since no calibration procedures are necessary.
- ☆ DPWs offer more flexibility in viewing and measuring several images simultaneously. This is a great advantage in identifying and measuring control points and tie points.
- ☆ Several persons can stereoscopically view a model. This is interesting for applications where design data is superimposed on a model. Free stereo viewing is also considered an advantage by many operators.
- ☆ DPW are more user friendly than analytical plotters. As more photogrammetric procedures will be automated, the operation of a DPW requires less specialized operators.

Among the many potentials of DPWs is the possibility to increase the user base. To illustrate this point, compare the skill level of an operator working on a stereoplotter, analytical plotter, and digital photogrammetry workstation. There is clearly a trend away from very special photogrammetric know-how to more generally available knowhow on how to use a computer. That stereo models can be viewed without optical mechanical devices and the possibility to embed photogrammetric processes in user friendly graphical user interfaces raises the chances that non-photogrammetrists can successfully use photogrammetric techniques.

It is possible to increase the roaming capabilities of DPWs beyond the stereomodel. Roaming should be performed in the entire project and it should include vector data, DEMs, databases, and design data. Such a generalized roaming scheme would further increase the efficiency and user friendliness of DPWs. One of the biggest advantages, however, lies in the potential to automate photogrammetric applications, such as aerial triangulation, DEM generation, and orthophoto production.

5. The Need for Integration

Many GIS application rely on accurate, up-to-date digital maps, but convectional techniques for producing these vital topographic resources are time-consuming and have become costly. New digital techniques under the umbrella of DP are emerging to provide cost-effective solutions (Schroth, 1999). Currently, digital technology is still in a development stage and not matured or robust enough to replace all analytical counterparts completely. A mixed analytical and digital photogrammetric mapping environment will continue to be used for the foreseeable future. A survey conducted in some mapping organizations shows that, both digital and analytical systems are at present being engaged in their production line. Refer to appendix A-2 for sample questionnaire of the survey.

Intermap Limited uses LH Systems DPWs and Zeiss $P_1/P_3/P_{33}$ analytical plotters. At Intermap, digital systems are mainly used for orthorectification and AT, while analytical systems are used for map compilation. DTM generation is performed on either system depending on the terrain; however, characteristic features such as breaklines, boundary areas and abrupt changes are still digitised manually on analytical plotters.

Swiss Federal office of Topography has 6 analytical plotters, BC3, from Leica and 2 digital stereoplotters from LH systems with Socet Set. Digital plotters are being used for automatic AT and orthophoto with mosaicking, however, every step has to be checked and sometimes improved manually. To the Swiss Federal office of Topography, road and building recognition and reconstruction to this day defy automation.

Hansa Luftbild, Germany Air Surveys, has 4 Zeiss P_3/P_{33} and 1 Imagestation II. In terms of digital photogrammetric production they can boast of 100 per cent orthophoto, 50 per cent DTM generation, 30 per cent AT and 5 per cent restitution. Analytical system is still used for AT right down to restitution depending on terrain type, cost and customer's requirements.

Besides, INPHO GmbH a producer (manufacturer) of digital photogrammetric systems comments that identifying natural ground control points seems to be rather difficult to automate. Although operators prefer digital photogrammetric systems for operational reason (flexibility), for map compilation, analytical systems have their own advantages. This is due to high-resolution optics hence reliable interpretation is achieved.

From this survey, it was realised that, DP technologies do not offer significant performance advantages for feature collection at the present time. Automatic or semi-automatic feature extraction is still in the research domain. Feature collection is still an interactive manual process performed on both the analytical plotter and DPW instruments. However, superimposition of vector data onto stereo view screen in a heads-up environment is a significant advantage for map updating/revision and quality control, which is very easy and less expensive on the DPWs. What is more, orthophoto can be performed more efficiently with digital devices due to the degree of automation.

Difficult terrain conditions and poor image quality makes it very difficult if not impossible to automatically generate DTM by auto-correlation1. The time spent on post-editing and quality assurance will negate the time saved by auto-correlation. An alternative way to automate DTM generation is to use semi-automatic methods where the digital system drives the cursor to pre-set locations for measurement.

Automatic breaklines extraction functions are not mature enough to be effective. The algorithms do not extract enough breaklines or the extracted lines are not accurate enough. Manual breakline collection is still necessary. In all cases visual verification and post-editing is still required for quality control and quality assurance. This process is the most time-consuming part of the whole DTM production process.

It is worth noting that while photogrammetric mapping is moving to the digital domain, not all related firms and agencies are using digital AT. Instead, many of them still apply analytical AT for image orientation and move the rest of the processing to DPW equipment. Also, most Mapping Companies have invested so much in the analytical systems, they still have competent operators, the systems are still giving good results hence cannot completely throw it away. There must be a transition period in which the digital systems will gradually replace the analytical systems. Consequently, there is the need to run both systems and find an ideal way to integrate them.

Close-Range Photogrammetry: Applications

The ever-expanding areas of application of close-range photogrammetry can be grouped into three major areas: architectural photogrammetry, biomedical and bioengineering photogrammetry (biostereometrics) and industrial photogrammetry.

1. Architecture

It is noteworthy that the very first measurements ever made by photogrammetry (in the middle of the 19th century) had to do with monuments. It is also a fact that an architect introduced the term "photogrammetry". Albrecht Meydenbauer, who made his first photogrammetric surveys in 1867. For over century, photogrammetric methods and equipment have continued to evolve. More recently, the field of architectural application of photogrammetry has undergone considerable expansion both in scope and diversity.

1.1 Surveys of Historical Monuments

Photogrammetric surveys of historic monuments can be grouped in three major categories: rapid and relatively simple surveys, accurate and complete surveys, and very accurate surveys.

(a) Rapid and Relatively Simple Surveys

These are used in preliminary studies for restoration and improvement, in inventory work, and in the study of the history of art. Stereometric cameras and other small format photogrammetric cameras are used extensively, together with "normal case stereo plotters". Plotting is generally at a scale of 1:100. To simplify the operations, inclined photography is taken at standard angles and slope calculators are used.

(b) Accurate and Complete Surveys

These are used for systematic documentation of architectural heritage. Plotting scale is generally 1:50, while the details are mapped at 1:20 or 1:10. Large-format metric cameras with long focal lengths are preferred in this type of work in view of the accuracy requirements and the sizes of buildings surveyed. The recently developed wide-angle cameras having local lengths ranging between 100 mm to 150 mm are particularly suitable for this class of photogrammetric surveys.

Accurate surveys are used to document the technical history of the construction of the monument and its evolution as time passes, also to analyse its structural lines and to document its condition and its need for conservation and restoration. This is why one needs high accuracy and precision and as detailed a survey as possible. The use of "first order" stereo plotters is, therefore, essential. Furthermore, normal case photography is often not possible due to the difficult conditions frequently encountered. In some countries, precision photogrammetric surveys have been made for "technical monuments" such as ancient bridges and viaducts of artistic value.

A special case of accurate photogrammetric surveys is the survey of building exteriors (facades). Such surveys are carried out, particularly in central Europe, for the systematic documentation of harmonious architectural groups formed by series of houses in a street or on a square in ancient urban centers in towns and villages. Because of space limitations, facade photography is often taken at an upward inclination (*e.g.*, 30g or 70g) or from an elevatable platform on a special truck.

A second special case of accurate photogrammetric survey is the partial detailed survey of particular parts of monuments. Such surveys are conducted in conjunction with restoration and consolidation projects. The highest possible accuracy is needed for these purposes. Depending on the needs, the final outputs of the survey can be in the form of plans, cross-section s, elevations, profiles (for arches), contours (for vaults and cupolas), and/or numerical data giving accurate dimensions between the main elements of the building or distances between these elements.

The photogrammetric surveys conducted in the framework of UNESCO campaigns to salvage prestigious monuments such as About Simbel, Philae, Petra, Borobudur, are good examples of accurate and complete photogrammetric survey.

(c) Very Accurate Photogrammetric Surveys

These are needed for highly refined studies. Accuracy requirement is generally in the order of 1 mm and in some cases 0.1 mm. The study of sculptures in monuments and the assessment of the evolution in the surface of defaced stones (in support of chemical and physical investigations into the "disease of the stone") require this very high accuracy. The principal difficulty in such cases is encountered in photography. Metric cameras permitting short object distances (*e.g.*, by having variable principal distance or through the use of additional lenses) are of great help in this type of work.

1.2 Operational Procedures

Procedures for all of the above-discussed types of photogrammetric surveys are well established and documented. Independent stereopairs of photographs are taken either horizontally, vertically or at some inclination using the camera(s) most suitable for the individual project. Base-to-distance ratio is kept rather small (1/5 to 1/15). External controls are kept as simple as possible (such as number of distances and checks on the levelling bubbles of the camera). In case of complex object, however, a network of reference points is necessary. Camera stations are normally located on the ground, on scaffoldings, on nearby buildings, on a hydraulic lift truck or even in helicopters, which are sometimes used to take horizontal photographs of the upper portions of tall buildings.

The photographs are catalogued and stored in "photogrammetric archives" and are plotted only when the need arises. Plotting is mostly done using analogue instruments.

In some photogrammetric surveys, rectifying and assembling a ground of photographs produce photo-plans. This technique is particularly suitable for plane surface of murals, for mosaics, for windows and for facades, particularly when the streets are narrow. In this case, photography is systematically taken at a given inclination and rectified in simple instruments. This approach is appealing both from the technical and economical view. Orthophotography has recently been experimentally applied in Italy on photography of interiors of cupolas and in Germany and Poland for decorated surfaces with some relief.

Some architectural surveys are made by analytical photogrammetry. In this approach, a certain number of points are accurately determined and then connected by architectural lines. Such a method is not applicable to complex and important monuments because it often involves too many assumptions on the course of the lines to be drawn between points thus giving a theoretical rather than a real representation. On the other hand, analytical methods can be advantageously applied in schematically treating groups of simple constructions, as has been done in Scandinavian countries.

The analytical approach is particularly suitable in studying the structure of monuments and in checking on their stability through the use of digital models. By forming digital models encompassing the monument's fundamental points and the skeleton of its structure, one can study the proportions, define the volumes, compare the form etc. By repeating these operations at intervals of time, one can follow and measure eventual deformations in the building and thus check on its stability. By targeting the points involved in the analytical increases the precision of the observations and the accuracy of the solution.

Both the analogue and the analytical approach lead to numerical data, which is used to determine architectural forms. Using a computer, one can determine the curve of surface that best fits the group of points measured, according to the method of least squares.

2. Biostereometrics (Biomedical and Bioenginnering)

The study of biological form is one of the most engaging subjects in the history of human thought, which is hardly surprising considering the immense variety of living things. As new measurement techniques and experimental strategies have appeared, new fields of inquiry have been launched and more minds have become absorbed with the riddle of biological form. Discovery of the microscope and X-rays prompted the development of microbiology and radiology, respectively. More recently advances in electronics; photo optics, computers and related technologies have helped to expand the frontiers of morphological research. Growing interest in the stereometric analysis of biological form typifies this trend.

Measurements of biological form and function were made from stereophotographs in the middle of the 19th century, shortly after the invention of photography. Why has it taken so long to establish a real place for photogrammetry in the biomedical world? Limited Space does not permit a detailed discussion of this questions: suffice it to say that the problem of bringing photogrammetrist and biomedical specialist together is a bit like trying to unite two tribes who speak different languages and are separated by uncharted territory. In this metaphorical setting, biostereometrics can provide the interpreter-guides needed to negotiate the no-mans land and make more durable connections than those, which have occurred by serendipity alone.

Over the years, contacts between photogrammetrists and biomedical specialists have been quite numerous but most of the contacts involved trying to tie photogrammetry to a particular biomedical speciality. Unfortunately, these efforts generated surprisingly little sustained interaction. Recently, the more wide-ranging approach of systematically relating stereometric analysis to biology and medicine in general has proved to be a more fruitful strategy.

If biological structures were regular geometric shapes, there would be no great problem measuring them because simple lengths, breadths, and circumferences would be entirely adequate. But as well all known, organism have irregular three-dimensional components and linear "atomistic" measures such as are produced by tapes and callipers cannot give an unambiguous, comprehensive spatial qualification of a part of an organism as a whole.

Biologist and medical specialists are showing renewed interests in the stereometric analysis of biological form. Recent advances in computer technology and a growing range of stereometric sensing techniques have helped to expose the potential of biostereometrics. As a result, the use of photogrammetry is growing in such fields as: aerospace medicine, anthropometry, child growth and development, dentistry, marine biology, neurology, orthodontics, orthopedics, pediatrics, physiology, prosthetics, radiology, and zoology, to mention a few.

The need for biostereometrics stems for from the fact that linear tape and calliper measurements of inherently irregular three-dimensional biological structures are inadequate for many purposes. When stereometric data are used to fill this information gap the potentials for achieving more realistic models and making a more thorough analysis of biological form and function are far reaching. However, the best tools in the world confer no advantages unless they are used wisely. A petroleum geologist is expected to have the necessary training to select promising sites for oil exploration. Similarly, training in biostereometrics can be helpful in making decisions about "when" and "where" to use photogrammetry in biomedical research and clinical practise.

As the potentials of stereometric analysis in biology and medicine become more widely recognized, the role of biostereometrics in helping to unravel the complexities of organic form and function should continue to grow. Already, several photogrammetrists have chosen careers in biology and medicine and this number is expected to increase over the next few years.

3. Industrial Photogrammetry

Photogrammetry has been applied in numerous industrial fields and the potentially for further expansion and growth is seemingly limitless. Industrial photogrammetry has been described as "application of photogrammetry in building construction, civil engineering, mining, vehicle and machine construction, metallurgy, ship building and traffic, with their fundamentals and border subjects, including the phases of research, planning, production engineering, manufacture testing, monitoring, repair and reconstruction. Objects measured by photogrammetric techniques may be solid, liquid or gaseous bodies or physical phenomena, whether stationary or moving, that allow of being photographed" by Meyer (1973).

The experiences in the fields of architectural photogrammetry and biostereometrics clearly indicate the effectiveness of this strategy. The consistent use of term "industrial photogrammetry" should be instrumental in drawing the attention of photogrammetrist and equipment manufacturers to this fertile field of application, and should be helpful in bringing the capabilities of photogrammetry to the attention of the various industries. Tis way, it is hoped that more and more industrial concerns would make full use of the economical and technical advantage of photogrammetry. Economic benefits of derived from photogrammetric approach has been stated as;

- ☆ Measurement time on the object is reduced by per cent 90–per cent 95,
- ☆ Saving in manpower,
- ☆ Reduce machine and time for blade machining through optimisation of the metal removal rate,
- ☆ Reduced material expenditure in the propeller casting manufacture through optimised molds,
- ☆ A cut in recycling time for non-ferrous metals,
- ☆ Shorter production time for propeller manufacture.

Photogrammetric technique is equally suited for other industries where work-pieces of a complex surface configuration are to be manufactured, which would be very time consuming to measure with conventional measuring tools. A review of selected application in metal working industry has confirmed that photogrammetric techniques can be both practical and economically feasible for industrial measurements inspection tasks. The development of a systematic approach to implementing such application is necessary to investigate the reduction of start-up costs, operating costs and equipment costs. Examples of Industrial applications are:

- ☆ Automobile Construction
- ☆ Mining Engineer
- ☆ Machine Constructions

- Objects in Motion
- Shipbuilding
- Structures and Buildings
- Traffic Engineering